D1073267

ALLE·ZEIT·WACH
1842

Hermann Remmert

ECOLOGY

A Textbook

With 189 Figures

Springer-Verlag
Berlin Heidelberg New York 1980

Professor Dr. Hermann Remmert
Fachbereich Biologie
der Universität
Lahnberge, Karl-von-Frisch-Straße
3550 Marburg/Lahn, FRG

Translated by:
Marguerite A. Biederman-Thorson Ph. D.
The Old Marlborough Arms
Combe, Oxford, Great Britain

2nd German Edition: *Ökologie*

ISBN 3-540-10059-8 Springer-Verlag Berlin Heidelberg New York
ISBN 0-387-10059-8 Springer-Verlag New York Heidelberg Berlin

Library of Congress Cataloging in Publication Data. Remmert, Hermann. Ecology, a textbook. Translation of Ökologie. Bibliography: p. Includes index. 1. Ecology. I. Title. QH 541.R4313.574.5 80-13091.

Cover design: W. Eisenschink, Heidelberg.

Typesetting, printing, and binding: Brühlsche Universitätsdruckerei, Giessen.

2131/3130—543210

Preface

> The wealth of the natural sciences
> no longer consists in the abundance of facts,
> but in the way they are linked together.
>
> ALEXANDER VON HUMBOLDT

There is no dearth of books on ecology. Why write yet another?

Each person is different, and each views the problems in a different way. Each emphasizes different aspects and describes them in a different style. When I was a student I often found certain books more helpful than others, and I still think it is useful to have a variety of presentations from which to choose. This variety also allows the student to appreciate the diversity within the field of ecology.

I have devoted considerable effort to making this book readable. Throughout I have refrained from using specialized terminology – thus also avoiding the problem that many terms are used differently in the various areas of ecology. Straightforward English is quite sufficient to describe complicated situations. Furthermore, precisely defined terms are usually associated with detailed quantitative descriptions, whereas we are concerned with a general understanding of the dynamics of ecology. For similar reasons I have tried to rely as little as possible on mathematical discussions. All too often, in recent years, people have overlooked the fact that mathematics – like language – can give only a description, albeit an especially precise one.

Further, I have made a point of presenting not only well-established results, but also current hypotheses that give expression to the dynamic development and orientation of present-day ecology. In so doing I hope to stimulate further work in the field, to arouse the reader's curiosity; hypotheses are the salt and pepper of research.

And finally, I have stressed the functional relationships within the complex that is ecology. Every phenomenon has its causes and its effects – a dualism that frequently goes unnoticed.

The first German edition was soon sold out. In preparing the second, I am aiming at a simultaneous English edition to make the book available to readers outside Germany.

Marburg/Lahn, July 1980 HERMANN REMMERT

Contents

A. Ecology: the Basic Concept

The term "ecology" was introduced more than a hundred years ago by Ernst Haeckel; "biocenosis" was introduced in 1877 by Möbius, and "ecosystem" in the 1920's by Woltereck. Now, after 100 years of ecological research, these words are being discovered anew. They have become modern, and echo on all sides. But ecology is not a doctrine of salvation. As Ernst Haeckel put it, ecology is the *Haushaltslehre der Natur* – the study of the economics of nature. It is a strict natural science, but must overcome considerably greater difficulties than physiology, genetics or biochemistry; it must accommodate a vast array of different parameters, so that predictions become infinitely laborious. With the organism most thoroughly studied physiologically – the human – it is still impossible to predict responses reliably. Ecologists face the problem of judging in advance the reactions and developments of complex systems within which an extremely large number of genetically distinct microorganisms, plants, and animals live. Even to attempt such a task is a daunting prospect. But the attempt must be made.

We are also confronted with a special modern dilemma of ecology. Today this old, strict natural science has suddenly become a focus of public interest. It is called upon for assistance in making political decisions, and in so doing it is forced to move out of the realm of pure science. This necessity represents a notable danger to ecology. Many have come to regard it as a method of generating results which, if assiduously applied, will help mankind to achieve steadily increasing well-being and happiness. Nothing could be further from the truth. Ecological systems are the result of long development, and the only systems in which man can live at all are those existing at present.

For example, consider the air we breathe. We know that the oxygen in the earth's atmosphere, on which all animal life depends, is derived from photosynthesis by plants, which is usually written

$$6\,CO_2 + 12H_2O\ (+\text{light}) \rightarrow C_6H_{12}O_6 + 6O_2 + 6H_2O\,.$$

If this formula were a complete description of the situation, the production of oxygen would continue until all the carbon dioxide or water had been used up. Of course, this has not occurred – the reaction can also proceed in the opposite direction:

$$C_6H_{12}O_6 + 6O_2 + 6H_2O \rightarrow 6CO_2 + 12H_2O\,.$$

This formula describes the overall process of respiration in microorganisms, plants, and animals. Because the conversion of CO_2–O_2 by photosynthesis has, on the average, exceeded the reconversion of O_2–CO_2, the amount of oxygen on earth built up in the course of evolution. If the rates of conversion and reconversion had been identical, so that the organisms used up all the oxygen the plants produced, neither we nor the other animals now inhabiting the earth would be alive.

Green plants, in fact, can be regarded as the first great polluters of the environment. While their photosynthesis was creating the present-day oxygen atmosphere of the earth, and the earth's surface was being oxidized, many organisms that had existed before the plants – adapted as they were to life without oxygen – were doomed to perish.

An amount of organically bound or oxidizable carbon equal to the amount of oxygen produced by photosynthesis must be present on the earth. Some of it is in a form familiar to everyone, the fossil fuels (anthracite and bituminous coal, petroleum, natural gas, graphite). But these represent only a minute fraction of the earth's reserves. The remainder is present in all sediments, in a finely subdivided form – so finely that even in future it will not be made available for use as a source of energy. (This is a reassuring thought: even burning of all the fossil reserves will cause no appreciable change in the oxygen content of the atmosphere. However, by such burning we raise the carbon dioxide content of the atmosphere; the consequences to the earth's climate are incalculable.)

We live on this earth, then, because of the way natural ecosystems have evolved over millennia, and we can hardly expect the insights of ecological scientists to generate a utopian world. But ecological theory may be crucial to our continued existence. We can live on earth under the conditions now prevailing, and under no others. As a pure science, ecology is concerned with understanding the balance and turnover of matter and energy in nature. In its applied form, ecology faces the problem of discovering how the conditions essential for present-day life can be maintained.

The field of ecology is generally subdivided into three areas: autecology, the study of the requirements of particular organisms for particular conditions; population ecology, in which the primary question is why populations of microorganisms, plants, and animals do not reproduce without limit, but rather maintain certain approximately constant numbers; and research on ecosystems, which is concerned with the cycling of matter and the flow of energy, with the way ecosystems function, and with questions of the stability and elasticity of ecosystems.

In all three areas various approaches – microbiological ecology, botanical ecology, and zoological ecology of the ocean, fresh water, and land – ought to be combined to provide general insight. But because of the extent of the field, and for historical reasons, this collaborative ideal has not been realized. As a result, the terminology has become utterly confused. Even the word "ecology" is used with quite different meanings by botanists and zoologists. (In botany it refers only to experimental, chiefly physiologically oriented research, whereas in zoology pure field research is taken for granted as part of the science. Botanical field studies, together with historical biogeography, are considered to lie in the area of geobotany.) Moreover, the terms used in the terrestrial domain are quite different from those in limnology and oceanography. In this book, therefore, we shall avoid technical terms almost entirely (for a review of ecological terminology see Tischler, 1975).

B. Autecology

B. Autecology

I. Theory of Autecology

Two organisms that require the same resource compete for this resource if they occupy the same area. The competition between the two is stronger, the greater the number of requirements they have in common. It becomes overpowering when the ecological demands of the two organisms are identical in all details. Two different species with identical ecological requirements can live together only if the number of individuals is so small that they never actually enter into competition – that is, if they do not utilize their habitat to full capacity. (The circumstances under which this situation can arise are discussed in connection with the regulation of population density, in the section on population ecology.) In general, however, it is possible for two species to coexist in the same space while competing for the same resources if they differ in various aspects of their biology or their ecological requirements.
Competition is most vigorous, then, between members of a single species. This extremely strong intraspecific (as opposed to "interspecific" – between different species) competition is the driving force for evolution, as Darwin described it more than a century ago. The best-adapted genotypes in each situation are continually selected, and those less suited are eliminated from the contest. Organisms can escape this competition in two different ways. As a result of selection they either become adapted to the conditions in a different habitat, or they evolve a different way of life, with different ecological requirements in the same habitat. The new features acquired as adaptations must become genetically fixed. If permanent changes are to be made in the gene pool, sexual isolation from the original population is necessary. This separation, the process by which new species are formed, is called speciation. Speciation – sexual isolation from neighbors – is thus an adaptive process which offers selective advantages in that it enables specialization and alleviates competition. The genotypes selected in the struggle for existence, then, are those best suited to the original situation as well as those able to colonize other habitats because of specific adaptations and those which, by virtue of other adaptations, are least exposed to competition from the original species in the original habitat.
The fact is often overlooked that acquisition of new adaptive features is never an unalloyed benefit – in all cases, adaptation takes its toll. Side effects can include slower development, a lower rate of reproduction, reduced mobility, impaired resistance to changes in the environment, or a combination of these and other symptoms. Another possible side effect, a high rate of energy consumption, is found only in exceptional cases, and then is usually temporary. Because high energy consumption must always be coupled with increased food intake, and there is hardly ever a surplus of food available, organisms must adopt other strategies for adaptation. In terms of the sort of cost-benefit analysis made by economists, the price of an adaptation would be set against its value to the organism – the advantage of escaping intraspecific competition.
So far no such cost-benefit analyses have been carried out to completion. In mutants of flowering plants that thrive on heavy-metal soils, dry-matter production is generally reduced by 20–50% (Ernst, 1975).

Organisms in cool habitats pay for the opportunity to live there with a lowered rate of development. Fresh-water and land organisms have become able to colonize these habitats at the expense of developing complicated and elaborate mechanisms for ionic and osmoregulation.

The recent attempt to incorporate the genetic information in microorganisms that governs fixation of aerial nitrogen into the genome of crop plants is to a great extent based on a mistaken estimate of the cost of ecological adaptation. Fixation of nitrogen "costs" much more than uptake of nitrogenous salts from the soil. Thus it must follow that if the experiment succeeds the productivity of the new crops will be distinctly lower than that of the original forms. And fertilization would still not become unnecessary, for phosphate and potassium – for example – would have to be supplied. No one has ever really calculated in detail whether, under these circumstances, this heavily-funded area of research is at all reasonable economically.

Once an adaptation has been acquired, further selection is directed toward optimizing it – that is, toward minimizing the price paid. On basic principles, the cost can never be eliminated altogether.

Optimization, however, does not imply optimal adaptation. This state may well never be attained. Adaptation occurs under a particular set of prevailing conditions; if the conditions should change it might turn out that still greater adaptation becomes possible. For instance, the fact that the American gray squirrel has supplanted the native squirrel in many parts of England must be regarded as the result of suboptimal adaptation of the native animal.

Where organisms are concerned – in contrast to abiotic systems – there is probably never a linear dose-response relationship. That is, the uniformly graded enhancement of one factor does not lead to a uniformly graded reaction of the organism. Everywhere we encounter optimal ranges; if their upper or lower limits are exceeded lethal effects are suddenly apparent. We shall discuss this principle further in connection with salt content, water relations, and temperature. It is especially important in the context of the current human situation. If increasing the water temperature in a river by 10 °C proves to have no dramatic consequences, it does not follow that a further one-degree increase would be harmless. If we can be exposed to the natural level of radioactive radiation without problems, we cannot therefore conclude that a further increase of the radiation dose by 1% or 1‰ would be tolerable.

II. The Range of Ecological Factors

The number of ecological factors is vast, as is the multitude of adaptations of organisms. Here no attempt will be made to survey them all; certain adaptive characteristics and particular factors and combinations of factors will be selected for discussion, with the aim of revealing various basic strategies of adaptation and principles of interaction. I have made an effort to consider each complex of factors from a different point of view, without giving an encyclopedic presentation.

III. Life-Form Types

In the course of evolution, organisms adapt to physiographically similar conditions in similar ways. Accordingly, animals of entirely different phylogenetic groups can closely resemble one another if they live under similar conditions. Famous examples of such resemblance include the cacti of the New World as compared with the euphorbias of the Old, both of which have evolved a "cactus-like" habit. The murres and auks (birds of the family Alcidae) in the northern hemisphere are very like the penguins (Impennes, named from the Irish word for murre) in the southern hemisphere. The plumage patterns of the northern-hemisphere creepers resemble in detail those of the South American Dendrocolaptidae, and those of the European lark (Alaudidae) and of the North American meadow lark (Icteridae) are just as similar to one another. We could go on listing examples indefinitely. It is clear, then, that members of quite different systematic groups take on a similar appearance under physiographically similar conditions – and indeed, they actually occupy the same position on the ecological map. They serve the same function in the system; they have the same ecological niche as defined earlier by Elton, though the term is no longer applied in this clear-cut sense (cf. p. 68). Examples of such "equivalent occupations" are given in Fig. 1. It is no new discovery – Cuvier himself pointed it out – that one can infer fairly accurately, from the structure of an organism, its ecological requirements. It is common to draw such conclusions about trees on the basis of their leaves. Normally we think of the sun's rays as being nearly parallel when they reach the earth, but this is of course not correct. Because of their divergence, a circular object casts a "deeper" shadow than a narrow or much subdivided object with the same projection area. If one considers the leaves of trees with this in mind, it becomes clear

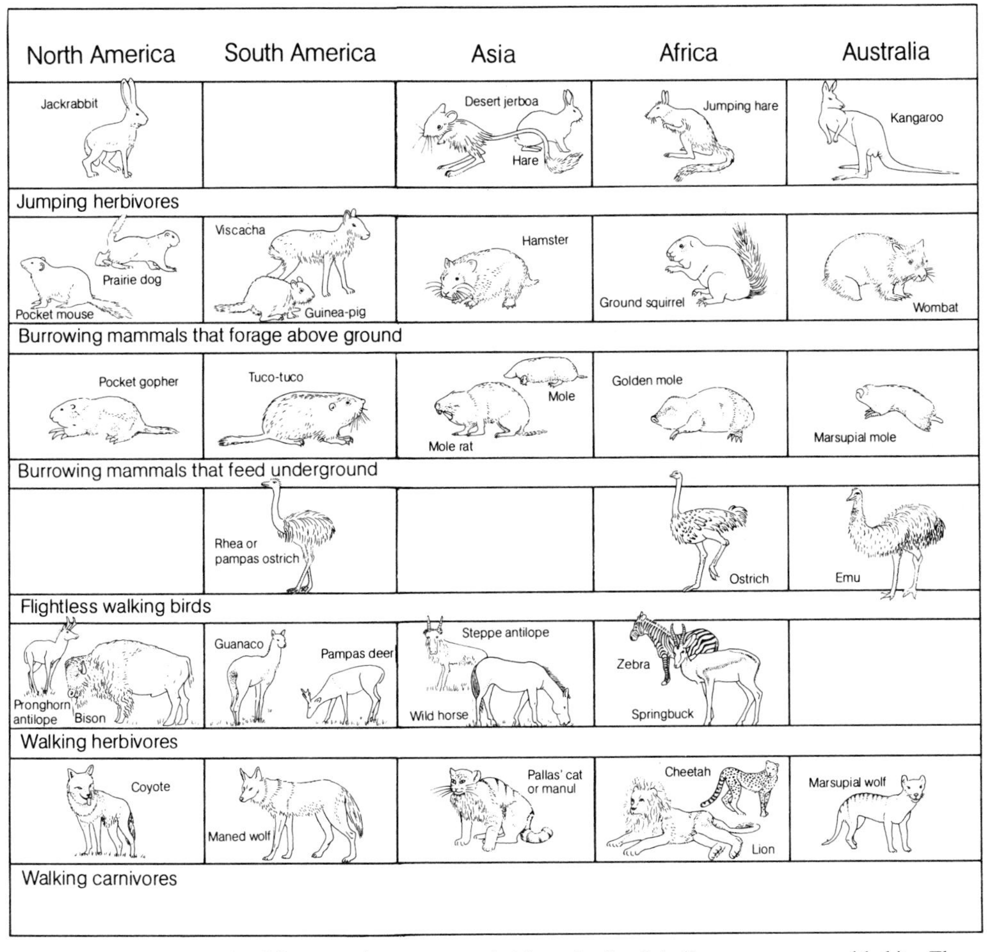

Fig. 1. Similar habitats in different regions are occupied by animals of similar appearance and habits. They have the same "professions" within the system. Elton applied the term "niche" to this equivalence in occupation (cf. Fig. 5; modified from Farb, 1965)

that species having narrow leaves (pine, willow, eucalyptus) or leaves with marked indentations (ash, maple, oak) give relatively little shade. There is enough light beneath these trees for other plants to thrive, whereas under trees with more rounded leaves (such as are abundant in the tropics, and are represented in temperate regions by the beech, Fagus, and linden, Tilia) hardly any other plants can grow.

A particularly well-known classification was proposed by Raunkiaer, to describe life-form types in terms of adaptation to various winter conditions (Fig. 2). In this scheme the following types are distinguished:

1. The phanerophytes (P) have perennial, erect shoots with buds at their tips. These buds pass the unfavorable season at a considerable height above ground. The megaphanerophytes (MP) comprise all the trees, and the nanophanerophytes (NP) the shrubs. In the case of evergreens, the leaves remain alive through the unfavorable season; in other cases they are shed, leaving behind only the sites of origin of new growth. These buds are an important source of winter food for warm-blooded animals – a fact which, as we shall see (p. 40f.), explains the pronounced effect of even small populations of such animals.

2. The chamaephytes (Ch) raise their buds only about 25 cm above the soil surface. Their evergreen leaves or buds are better protected in winter, in regions where snow cover is ordinarily present. This group includes all the dwarf shrubs with woody stems (Calluna, Vaccinium) and the cushion plants, but also species with prostrate or creeping stems lying close to the ground (Sedum, Stellaria holostea, Thymus, Helianthemum, Veronica officinalis, Vinca minor, Lysimachia nummularia).

3. The hemicryptophytes (H) have aboveground shoots that die back completely during the winter. Just at the surface of the soil there remain living buds, which send out shoots the next year. Nearly half of all plant species in the temperate zone are in this group. In the winter, the living buds can be detected at the bases of the dead stems. The entire root system survives the winter and serves as a storage organ. The most shallow snow cover, or even the layer of litter on the ground, provides protection

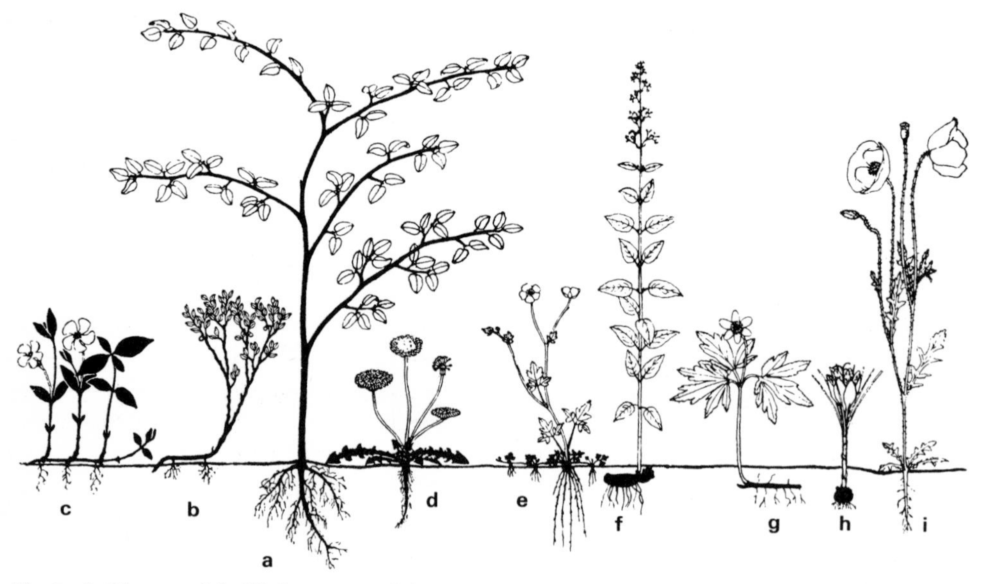

Fig. 2a–i. Diagram of the life-form types of plants. **a** phanerophytes, **b** and **c** chamaephytes, **d–f** hemicryptophytes, **g** and **h** cryptophytes, **i** therophytes. The overwintering parts are shown in black. (Walter, 1949; see text)

to keep the overwintering parts of the hemicryptophytes from drying out. In warm winters these plants can lose a great deal of energy by respiration; then the new growth in the spring is delayed and weak. The group is extremely diverse. Its subgroups are the plants without rosettes, plants with semirosettes, and rosette plants.

4. Cryptophytes (Cr) are plants with above-ground organs that die back completely and buds that either are deep underground or pass the unfavorable season underwater. The overwintering parts of the plants have the form of rhizomes (Anemone nemorosa, Iris), corms (Colchicum), tubers (Orchidaceae), or bulbs (Allium).

5. The therophytes (T) or annuals are the last group, and the best adapted to survive an unfavorable season, for the only part of the plant that remains alive is the seed. The disadvantage is that relatively scant reserves are available for sprouting. The young plant must provide for itself all the materials it needs to flower and develop fruit. Where the growing season is short such plants cannot become established (Tables 1 and 2).

Returning to tree forms, we find that the distribution of the leaves is also characteristic of particular types. In some species they tend to be restricted to the periphery, while in others they occupy the entire twig from base to tip. Both types can be found within a single genus. But the optimal light conditions for the two are quite different; species with peripherally situated leaves are superior in weak light, and can grow tall even if they start out in the dense shade of the forest floor. On the other hand, if there is more light the species with leaves filling the crown have the advantage. Because of their larger photosynthetically active leaf area they are capable of more rapid growth and can thus quickly colonize newly available sites.

It is already evident from these examples that several life-form types can coexist in a single habitat. To a great extent, life-form type is dictated by the size of the organism. Where the ocean floor is sandy there are animals that live either on the substrate or burrowed a short distance into it. Frequently these are flattened forms (the flounder Pleuronectes; the ray Raja; the sea star Asterias; the brittle star Ophiura). These are accompanied by forms that burrow deeply into the sand but extend far enough into the open water to filter food from it (the clam Cardium; the sea urchin Echinocardium; the chaetopod Lanice; the lancelet Branchiostoma). Finally, there are a number of forms that burrow in completely, like earthworms. Among these are a few widely distributed polychaetes such as Nephthys, Scoloplos, Ophelia, und the gastropod Natica.

Table 1. Distribution of the plants (see text for abbreviations) in various geographical regions among the life-form types of Raunkiaer, a convenient representation of the biospectrum of a zone. (Walter, 1949)

	P	Ch	H	Cr	T
Tropical zone:					
Seychelles	**61**	6	12	5	16
Desert zone:					
Lybian desert	12	21	20	5	**42**
Cyrenaika	9	14	19	8	**50**
Mediterranean zone:					
Italy	12	6	29	11	**42**
Temperate zone:					
Paris basin	8	6.5	**51.5**	25	9
Central Switzerland	10	5	**50**	15	20
Denmark	7	3	**50**	22	18
Arctic zone:					
Spitsbergen	1	22	**60**	15	2
Nival altitude zone:					
Alps	—	24.5	**68**	4	3.5

Table 2. Distribution of the plants in the high Alps among the life-form types of Raunkiaer. (Walter, 1949)

Altitude	No. of species	P	Ch	H	Cr	T
3,050–3,150 m	82	—	40.3	**52.5**	2.4	4.8
3,150–3,260 m	42	—	**53.3**	34.9	2.3	9.3
3,260–3,350 m	31	—	**64.6**	29.0	3.2	3.2
Above 3,350 m	16	—	**69,0**	31.0	—	—

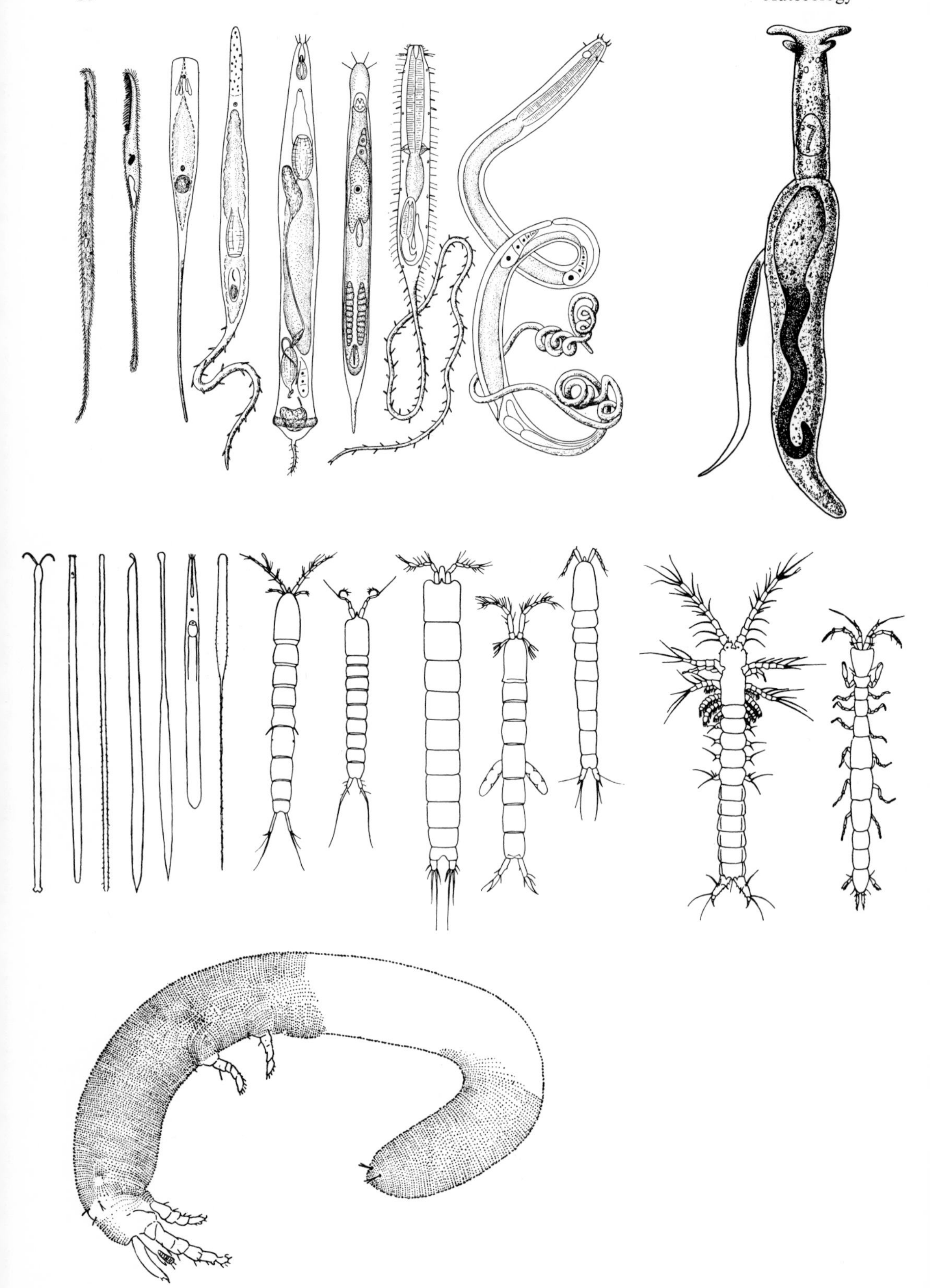

For a long time yet another sandy-bottom life-form type has attracted special interest; these are the animals that live in the system of spaces between the sand grains. Sand grains of a particular size are enclosed in water by adhesive forces; a mass of such grains forms a relatively stable structure through which slender animals can move without difficulty. A very rich fauna from a large number of animal groups has colonized this habitat (Fig. 3). At first glance one notes an astonishing similarity among the representatives of quite different groups – crustaceans, polychaetes, and archiannelids, protozoans (ciliates), turbellarians, gastrotrichs, gnathostomulids, nematodes, and even mites. This striking resemblance is not just superficial; it extends to many aspects of anatomy and habit. Common to all of these interstitial animals is the presence of an adhesive apparatus by which they can attach themselves to sand grains when the sand is in motion. These adhesive organs are very often located on filaments, and frequently there are cilia – sense organs – at their bases. A number of forms have developed tissue resembling the vertebrate notochord. A very common type of locomotion involves a serpentine movement, with which the animals push themselves from one interstitial space to another. Fertilization is direct in all cases, and often by way of spermatophores. This method has been adopted even by representatives of groups that ordinarily employ quite different mechanisms (for instance, the polychaetes and opisthobranchs). The planktonic larval stage otherwise so typical of marine animals is entirely lacking; development is direct. In a number of groups the genital apparatus is multiple.

In addition to the animals that wander through the interstitial system, there are others of comparable size but more or less sessile habit, attached to the sand grains. Among these are a number of tunicates and the only known bryozoan that lives as an individual. Finally, each sand grain is covered by a film of bacteria and by very small algae which, together with the rain of detritus, serve as food for the animals.

Particularly good examples illustrating life-form types can be found among the parasites. Long-term parasites on the bodies of other animals undergo parallel development of a flattened body with clasping legs (Fig. 4; see also Fig. 5). Examples include the lice (Anoplura) and members of various fly families (Hippoboscidae, Nycteribiidae); the arachnids are represented by the ticks (Ixodidae) and the crustaceans by copepods (the carp louse Argulus and many others) and amphipods (Cyamidae). As evolution progressed members of very divergent groups developed a rootlike projection into the host, by which food is absorbed; this innovation is accompanied by reduction of the gut. Such animals include mites parasitic on insects, many copepods, and the rhizocephalans (genus Sacculina), closely related to the cirripedes. Loss of the intestinal canal and intake of food by way of the skin is common to a large number of tissue parasites (the tissue-parasitic stages of trematodes and cestodes, and tissue-parasitic protozoans), and the same modification is found in intestinal parasites. In general, parasitic

◄ **Fig. 3.** Life-form types in the meiofauna inhabiting the interstitial system of the seashore. (Ax, 1966, 1968; Strenzke, 1954) Reading left to right, these are: *Top:* Remanella caudata (Ciliata), Spirostomus filum (Ciliata), Mecynostomum filiferum (Turbellaria), Boreocelis urodasyoides (Turbellaria), Cheliplanilla caudata (Turbellaria), Gnathostomula paradoxa (Gnathostomulida), Urodasys viviparus (Gastrotricha), Trefusia longicauda (Nematoda), and Microhedyle lactea (Gastropoda Opisthobranchia; the structure attached at the left is a spermatophore). *Middle:* vermiform animals on the left: Protodrilus (Archiannelid), Coelogynopora (Turbellaria), Michaelsena (Oligochaeta); two ciliates; Proschizorhynchus (Turbellaria), Urodasys (Gastrotricha); crustaceans on the right: the copepods Leptastacus macronyx, Evansula incerta, Parastenosella leptoderma, Stenocaris minor, and Arenopontia, and finally Derocheilocaris remanei (Mystacocarida) and Microcerberus stygius (Isopoda). *Bottom:* Nematalycus nematoides (Acari)

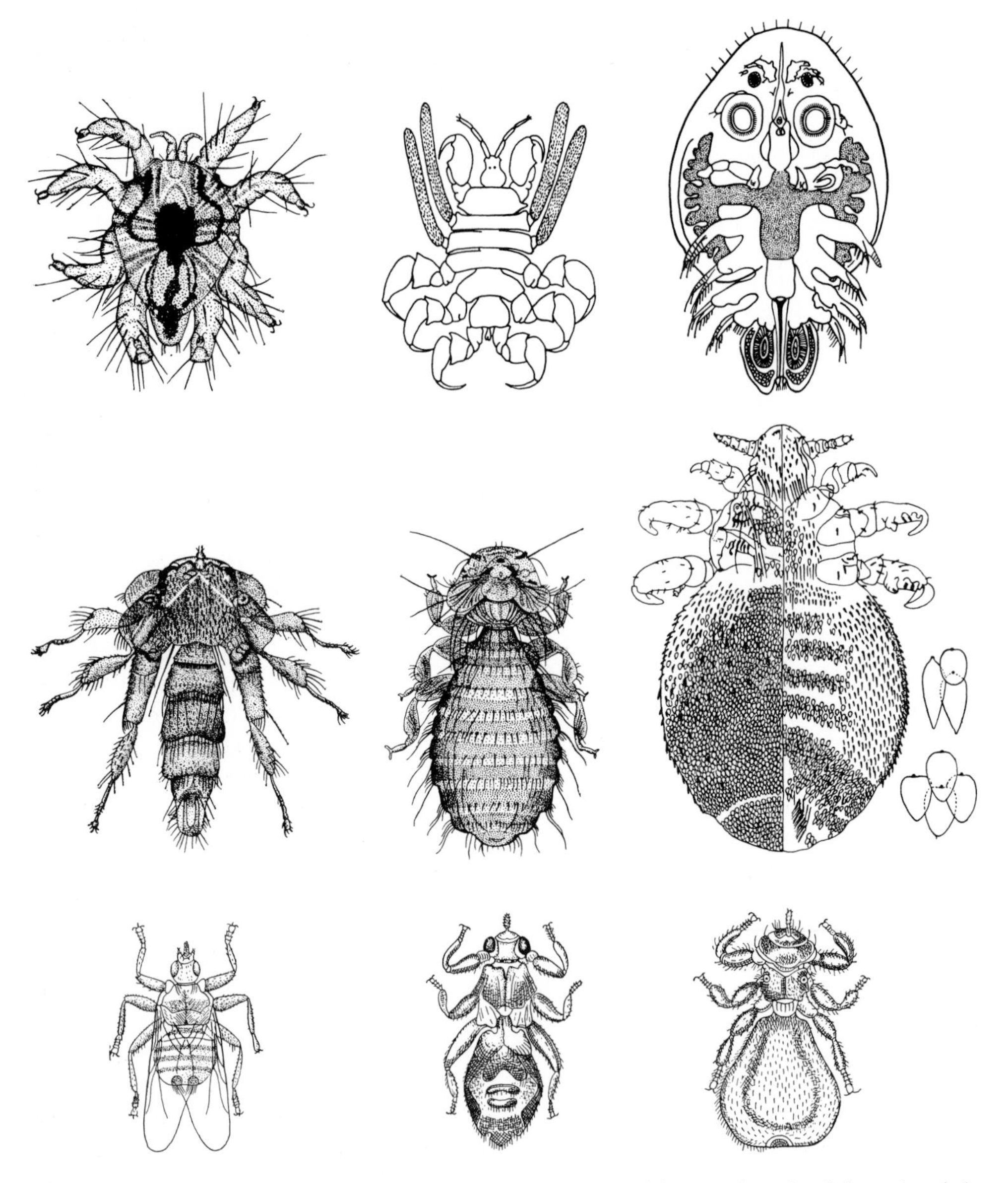

Fig. 4. Life-form types of ectoparasites. (Dogiel, 1963). From left to right: *Top:* bat mite, Spinturnix; whale louse, Cyamus (Amphipoda); carp louse, Argulus (Branchiura). *Middle:* bat fly, Nycteribia; cuckoo biting louse, Cuculiphilus; sea-lion louse, Antarctophthirius. *Bottom:* louse flies (Lipoptena, Lynchia, Melophagus)

stages exhibit drastic reduction of nervous system and sense organs, and an equally dramatic increase in rate of reproduction. Extraordinarily aberrant parasitic forms exist in a wide variety of marine groups – isopods, amphipods, gastropods, bivalves, and ctenophores. (A major source of information about the vast range of adaptations found here is the book by Dogiel, 1963.)

A modification at a different level – externally almost indetectable, but physiologically of great significance – is the development of polyploidy among land plants. This multiplication of the chromosome complement accompanies migration into

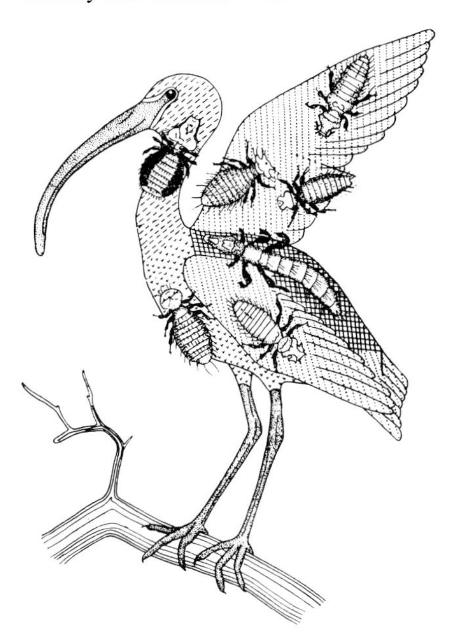

Fig. 5. Various closely related species of biting lice parasitize the body of an animal at strictly separate locations. (Dogiel, 1963)

extreme habitats such as deserts, saline environments and the arctic. Polyploidy is evidently the first step in adaptation to extreme conditions; the increased redundancy in genetic information appears to permit initial survival of the plants. In the long run, the polyploids are replaced by specially adapted forms. This development is apparent when one compares very old desert regions with geologically younger deserts. In animals polyploidy is exceedingly rare, and it is almost always associated with parthenogenesis.

We have seen that organisms from very different phylogenetic groups can occupy very similar ecological niches. This multiplicity of origin is not only found when different regions are compared, but is just as evident within a single habitat. Strictly speaking, then, it is inappropriate to work on the ecology of a single group of plants or animals in isolation, to consider the niches it occupies, their overlap and their diversity (cf. pp. 68) without reference to other groups. It is entirely possible that organisms from a quite different group are present and in competition for some of the same important resources – that they may actually be superior competitors. This possibility has often been neglected, particularly in research on bird communities; but it has been demonstrated, by Reichholf (1975) for the reservoirs along the Inn River and also by researchers in Sweden, that there is a vigorous competition for food between fish and water birds. As a reminder of the need for caution, think of the interstitial fauna of the ocean floor.

It is also dangerous to draw inferences about habitat from the anatomy of an organism. Flies of the genus Coelopa closely resemble birdflies (Hippoboscidae) and even behave quite similarly – but they live in the stratified debris on ocean beaches. Many species regularly alternate biotopes. In particular, such behavior is known among arctic birds. The sandpiper Calidris maritima breeds in the tundra; it often leads its young to small ponds with a variety of plant life, though this is not obligatory behavior. If the seacoast is nearby, these birds can be seen looking for food on sandbanks. When migration begins, they are found only on coastal cliffs. In the far north this change in habitat takes place within a few days. The phalarope (Phalaropus) breeds near arctic ponds with an abundance of mosquitoes; the birds peck this food from the surface of the water while swimming. One can see them feeding in this way in the Reykjavik City Park. But they spend the winter on the open ocean, far from the coast. In summer an inland bird on small bodies of water, and in winter a denizen of the high seas!

IV. Ecological Factors

1. Salinity and Osmotic Pressure

The most deterring ecological barrier we know is the salt content of sea water (Fig. 6). Only very few plant and animal species can thrive in a range of salinity extending from sea water to fresh water. Moreover,

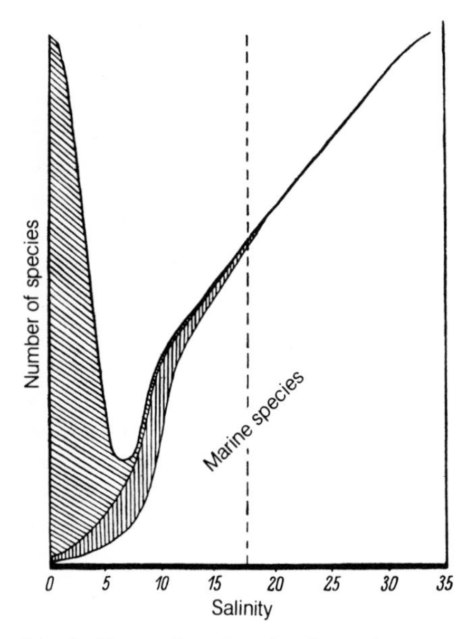

Fig. 6. The number of species of organisms present in waters of different salinity; reconstructed from many reports. *Diagonal hatching:* fresh-water species; *vertical hatching:* specific brackish-water species; *unshaded:* marine species; *black* (bottom of graph): species that live in both fresh and salt water. In each case the number of species is represented by the vertical extent of the surface. (Remane, 1954)

within this continuum of salinity, a particular degree of change is not necessarily accompanied by a corresponding change in the organisms. In a certain range a relatively pronounced difference in salinity (for example, 25‰ vs 15‰) produces only a slight difference in number of species, whereas in the vicinity of 8‰ and at the boundary between fresh and brackish water there is a sudden dramatic effect. These effects of water salinity and the associated osmotic pressure deserve closer attention. We shall take animals as an example.

All the animals and plants alive today originated in the ocean. In order for them to invade and conquer terrestrial and fresh-water habitats, they had to undergo adaptive modifications of their original states – at a price, of course. We must keep this fact in mind whenever we are discussing plants and animals of fresh water and land.

To begin, let us consider the primary marine animals – species which throughout their phylogenesis have occupied no environment other than the ocean. The osmotic pressure of their intercellular fluid is very slightly higher than that of the open sea; it corresponds to 35‰ total salinity. The fluid in the cells has the same osmotic potential, and differs from that outside the cells only in ionic composition. Whereas intercellular fluid – in the water-vascular system of echinoderms, for instance, or in the blood vessels of annelids – closely resembles sea water, the cells contain considerably higher amounts of potassium ions. Even in these animals, then, ionic regulation occurs; it leads to the well-known electrical phenomena at cell surfaces. Moreover, the cells also contain organic substances (in particular amino acids, and sugar in many plant cells) which contribute to maintenance of osmotic pressure.

Animals of this type can endure only minimal fluctations in the osmotic pressure (that is, the salinity) of their environment. Their distribution is thus restricted to the oceanic plankton or to deep regions of the sea. The majority of benthonic animals – echinoderms, tunicates, crustaceans, annelids, molluscs, and cnidarians – are in this category. Animals living in shallow water, tidal zones and estuaries must be able to tolerate fluctuations in salinity. Changes in the osmotic pressure of the milieu, which have an immediate effect on the intercellular fluid and thence on the cells, must be compensated. Without such compensation, reduction of environmental salinity would cause water to flow into the animal and its cells, which would swell and burst. Conversely, increase in salinity results in shrinking of the cells.

These processes can be counteracted only by active alteration of the osmotic pressure in the cells and in the intercellular fluid (Fig. 7). Water is expelled and ions are actively taken up. Within the cells osmotically effective amino acids are synthesized and broken down. "Poikilosmotic" behavior of this sort, effective over a wide range, is thus by no means a passive phenomenon; it requires energy-consuming activity

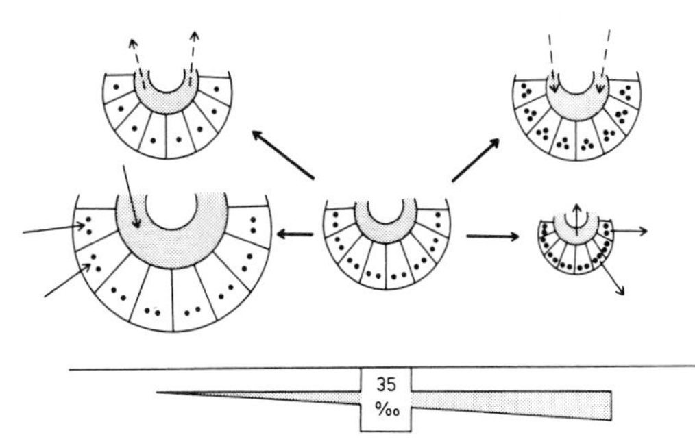

Fig. 7. Response of primary marine animals to rapid change in the osmotic pressure of the milieu. The central drawing represents the animal in water of normal marine salinity. *Below:* response of a species strictly limited to a small range of salinity. *Above:* response of a less restricted species. *Solid arrows:* passive water influx; *dashed arrows:* active ion transport; *dots:* osmotically effective organic molecules in the cells, which can be synthesized or broken down to adjust to the prevailing external osmotic pressure; *shaded region:* intercellular fluid; *abscissa:* salinity of the milieu. (Remmert, 1969a)

of single cells and of the entire body. Animals in this category are characteristic of regions with moderate salinity fluctuations; they invade the tidal zone, the mouths of rivers, and bodies of brackish water. This type represents the point of origin of land and fresh-water animals. But purely poikilosmotic behavior is potentially lethal for both lines of development. No animal can exist with the low osmotic pressure of fresh water in its cells or blood. In addition to the ionic regulation we have just discussed, then, a further mechanism is necessary – osmoregulation. First let us consider the evolution of the terrestrial habit (Fig. 8). Animals living in an air milieu are less exposed to the osmotic pressure of the substrate fluid, because only small areas of the body contact the ground. In the long term, though, they are of course affected in the same way as water animals, by way of the water they drink and osmotic processes in the parts of the body exposed to the water in the soil. The results we shall now consider are based entirely on prolonged culture experiments. When, during the course of evolution, many groups of animals migrated from the sea to the land, they moved from a milieu in which the salinity varied hardly at all through one with highly unconstant conditions – the marine supralittoral – before they again reached a stable, though quite different, situation inland. Most crustaceans initially protect themselves from the wide fluctuations in the coastal region by retaining an osmotic pressure higher than that of the external medium (hypertonicity regulation). Then the beach forms acquire an additional capacity for

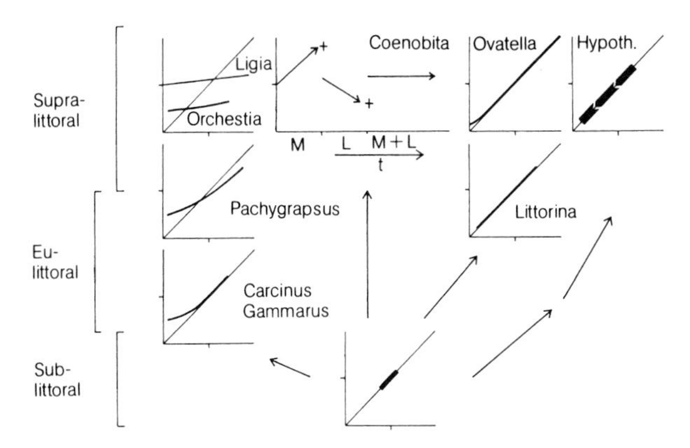

Fig. 8. The migration, during phylogenesis, of primary marine animals to land. Abscissa: osmotic pressure of the milieu; ordinate: osmotic pressure of the hemolymph. The scale mark on each axis represents a pressure corresponding to about $30/10^3$ (salinity of the open sea). Along the 45° line internal and external pressures are identical. In the graph for Coenobita, the abscissa indicates the time during which the animals were offered fresh water alone *(L)*, sea water alone *(M)*, or both *(M+L)*. In this arrangement, the species represent particular physiological states; they are not phylogenetically related in the sequence shown here. (Remmert, 1969)

hypotonicity regulation; they are able to keep the internal milieu at an osmotic pressure lower than that of the environment. Such constancy of the internal milieu is typical of most of the coastal crustaceans living above the water line (Uca, Ocypode, Carcinus, Talitrus, Orchestia, Ligia).

The terrestrial hermit crabs (Coenobita) and some brachyurans also appear to keep their internal milieu constant. In this case, however, another method is used. Evidently these animals require both fresh and sea water; if only one kind is available, the internal osmotic pressure rises or falls and the animals eventually die. The time of death depends on many other factors (molting is particularly hazardous). When land hermit crabs drink, then, they sometimes choose fresh water and sometimes sea water. Their internal pressure can vary over a wide range, but can be continually readjusted if water of the appropriate salinity is consumed in time. An entirely different approach enabled gastropods to conquer the supralittoral zone. On flat soft beaches they too develop hypertonicity regulation; its acquisition can be followed within genera. For example, Limapontia capitata exhibits no osmoregulation at all, whereas a species living higher up the beach, Limapontia depressa, can regulate its osmotic pressure. Gastropods have responded to high salinity by developing a poikilosmotic mechanism effective over a wide range; in this case there is no hypotonicity regulation. By these means, gastropods of the ocean beaches can tolerate extremely wide fluctuations of the pressure in the internal medium and in the cell fluid. In Ovatella and Assiminea, pressures corresponding to a salinity between 6‰ and 90‰ (sometimes up to 100‰) have been measured. This situation requires highly specific adaptation of the enzymes, which must remain functional despite great differences in their intracellular milieu. By contrast, gastropods on rocky coasts apparently never perform osmoregulation. When conditions are unfavorable they shut themselves off from the outside world by closing the operculum.

Of course, adaptation to changing salinity conditions cannot be obtained free of cost. In the case of homoiosmotic animals such as Ligia or Orchestia platensis the price is obvious. In both low- and high-salinity water they must perform hard osmotic work, so that it comes as no surprise that with respect to growth and rate of reproduction they exhibit a distinct optimum in the approximately isotonic range. The same is true of poikilosmotic species. Synthesis and breakdown of the osmotically effective substances costs energy. Even the presence of enzymes that operate within a wide osmotic range (Sarkissian, 1974) evidently represents a drain on energy. Poikilosmotic species, then, also pay for their

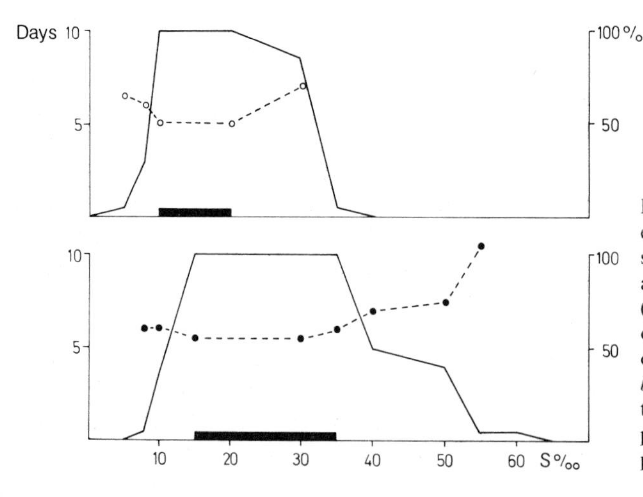

Fig. 9. Alderia modesta: response of eggs and embryos to different salinities. *Solid line:* percent of eggs and larvae that develop normally (right ordinate); *dashed lines:* duration of development, in days, to the stage of larval emergence (left ordinate); *black bar:* optimal range. *Above:* Baltic Sea population; *below:* North Sea population. Temperature during experiment: 10 °C. (Seelemann, 1968)

ability to tolerate a wide range of salinity with very different rates of development and reproduction (Fig. 9). It seems clear that when organisms are under such stress the strategy adopted is not to increase energy consumption (which would mean eating greater amounts of food, and this is normally not available) but to cover the cost by slower growth, a lower rate of reproduction, and less resistance to generally harmful environmental factors. The poikilosmotic mussels in the inner Baltic Sea, for example, are considerably less able to tolerate cold than those in the North Sea – although in the Baltic selection for high cold tolerance would be expected. Desert plants and lichens are particularly sensitive to chemicals in their environment.

Only two of the four types were able to evolve further, into true land animals. Terrestrial hermit crabs and snails on rocky coasts need at least occasional access to water of marine salinity. Their evolution therefore ends in the supralittoral. Once the other terrestrial crustaceans and gastropods had evolved into genuine land animals, the salinity at which body fluids could be maintained by maximum regulation fell significantly. For terrestrial arthropods it corresponds to about 10‰, and for soft-skinned land animals to about 6‰. At the same time, the ability of arthropods to tolerate high salinity was greatly restricted by loss of hypotonicity regulation. Phylogenetically young land animals (isopods) can still exhibit such regulation, and can thus tolerate high environmental osmotic pressure. This physiological feature is useless in the present habitat of these animals, and can only be understood in the historical context. In the evolutionary tree of soft-bodied land animals, the poikilosmotic branch is reduced; salinity tolerance is strictly limited. The pattern that thus developed – in arthropods and land vertebrates (which originated in fresh water), a fixed limit for regulation corresponding to 10‰ with no hypotonicity regulation, and in soft-bodied land animals (gastropods and annelids) a 6 ‰ limit

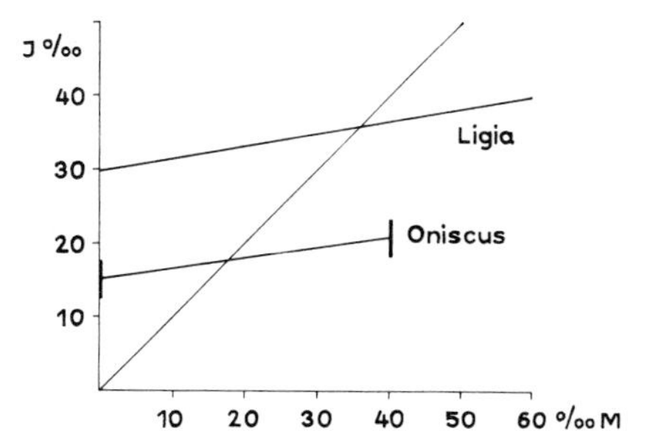

Fig. 10. The osmotic pressure of the hemolymph of the terrestrial isopod Oniscus is always much lower than that of the beach isopod Ligia. Ordinate and abscissa scales as in Fig. 8. (Remmert, 1969)

with poikilosmotic behavior limited to a very small branch – gives the starting point for the recolonization of the ocean by land animals (Figs. 10 and 11).

Certain animals were predisposed for this return to the sea – desert animals that drink from salt lakes and must excrete salt, and animals from environments with aberrant chemistry (such as liquid-manure pits). Another predisposing property, evidently, is alteration in the set level of tonicity during the course of the year, as a protection against cold. The comparatively massive advance of terrestrial animals (insects, mites) into arctic oceans argues for this interpretation. In the transition from land to sea arthropods redevelop hypotonicity regulation. Only the Collembola appear to extend the poikilosmotic branch of evolution, which otherwise includes soft-skinned forms – amphibians, gastropods (Succinea), and enchytraeids. The increase of osmotic pressure in the blood of the only truly marine frog (Rana cancrivora, eastern Asia) is brought about not by salts but by dissolved urea. By such modifications secondary marine animals are initially capable of living in either sea or fresh water. They need no salt; they can either tolerate or excrete it. A real dependence on marine salinity arises by loss of hypertonicity regulation, as has happened in the beach fly Coelopa. Once this has occurred, the animal is cut off from both land and fresh water; examples in this category

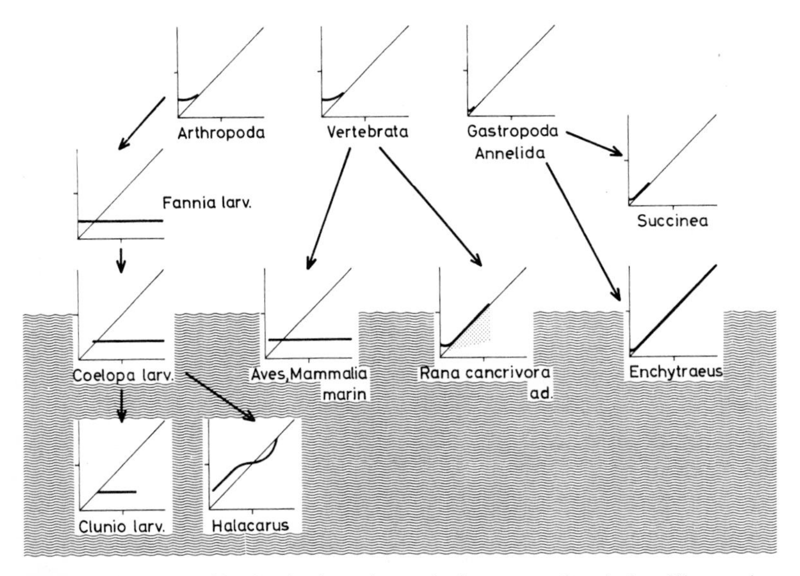

Fig. 11. The return of land animals to the sea in the course of evolution. The graphs correspond to those in Fig. 8; in that for Rana cancrivora, the shaded region indicates the contribution of urea to the osmotic pressure. (Remmert, 1969)

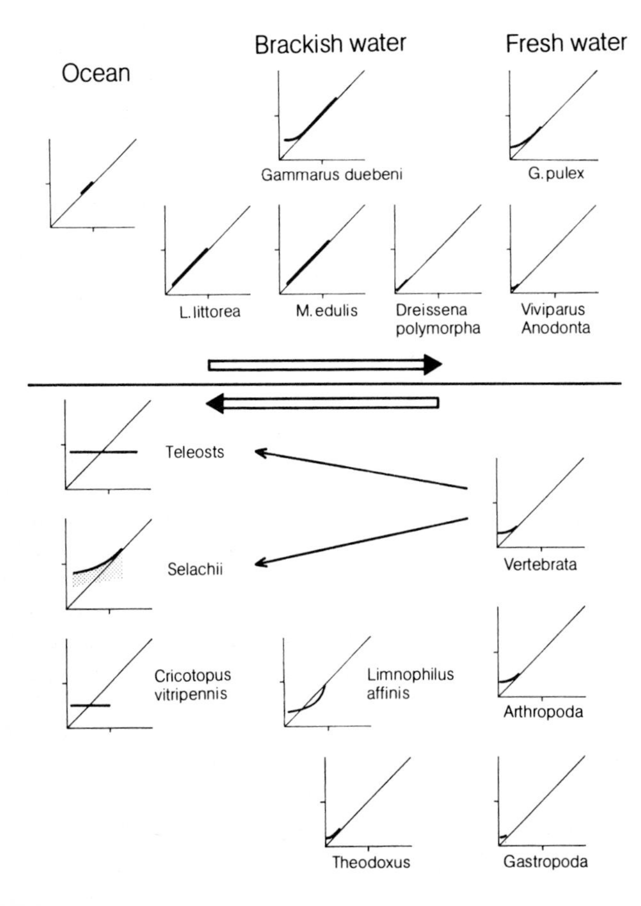

Fig. 12. Phylogenetic migrations between sea water and fresh water. *Above:* from the ocean to fresh water. *Below:* the opposite direction. The graphs correspond to those in Fig. 8; the shaded area in the selachian graph indicates the contribution of urea to the osmotic pressure. (Remmert, 1969)

include marine nematocerans (most species of Clunio) and mites (Halacaridae). These arthropods, which returned to the sea very long ago, exhibit hardly any trace of their old homoiosmotic characteristics; they have secondarily become almost poikilosmotic. Moreover, they have secondarily raised the regulated internal tonicity to such a level that they are isoosmotic in the water of the open ocean, and thus are exposed to only slight physiological stress.

Another route away from the ocean, simpler to understand and not associated with such a diversity of types, is that leading to fresh water by way of brackish water. Salinity does fluctuate in brackish-water regions, but these fluctuations are much smaller than those in the supralittoral. No species travelling this route has developed hypotonicity regulation (Fig. 12). Again, a typical feature is lowering of the internal tonicity; in fact, this is much more pronounced than in terrestrial animals. This progression can be readily followed in species of Gammarus from the ocean, brackish water and fresh water. Soft-skinned fresh-water animals (clams and snails) have the lowest internal osmotic pressures of all animals. Because of this low set tonicity level remigration into the ocean presents difficulties. Bony fish (Teleostei) in the ocean must continually excrete salt; they have developed hypotonicity regulation as an adaptation to marine salinity. They have also raised their regulated tonicity as compared with bony fish in fresh water, but not enough to make salt excretion unnecessary. And this regulation uses up energy, with severe consequences. Fish in the arctic freeze at higher temperatures than does the medium surrounding them. Selachians (sharks and rays) – like Rana cancrivora and Latimeria – have followed the poikilosmotic route. They raise their internal pressure to levels equivalent to those of the environment by retaining urea. This feature is an extremely strong indication of the fresh-water origin of selachians – if, indeed, the high urea production does not actually imply a semiterrestrial ancestry. Fresh-water arthropods adapted to marine life by hypotonicity regulation, acquired in a stepwise manner. A good example of a typical brackish-water arthropod on its way back to the sea is the caddis fly Limnophilus affinis. But only relatively few species returned from fresh water to the open ocean; examples include the bug Halobates and the nematoceran Cricotopus. No arthropod that originated in fresh water has been found to have secondarily raised its regulated tonicity level. The internal osmotic pressure of Clunio, high in comparison to that of other chironomids such as Chironomus and Cricotopus, may well be associated with the terrestrial origin of this genus (Chironomus is discussed further below). Soft-skinned fresh-water animals have achieved this return journey only to a slight extent; the chief examples to mention are the marine rotifers. Nothing is known about how they maintain water balance. The river gastropod Theodoxus has advanced into brackish water up to fairly high salinities; it exhibits a slight increase in poikilosmotic capacity as compared with its limnetic relatives.

The routes of migration, then, can lead from sea to land, land to sea, sea to fresh water and the reverse – and there are many others. We shall mention only one, which involves salt lakes and deserts. Salt lakes and highly saline lagoons in arid maritime regions are colonized from the land, from the ocean and from fresh water. It is not difficult to distinguish animals of marine and limnetic origin on the basis of differences in internal osmotic pressure. Insects that came from land or fresh water develop a marked hypotonicity regulation, though the maintained tonicity is significantly increased. This process can readily be followed within the genus Chironomus, which has progressively invaded small bodies of water of increasing salinity. There are many ways by which animals can migrate from these saline environments to brackish-water regions, where the inventory of species is very similar (Fig. 13).

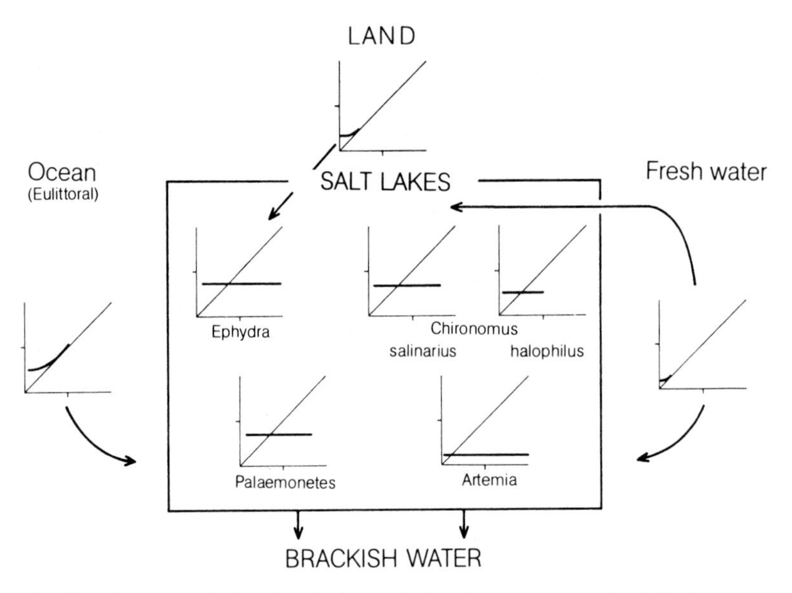

Fig. 13. Directions of migration during evolution, between ocean, land, fresh water, and brackish water via salt lakes. The graphs correspond to those in Fig. 8. (Remmert, 1969)

We have seen that a given environmental factor can elicit quite different physiological responses among animals coexisting in a particular habitat. These responses can be understood only in relation to the ecological history of the animals. The same considerations hold true for ecological factors other than salinity. There is evidently little connection between the development of physiological functions and the phylogenesis of the animals in the sense of evolution of higher forms of life. Rather, physiological functions appear to arise chiefly in relation to the changes of habitat that occur during phylogenesis. This interpretation is generally accepted in the case of excretion of nitrogenous metabolic waste. The complicated processes by which ammonia is converted to larger molecules in land animals are determined by the animals' ecological history. And the same is true of sensory functions. Dolphins evolved the capacity to orient by ultrasound, and it may be that this new adaptation, "economical" in terms of the price paid, made possible their great success as inhabitants of the sea – a habitat where other animals in a more favorable osmotic situation and/or with gill respiration are much better off energetically. One could say that the dolphins can "afford" higher energy consumption because of a "newly invented technological trick."

In this discussion we have emphasized the costs incurred by animals that avoid intraspecific competition by moving to different habitats. The uptake of dissolved organic matter, which plays a decisive role in the nutrition of a great number of marine invertebrates (Schlichter, 1975), is much more difficult under the osmotic conditions in fresh water; indeed, it is impossible (Siebers and Bulnheim, 1976). In migrating to fresh water, marine animals gave up an extremely abundant source of energy-rich and qualitatively valuable food, leaving it to the microorganisms. Furthermore, hypertonicity regulation requires that the water drawn in by osmosis be continually pumped out, and that there be continual active uptake of salts against a concentration gradient. Special organs for salt uptake and water elimination are necessary for existence on land and in fresh water. Energy is expended for the construction and operation of these organs. And there are still other problems. The organs for salt uptake are evidently not so specific that they can recognize precisely the correct ions in the water. They "confuse" heavy metals with sodium and potassium ions, and if uptake of the latter is blocked by heavy metals the animal dies. The same problem arises when animals return to the ocean. Fish and birds and mammals in the ocean must continually excrete salt; they must develop specific organs for the purpose, as well as others to take in water to replace that lost by osmosis. From the viewpoint of energy consumption, fish and birds are inferior to the primary marine animals. As Dollo's Law of the irreversibility of evolution states, animals returning to this intrinsically "economically favorable" environment do not re-evolve the original physiological constitution. It is worth making a cost-benefit analysis to learn how warm-blooded animals have managed, despite these difficulties, to be so successful in the ocean; but no such analysis has as yet been carefully done.

The best solution to the problem of minimizing costs on the land has been developed by the vascular plants. Their roots extend into the moist and mineral-rich soil, their leaves into the dry air. Water evaporates continually from the leaves, generating a flow of sap through the plant, initially always from the roots to the leaves. Sap flow in this direction makes it possible to maintain, without cost, an osmotic pressure within the plant that is relatively high in comparison with the soil, as well as a cost-free influx of water against the prevailing osmotic gradient. Both the movement of materials from the roots to the leaves and osmoregulation, then, are initially brought about in a purely physical way, by means of the differences in the vapor pressures of soil and air and in the structures of the plant below and above ground – all at no cost to the plant. Of course, water consumption is very high. Unlike animals, plants have no internal circulation of water.

In all our discussions of the effects of seawater salinity and osmotic pressure we have been thinking in terms of the normal composition of sea water. The proportions of the different salts are in fact uniform throughout the oceans and in all waters directly connected with the ocean. Saline habitats inland, by contrast, have an entirely different ionic composition. Thus they can usually not be colonized by primary marine forms, which are really dependent on salt and have correspondingly high concentrations of salt in their cells and body fluids. Only secondary marine organisms, not actually dependent on salt but simply capable of excreting it efficiently, can colonize such sites. This fact explains the great differences between the fauna und flora of inland saline habitats and those of marine habitats. A particularly striking characteristic is the absence of most of the true marine algae (only very pollution-tolerant forms can survive in some inland saline sites) and sensitive ocean animals such as all of the marine Cnidaria, echinoderms and molluscs.

The same considerations apply to plants. The primary marine species have tissue osmotic pressures identical to that of the surrounding medium. Plants in fresh water and on land must actively take in salts and protect themselves from the osmotic influx of water. Land plants have exploited this situation as described above; the water simply evaporates, and in the process sets up a current that simultaneously carries nutrients through the plant. If vascular plants return to saline habitats they are in the same predicament as animals, and must eliminate the excess salt. Salt glands develop in such plants, just as in animals (Fig. 14). Land and fresh-water plants are thus from the outset in a less favorable energetic position than marine plants. Adaptation to the new habitat exacts a price. Should these terrestrial plants return to the marine environment, they cannot simply realize the old advantages of life in the sea – they must now bear additional costs. If they are to maintain themselves in

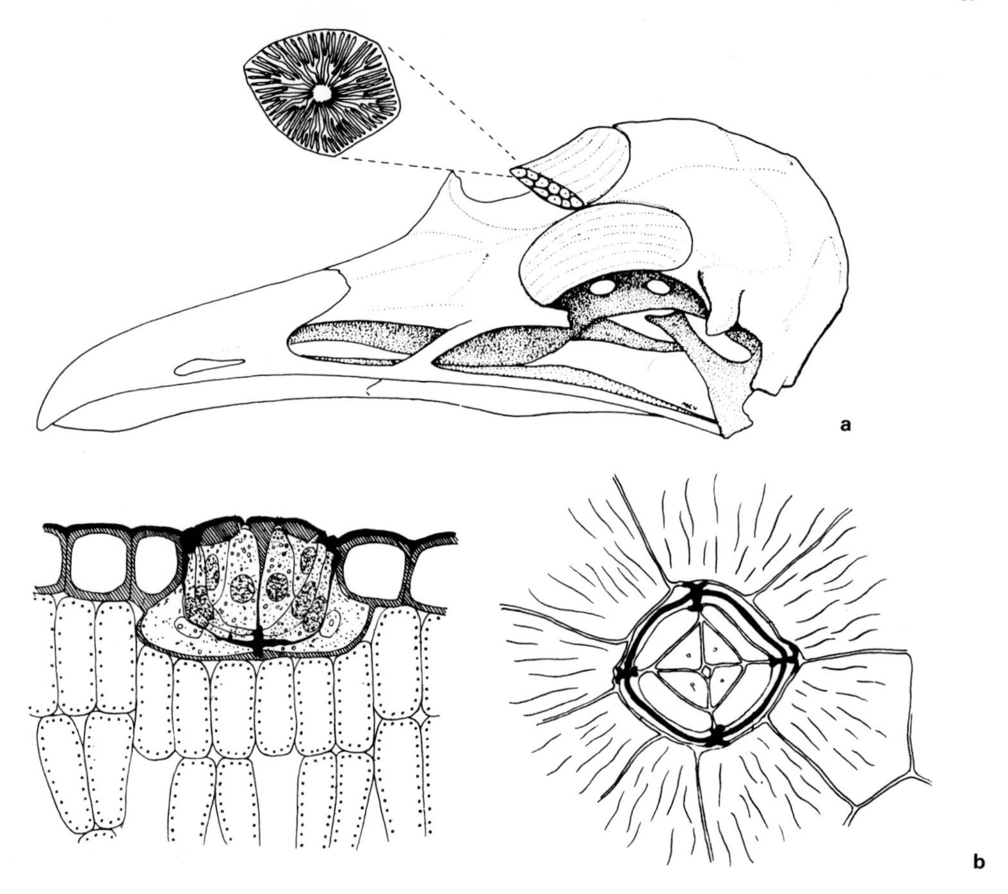

Fig. 14 a, b. Salt glands for the excretion of salt from secondary marine organisms. **a** Salt gland of a gull (Larus) (Schmidt-Nielsen, 1965), **b** salt gland of sea-lavender Statice gmelini (Ziegler and Lüttge, 1966)

competition with marine plants they must become or remain competitive in a different manner.

This sort of optimization in a new habitat can occur in surprising ways. A secondary marine animal which relies on the continual excretion of salt can reduce the amount that must be excreted by selective feeding; when marine fish or birds are taken as food (as they are by many seals and dolphins) relatively little salt is consumed. An animal with such a diet need not secrete salt as vigorously as one that eats primary marine animals. A desert animal that selects, from the very salty plants available to it, the parts containing little salt, is in a favorable energetic position. For example, the kangaroo rat Dipodomys microps of the North American deserts scrapes off the less salty tissues of plants with specially shaped teeth. It consumes only as much salt as a normal herbivorous land animal (Kenagy, 1973). Related rodents make use of the water produced in their bodies when food is metabolized – so efficiently that they never need to drink, despite the dryness of their food (they eat mainly the very dry but energy-rich seeds of plants; Fig. 15).

Land-dwelling plants and animals face special problems with regard to water balance. Because of the evaporation of water into the air, their milieu in some respects corresponds to a highly saline environment in which water is lost due to osmotic effects. We shall now discuss the methods by which water is conserved under these conditions, and consider the question whether

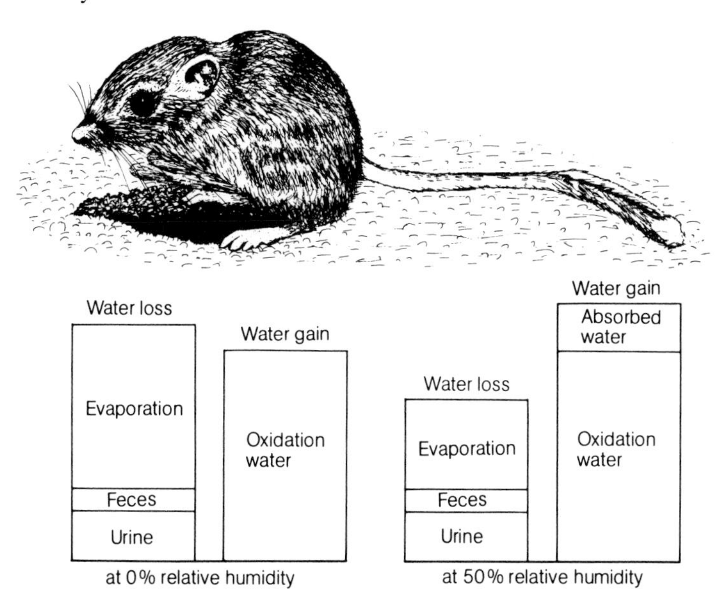

Fig. 15. Water balance in the kangaroo rat Dipodomys spectabilis at different relative humidities. *Left:* at zero humidity; *right:* at 50% humidity, 25 °C. In each case the left blocks indicate the water lost by transpiration and excretion, and the right blocks show the amount of water acquired. At 50% humidity the plant seeds on which the rat feeds take up water that can be utilized by the rat (absorbed water). The major fraction of the water required is supplied by oxidation within the body. European rats lose much more moisture in their urine and feces and by transpiration. (Schmidt-Nielsen in Tischler, 1977)

it is possible to make use of the moisture in the air by taking it in directly. It will become apparent that in the last analysis water balance is merely a special case of salt balance and osmoregulation. The problem of moisture, which occupies a central place in textbooks on terrestrial ecology, must be given less emphasis in a general presentation that includes marine ecology. It is therefore treated as a subtopic in this chapter.

First let us consider the strategies by which animals adapt to differences in relative humidity. Simple model experiments and theoretical considerations show that a small sphere of gelatine loses more moisture per unit time than a larger sphere under the same conditions. The small sphere soon dries up; the larger one retains moisture for a longer time. As a strategy of adaptation to low-humidity regions, then, we would expect to find increase in the size of animals. But this increase brings a disadvantage, in that larger animals take longer to reach sexual maturity. Their reproductive rate is fundamentally lower than that of a small animal. Thus in regions where there is little or no danger of desiccation, smaller animals have a selective advantage over larger ones. From the cold deserts of the arctic to the arid deserts of the tropics, there is a distinct increase in mean body size of the insects sampled (small insects, if present, are active only at night, when humidity is high). On the other hand, the number of individuals per unit area is greater in cold deserts than in hot deserts. What has been said of deserts applies similarly to other habitats. Humid-cool regions are occupied, on the average, by smaller insects than dry slopes; maritime regions, such as the North Sea coast and the British Isles, on the average have smaller insects than continental regions like eastern Prussia or Hungary.

But there is a catch in this arrangement – larger insects must pass through juvenile stages small enough to be in danger of desiccation. The danger is particularly great in arid regions. Many of the animals living here provide protection for their young or, by actual brood-care behavior, shift the emergence of the youngest stages to a time when the humidity is relatively high. They actively select particularly favorable (i.e., moist) elements in the habitat. The youngest and thus smallest stages of crickets stay in the more moist places, near the bases of the plants; mature crickets prefer the open areas between the plants. A similar shift of locale occurs during the bug life cycle; from their original post at the bases of plants, the young stages migrate to the exposed tips. Another way to escape desiccation is to adopt the nocturnal habit. Mammals lose more water by evaporation when the air temperature is higher than that of their bodies. Under such conditions, they must allow water to evaporate in order to keep the body cool. Some desert mammals save water by doing without such cooling; they are able to tolerate an increase in body temperature when the air temperature is very high. The same species preserve the water ordinarily lost in respiration, by retaining it in the nose with a specific cooling mechanism.

On the other hand, remarkably few organisms have developed the ability to withdraw water vapor from the air. Such behavior is known only among a very small number of animals (book lice) and lower plants. No vascular plants can do it, but many of them can utilize water suspended in the air as droplets – fog, for example. This occurs in the high deserts of the Andes.

Very high humidity is unfavorable to terrestrial organisms in all temperature zones. It is obvious that plants should suffer, for they depend on transpiration. But we still do not know why grasshoppers, crickets, butterflies, beetles, and their larvae show practically no growth or development at relative humidities above 80%.

2. Temperature

No environmental factor seems so easily measurable, and there is no factor of which we are so readily aware, as temperature. Because of its accessibility, a chaotic mass of temperature data has accumulated, most of which is unrelated to the ecological context within which it was published. In fact, it is difficult to measure temperature at the site of importance to the organism – in the organism itself; the temperature here (in the case of land organisms) bears little relationship to the meteorological temperature. The latter can serve only as a rough estimate. Accordingly, it is considerably harder to establish a relationship between the distribution of an animal or plant and the temperature than, for example, to learn how distribution depends on the salinity of the available water.

As an illustration of the difficulty of such judgements, consider the following example. In Bavaria, the part of West Germany with the most continental climate, the nightingale is found in only a very few places where the climate is especially mild. It is of common occurrence in many other parts of Germany. About April 20 the nightingale returns from its wintering grounds; it cannot tolerate nighttime frost, and hard frosts are generally to be expected in Bavaria until the beginning of May. By contrast, the hoopoe is a regular inhabitant of many locations in Bavaria. It requires considerably higher temperatures than does the nightingale. It arrives later (not before the beginning of May), when the night frosts in general have stopped and the approaching continental summer brings very warm days and very small amounts of rain. In most other parts of Germany (apart from Baden-Württemberg) the hoopoe is considerably less common, because the summer temperatures are lower and the precipitation greater. It is not clear whether the influence of temperature here is direct or is exerted by way of another factor such as food supply.

Temperature affects all chemical processes as formulated in Van t'Hoff's Law: a 10 °C increase in temperature accelerates a chemical reaction by a factor of 2–4. We say that this chemical reaction has a Q_{10} of

2–4. The biochemical reactions of organisms are naturally subject to this law. Given that the temperature can fluctuate over a wide range during the course of a day, it is understandable that during evolution all organisms have developed mechanisms that liberate them, to a greater or lesser extent, from this temperature dependence. The warm-blooded animals have gone furthest in this respect, and will be discussed separately. The ectothermic organisms, with body temperatures that can change over a wider range (microorganisms, plants, and poikilothermic animals), seem at first glance to be obliged to follow even the most erratic changes in environmental temperature. But in fact they too have developed a lavish array of regulatory mechanisms, and in many respects even these organisms are evidently unaffected by temperature.

The most illustrative example is the physiological clock of plants and animals, which runs with a period of about 24 h regardless of the prevailing temperature. It is self-evident that this should be so; a physiological clock that ran faster at high than at low temperatures would offer no advantage in the struggle for existence and would therefore never have become established in evolution. But this example demonstrates that temperature compensation is possible even in ectothermic organisms. If we want to know more about the general aspects of such compensation, we must ask how widespread these phenomena are and explore the underlying physiological mechanisms.

It is astonishing that true compensation seems never to have become a property of the processes of development. All ectothermic organisms appear to develop in dependence on Van t'Hoff's Law; an example is given in Fig. 16. And this dependence is even more extensive than is evident in the figure. Organisms adapted to low tempera-

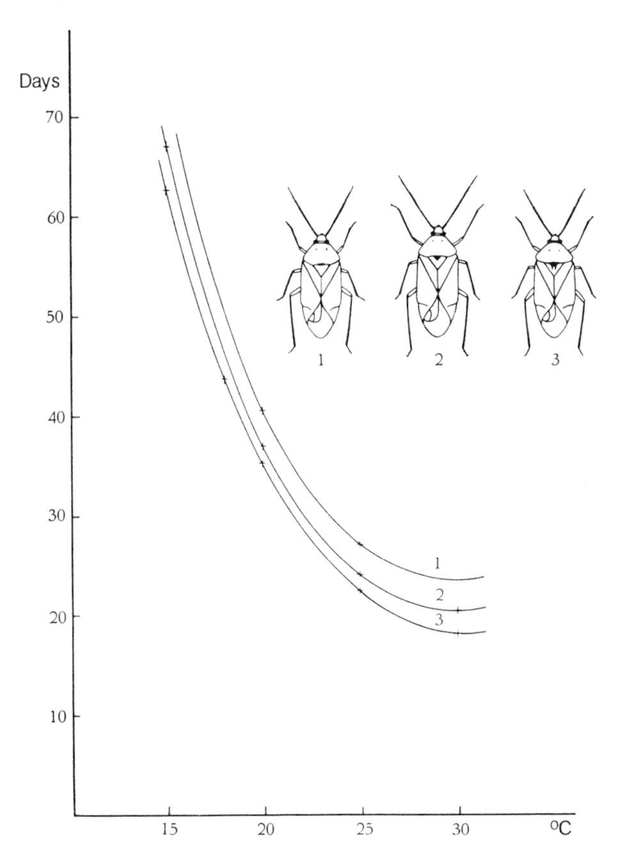

Fig. 16. Temperature-dependence of the duration of development of three species of the bug Lygus (*1* maritimus; *2* pratensis; *3* rugulipennes). (Boness in Tischler, 1965) Growth and temperature are related identically, in principle, in all heterotrophic ectothermic organisms. The product of duration of development (T) and the temperature (t in degrees above the null point t_0 for development of a species) is constant (the thermal constant K); $T(t-t_0)=K$. The null point for development can be computed from experiments at two temperatures, by the formula

$$t_0 = t_1 - \frac{T_2(t_2 - t_1)}{T_1 - T_2}$$

(Boness in Tischler, 1965)

tures can develop at low temperatures, but their development occurs no more rapidly than would that of a tropical organism if it were able to develop at that temperature. That is, adaptation to permanently low temperatures is bought at the cost of extending the period of development. The stoneflies and mayflies that live in mountain brooks need a whole year to grow to the size of a housefly; but a housefly could not grow at all at such temperatures. The antarctic fish Trematomus lives in an environment with a constant temperature of −1.6 °C and requires about 10 years to reach the size of a small trout. And there are still other costs involved. An animal that has adapted to low temperature and is paying for this adaptation with a prolonged period of development at the same time loses the ability to exist at higher temperatures. All the mayfly larvae, stoneflies, and amphipods indigenous to mountain brooks die if exposed to the temperature that any pond will reach during the summer, about 25 °C. Typical winter animals such as the snow scorpionfly Boreus cannot tolerate temperatures around 20 °C for very long. The price for this adaptation, then, is high. The best proof is given by the fact that we find refrigerators useful. If during their millions of years of evolution microorganisms had developed vital functions that could proceed as rapidly at low as at high temperatures, we would not be able to keep food in simple refrigerators. From the outset, then, we would hardly expect to find truly temperature-compensated processes of development in ectothermic organisms. It is still not entirely clear why organisms were able to achieve independence of temperature in many functions, but never in those of growth and development.

Comparative biochemical analyses have led to a current hypothesis that can be expressed in simplified form as follows:

The normal enzymes of an organism, as discussed on p. 27, are adjusted to operate best at temperatures in the range from 28 °C to just over 30 °C; here they are most effective and last for a long time.

There are enzymes that can operate just as effectively at higher or lower temperatures as the "normal" enzymes do at 28°–30 °C. However, these enzymes survive only briefly; they soon break down and must continually be resynthesized. The possession of such enzymes costs energy.

Functions involving these special enzymes can proceed at temperatures higher or lower than normal at the same rate as in the normal temperature range.

However, because of the high energy requirement, it is not possible to provide all functions with such enzymes; there must be a system of priorities. In general, sense organs and organs necessary for flight reactions (muscles) tend to take precedence over organs subserving metabolism (and hence growth).

This fundamental hypothesis has found wide acceptance, but cannot as yet be considered sufficiently well documented to be given the status of a theory.

Finally, by van t'Hoff's Law, the curves we find for temperature dependence are exponential (Figs. 16 and 21). That is, very slight temperature differences can have very large or small effects. We can imagine a species of which most individuals cannot reach maturity during a normal year; their number decreases from year to year, until eventually one of the rare very warm summers makes it possible for all the individuals to complete development. Now the size of the population leaps suddenly from very small to very large, so that the species can persist during the following normal years. We shall consider such an example in more detail on pp. 164ff.

Of course, one must bear in mind that it is not the temperature of the air but that of the organism itself that determines the rate of development. Body temperature can depart considerably from air temperature. In flight, the temperature of all the larger insects rises to more than 35 °C because of the activity of the flight musculature. Bees can warm up their hive in this way; bumble-bees, by beating their wings, keep their brood warm so effectively that the species can invade even arctic regions. Furthermore, a great many animals can heat themselves in the sun. Grasshoppers, in the cool of the morning, position themselves broadside to the sun's rays and as a result quickly become very warm. During the midday heat they turn their heads to the sun so as to present the smallest possible surface, and absorb relatively little heat. This sunbathing behavior may well be an absolute necessity for the survival of many insects in temperate and cool regions. It is characteristic of red ants when they first leave the nest in spring, of many lepidopterans in arctic and subarctic regions, and probably of a number of insect larvae. Redbugs (Pyrrhocoridae) in spring, and later their larvae, use the sun's radiation to regulate their body temperature. Usually they take up a position at the foot of a tree to sun themselves. But if forced to stay there permanently, so that their body temperature is always relatively high, they develop too rapidly; the synchronization between stage of development and time of year is lost, and the animals die. Under natural conditions the bugs move back and forth between sunny and shady places, adjusting their movements to the changing conditions through the year.

On the forest floor in Denmark, the development of the large brown weevil Hylobius abieties generally takes three years. But in clearings, where the ground is heated by the sunshine, development is complete in two years. As a consequence, the damage this beetle does to coniferous forests is considerably greater where clear-cutting is practised. The difference in duration of development corresponds to a southward shift of about 1,500 km (Bejer-Petersen, 1975).

It appears that all ectothermic land animals begin to lose water when their body temperatures reach 34 °–35 °C, and water loss increases sharply as the temperature continues to rise. This is probably a cooling mechanism (Jakovlev and Krüger, 1954; Jakovlev, 1956). In experiments on Mediterranean crickets the body temperature could not be raised above this level even by irradiation, whereas similarly irradiated dead animals reached a temperature of about 45 °C within two minutes. Many of the physiological parameters that have been measured (transpiration, respiration, the enzyme activity of various organs) indicate that in terms of the biochemistry of their enzyme complement all the eurythermic animals are adapted to an optimal temperature of around 27 °C (Fig. 17). In many cases reptiles, too, can invade cold regions only because they use the sun's radiation to accelerate development. The adder and the lizard Lacerta vivipara have

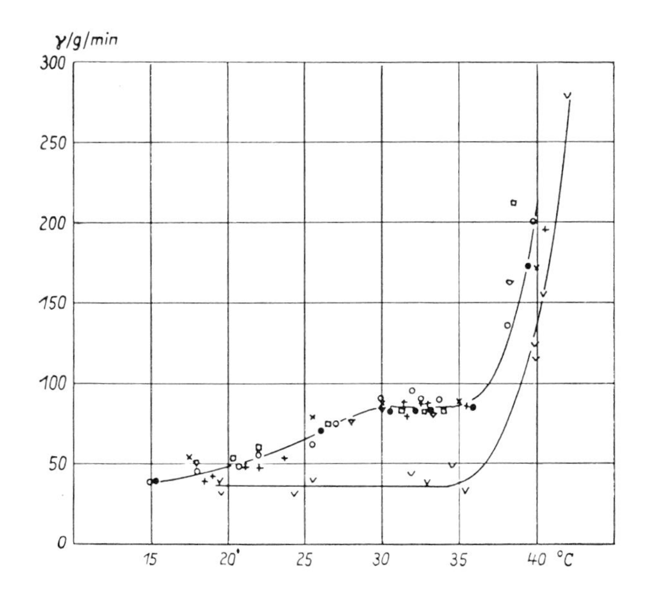

Fig. 17. The relationship between transpiration and temperature in grasshoppers. ∨ = Oedipoda coerulescens; the other symbols indicate other species. Ordinate: water vapor released per unit time; abscissa: temperature. (Jakovlev, 1956)

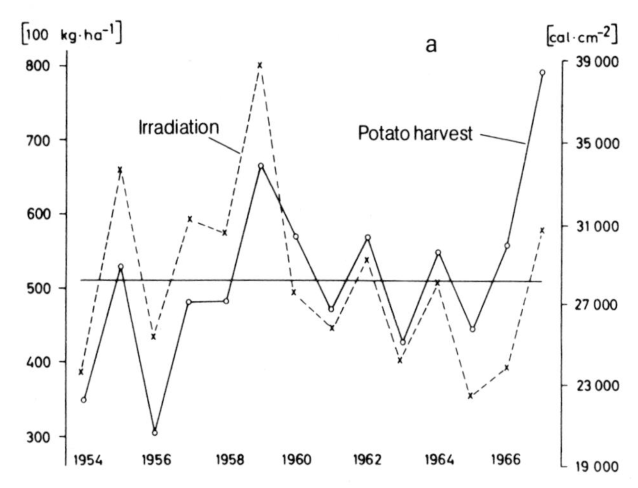

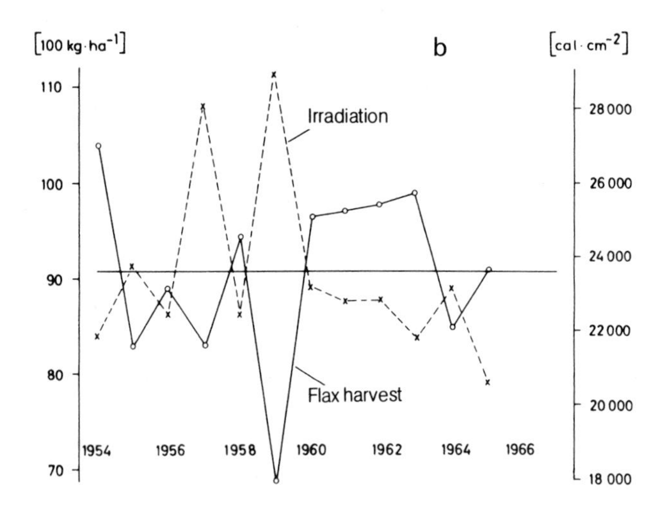

Fig. 18 a, b. Relationships between yield and irradiation during the growing seasons at an experimental station in the Netherlands. **a** Potatoes, **b** flax. Growing period: 3 weeks after sprouting until harvest. (Sibma in Baeumer, 1971)

advanced further north than any other reptiles. Neither lays eggs; they bear living young. During the day they let their bodies get very hot in the sunshine, and the unborn young are heated as well.

Plants can be injured if they are overheated by the sun (Fig. 18). Under intense irradiation the dark bark of a tree can become very warm – a danger both in winter, when a sharp temperature gradient is produced between the lighted and shaded sides of the tree, and in summer, when the heating can be sufficient to interrupt sap flow. Injury so caused is evident at the edges of clearings and along roads, especially in beeches. Green leaves are not heated very much by radiation; furthermore, their temperature can be brought appreciably below that of their surroundings by transpiration (Figs. 19 and 20). The same is of course true of animals; by high transpiration rates and specific behavioral mechanisms ectothermic animals can also become cooler than their surroundings. When desert birds bury their eggs in sand, in many cases it is done to protect them from heat. For the same reason insects, spiders, and lizards seek out shady places.

It is evident, then, that the temperature of an organism can fluctuate even more widely than the meteorological temperature. On a sunny day it can mount very rapidly, to plunge within minutes when a cloud passes. Even on cool spring days

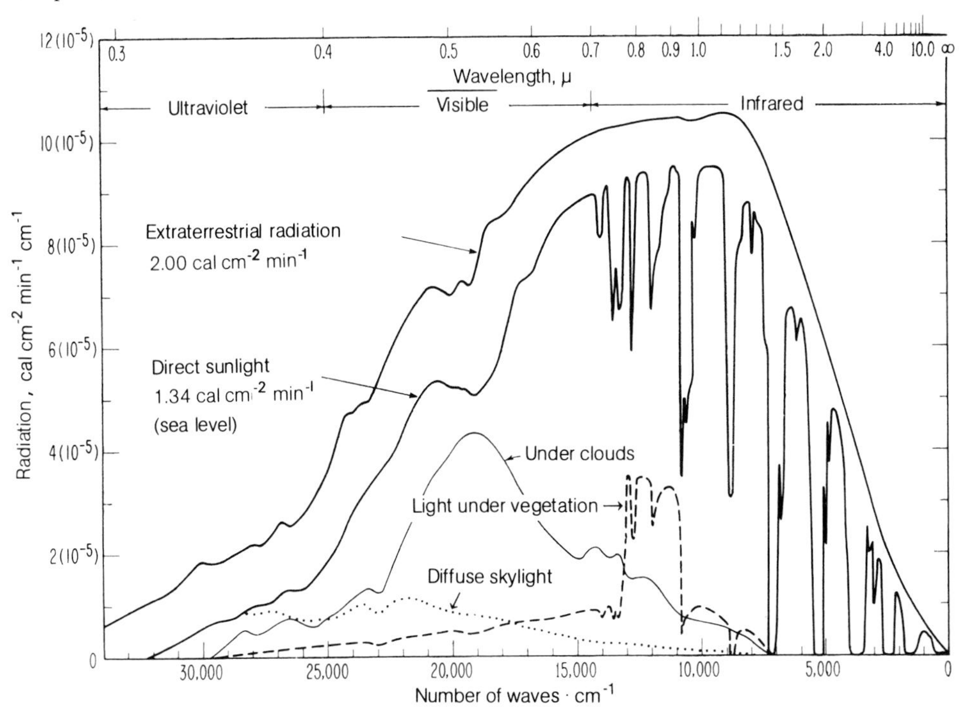

Fig. 19a. Spectral distribution of the extraterrestrial solar radiation and the solar radiation on earth during clear weather, when the sky is overcast, and under plant cover. (Gates, 1965)

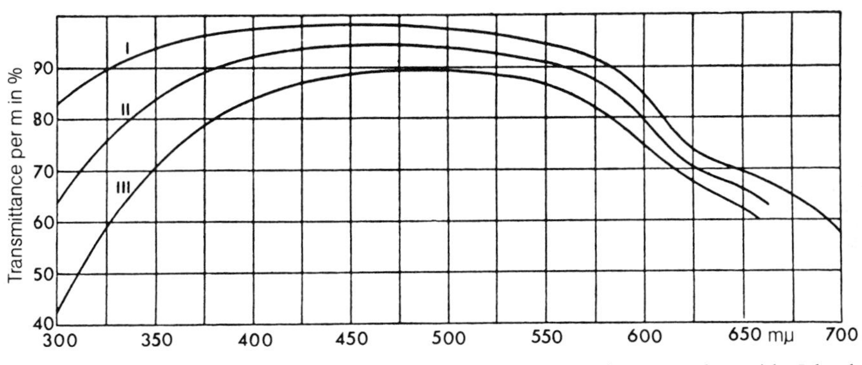

Fig. 19b. Transmittance of sea water to light of different wavelengths; sun at the zenith. *I* the clearest ocean water; *II* subtropical and tropical ocean water of high turbulence; *III* ocean water at temperate latitudes. In arctic waters the transmittance is probably still lower. (Hedgpeth, 1957)

sunbathing butterflies and lizards can achieve body temperatures above 35 °C, whereas at night they are chilled by frost. Can we make direct comparisons between such changing temperatures and what we know about the effects of constant temperatures? If we know the time (in hours or minutes) during which an organism is exposed to a particular temperature, and its rate of development under constant temperatures, we should be able to calculate how long it would need for development under a particular changing-temperature regime. As early as 1928 Kaufmann made such calculations. His conclusions have been confirmed recently by a large number of experiments. The calculations are complicated by the exponential relationship

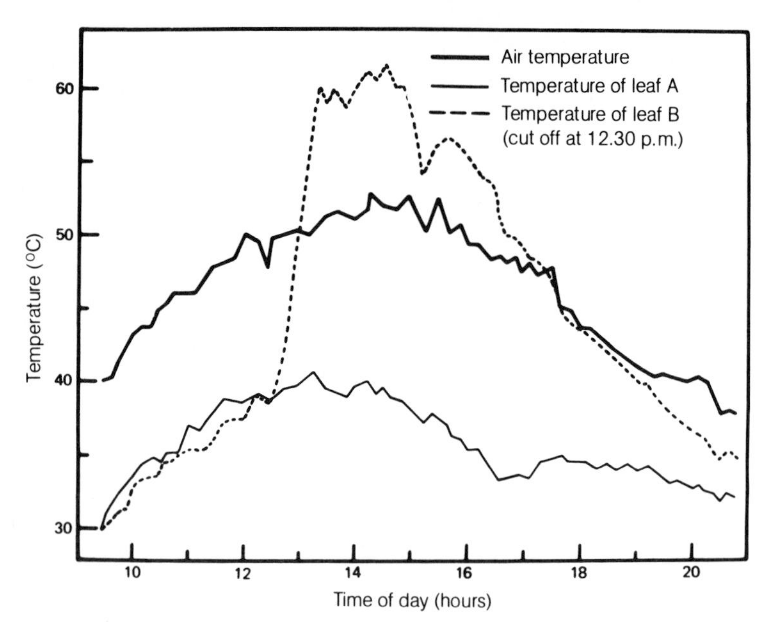

Fig. 20. The temperature of a leaf can be well below that of the air, as a result of transpiration and the associated cooling. A cut-off leaf is rapidly heated by sunlight, because of lack of water. (Lange in Gates, 1975)

between growth and temperature, and by the existence of a threshold – a null point for development. Because of these features, growth and development in general proceed more rapidly under fluctuating temperatures than in a constant temperature equal to the mean, and the difference is most pronounced at low temperatures (Fig. 21). However, there are a few notable phenomena that have not yet been satisfactorily explained.

1. The productivity of many plants is increased when the temperature fluctuates (Fig. 22; of course, only the parts above ground are exposed to the full fluctuations). Perhaps low nighttime temperatures act to prevent high respiratory losses. Such an explanation could also apply to the diurnal vertical migrations of planktonic animals, which when they are not feeding sink to the colder levels in the water.

2. In many of the ectothermic animals that have been studied the rate of reproduction is greatly increased in fluctuating temperatures (these include the rotifer Brachionus and various terrestrial insects; Fig. 24). Parasitic wasps of the genus Trichogramma raised under fluctuating-temperature conditions have appreciably greater effects, when they are set free for the control of the host insects, than those raised in constant temperature (Stein, 1960 a, b).

So far it has not been possible to decide whether such examples represent a general rule, nor do we know enough about their physiological bases. In view of the ecological significance of such an increase in the production of organic matter or of eggs, this phenomenon definitely deserves further study.

Finally, temperature almost always acts in concert with wind and moisture (rain). Their effects are practically inseparable. Indeed, it seems almost miraculous that we are able to say anything at all about the influence of temperature on animal distribution. In the area of forest entomology it has proved useful to construct climate curves based on both precipitation and temperature. Merkel (1977), for example, used such climate curves to study a bark beetle. Temperature was automatically recorded on an hourly basis; the times when

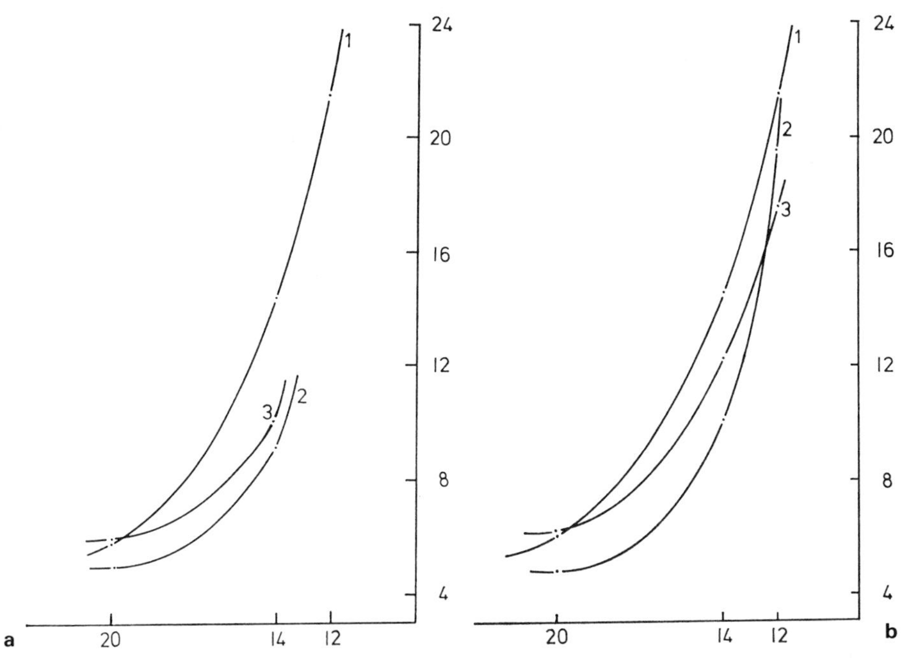

Fig. 21 a, b. Time elapsed from emergence of the imago to the first egg-laying by Drosophila sp., as a function of temperature. Ordinate: time in days; abscissa: temperature. *1* constant temperature; *2* temperature fluctuating about the corresponding mean during the day; *3* computed curve for the fluctuating temperature. **a** Amplitude of the temperature cycle 8 °C, **b** amplitude 5 °C. At the lower temperatures development takes less time when the temperature fluctuates than at the corresponding constant temperature. (Remmert and Wünderling, 1970)

it exceeded 7 °C, the null point for movement and feeding by the beetle, were added and the effective temperature sum per day and month was calculated. Plotting these sums on the x axis of a coordinate system and the sums for precipitation on the y axis

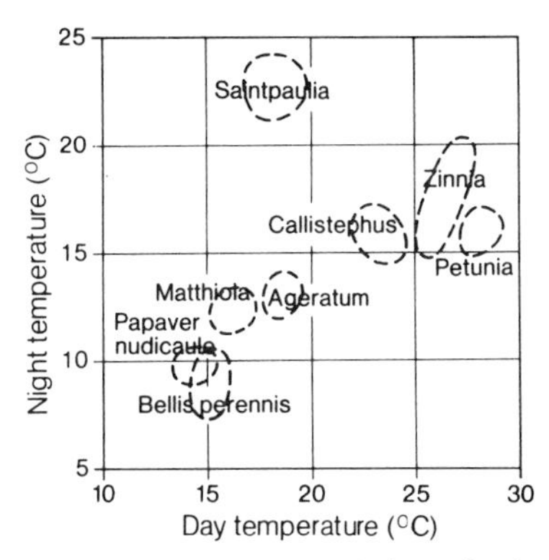

Fig. 22. Many ornamental plants develop best when the night-time temperature is below that during the day. The most favorable temperature difference is about 5°–10 °C. The African violet is an unusual exception. (Geisler, 1971)

gives the climate curve (Fig. 23). When there is heavy rainfall the curve rises steeply, and when the weather is dry it is nearly parallel to the x axis. By entering observations of stages in the life of the beetle in this graph, one obtains data specific to the measurement site which can be compared with results similarly obtained at other locations or in other years. Using this method, researchers have developed considerable insight into the dependence of insects on the climate in their habitats. But even this procedure is not entirely satisfactory, because depending on the general state of the weather the beetles can require different integrated temperatures. For example, the large bark beetle Ips typographus does not appear until the temperature at the site where it has spent the winter rises above 7 °C. By this time the surface of the soil has reached 10°–20 °C. The beetles then begin feeding under the bark; to do so, they require temperatures

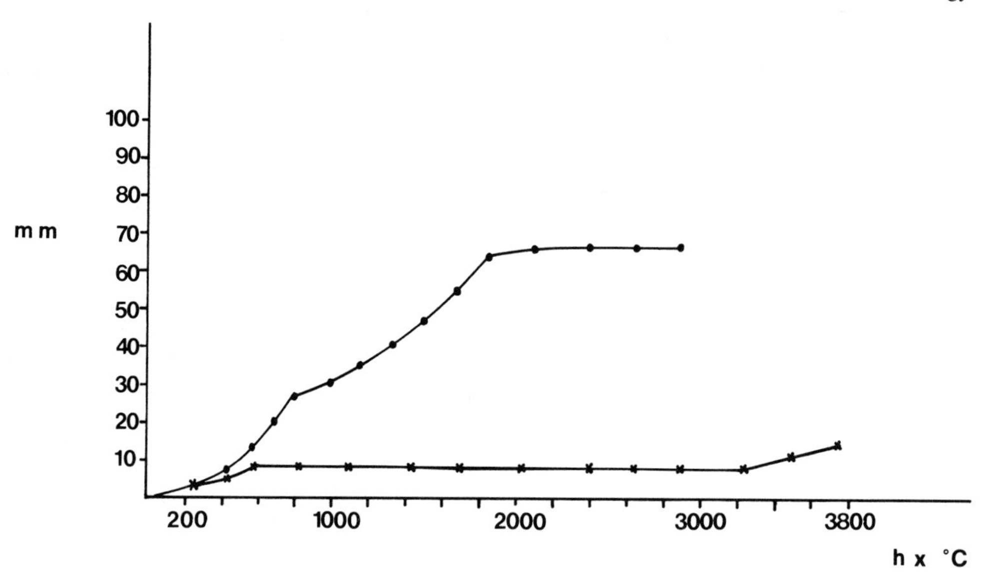

Fig. 23. Example of a way in which the climate at two different sites can be described. Ordinate: precipitation in mm; abscissa: integrated temperature (hours × °C). The curves are cumulative, with the months plotted in succession. In the first month the two curves coincide (the weather was the same at the two sites); during the following months one curve rises sharply (rain and low temperatures) whereas the other is flat (dryness and warmth). To make the differences more distinct, one can take as a reference level the null point for development of a species

between 12 ° and 19 °C. The temperature sums necessary for maturation of the gonads vary depending on the weather. The breeding flight of the mature beetle can occur only if the air temperature is at least 20 °C. The body of the animal must have reached at least 23 °C, which can happen at an air temperature of 20 °C if the weather is sunny but requires 23 °C air temperature under overcast skies.

Another approach to the difficult problem of temperature in terrestrial habitats is to determine the effective mean temperature (eT) by the method of Pallmann et al. (1940). This procedure is based on the fact that sucrose in aqueous solution is broken

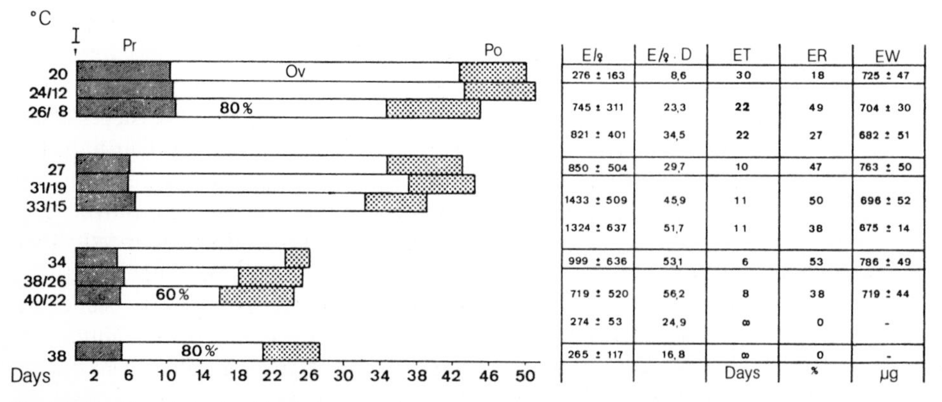

E/♀	E/♀ · D	ET	ER	EW
276 ± 163	8,6	30	18	725 ± 47
745 ± 311	23,3	22	49	704 ± 30
821 ± 401	34,5	22	27	682 ± 51
850 ± 504	29,7	10	47	763 ± 50
1433 ± 509	45,9	11	50	696 ± 52
1324 ± 637	51,7	11	38	675 ± 14
999 ± 636	53,1	6	53	786 ± 49
719 ± 520	56,2	8	38	719 ± 44
274 ± 53	24,9	∞	0	-
265 ± 117	16,8	∞	0	-
		Days	%	µg

Fig. 24. Fertility of Gryllus bimaculatus under conditions of constant and diurnally fluctuating temperatures. *I* imaginal molt; *Pr* pre-oviposition period; *Ov* oviposition period; *Po* post-oviposition period; *E* number of eggs per female; *E/D* number of eggs per animal per day; *ET* time until emergence of the larvae; *ER* emergence rate; *EW* weight at emergence. (Hoffmann, 1974)

down by hydrogen ions to form glucose and fructose. In a buffered solution, with constant hydrogen-ion concentration, the reaction is temperature-dependent. The degree of sucrose inversion can readily be monitored polarimetrically, providing a convenient measure of the temperature situation in a particular period of time. The advantage of the method lies in the fact that the reaction rate, like the growth processes of ectotherms, rises exponentially with temperature. Sucrose inversion ought therefore to give a very precise measure of temperature-dependent biological processes. But in spite of this obvious advantage the method has not been widely adopted, no doubt for fear of overgeneralization. That is, there are no really well documented relationships between organic life and the temperatures recorded in this way. Exemplary studies in which this method is compared with others would be most informative.

In water things are simpler. A general heating by the sun's radiation is impossible because infrared light does not penetrate the water far enough. It is almost completely absorbed in the upper millimeters, which may become distinctly warmer than the rest of the body of water. If black bodies (heat-absorbing objects) are present in this uppermost layer they become quite warm. By this means Aedes larvae can develop very rapidly in the spring, even in ponds still partly covered by ice.

Only one ecologically fundamental process is clearly not covered by the above considerations – photosynthesis in green plants. Photosynthesis is not a simple biochemical process; to a considerable extent it involves photochemical events and is thus not so dependent on temperature. The Q_{10} of photosynthesis is less than 2, distinctly lower than that of biochemical reactions in general (with a Q_{10} of 2–4). Hence photosynthesis can operate at low temperatures, and at higher temperatures does not increase to the same degree as respiration or any of the functions of animals and microorganisms. Figure 25 illustrates a

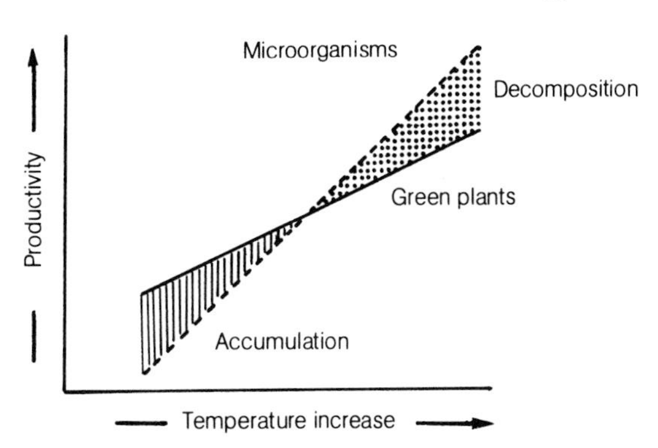

Fig. 25. The influence of temperature on production by green plants and on the processes of decomposition by respiration in animals and microorganisms. In a tropical warm-moist climate plant matter is decomposed more rapidly than it is synthesized; the layer of humus on the ground is but a thin film covering the mineral soil, and in a few months it can be completely eroded. In cool regions the humus layer is potentially much thicker. (Beck in Remmert, 1973; for documentation of the relatively low temperature dependence of photosynthesis see, e. g., Ehleringer, 1978)

basic ecological phenomenon: in the moist, warm tropics dead organic matter is rapidly decomposed, whereas in cool regions decomposition occurs slowly. However, the production of organic matter by the green plants is not very different in the two locations. This fact was remarked upon by Darwin in his book about the voyage of the Beagle. In comparing the jungles of the Amazon and of Tierra del Fuego, he writes that the tropical rainforests appeared to him as a symbol of life, full of power and growth; the forests at the southern tip of the continent seemed like a symbol of death, full of dead tree trunks, branches and twigs. Forests with dead trees are characteristic of temperate and cool zones, because at low temperatures the dead wood remains intact for a long time. In a tropical rainforest fallen trees decay so rapidly that they are hardly noticeable. From the different slopes of the two lines in Fig. 25 we would also predict that plant productivity at consistently high temperatures cannot be much higher – and in certain circumstances may be even less – than at temperatures in the intermediate range. At high temperatures the steady losses due to respiration are so large that the balance between photosynthesis and respiration may show a deficit. Again, of course, the temperatures of interest are not those in the meteorological reports. If the plants have access to enough water, massive transpiration can hold the leaf temperature well below that of the surrounding air. Warming by heat radiation, such as occurs in animals, plays no role in the leaves of plants; essentially all the infrared radiation passes through the leaf (Fig. 19) or is reflected by the chlorophyll.

So far we have been considering only the period in which plants and animals are active. In climates with a cold winter the organisms face the problem of surviving freezing temperatures – and sometimes temperatures well below the freezing point. Many animals can avoid exposure by moving to parts of the water or soil that do not freeze, but many others cannot do this; in the permafrost regions of the arctic such a strategy is entirely impossible.

Very few organisms can tolerate freezing of their body fluid or of their cells. At the lowest temperatures which they can survive, nearly all organisms manage to keep their internal milieu in the liquid state. This fact, together with the photochemical nature of photosynthesis, explains the ability of green plants to assimilate carbon dioxide with a positive balance even in winter. Cold-weather assimilation has been demonstrated in the coniferous forests of the taiga and even more strikingly in antarctic lichens, but it also occurs elsewhere – for example, in the high-mountain plant Ranunculus glacialis. How is this prevention of freezing achieved? The organisms produce substances that lower the freezing point, and accumulate them in their body fluids and cells. A particularly well-known antifreeze agent in animals is glycerol, those of plants include the sugar hamamelose. There are a number of chemically similar substances that tend to produce the same effect. Because of these, some beetles can tolerate temperatures of −80 °C or lower. When an organism is so well adapted it would seem irrelevant whether the winter temperatures stay above zero or fall to −20° or −30 °C. In fact there is an additional complication; at temperatures between the freezing point and about +10 °C ectothermic animals and plants lacking green parts use up a great deal of energy in respiration; in the temperature range 6°–10 °C many animals can even more actively about. But only predatory organisms can, under certain conditions, cover these losses, since the nutrients in animal tissue are readily utilizable. Herbivores and detritus feeders cannot eat enough to replace the lost energy at those temperatures. It is considerably better for them if the winter temperatures remain well below the freezing point, so that none of their stored energy is used up. The paucity of species in maritime regions (for instance, Schleswig-Holstein or the British Isles) is in part explainable on this basis; the winter is not warm enough for feeding, but not so cold that animals can retain all of their stored energy. In this regard continental climates are more favorable. The great variety of insects in the northern United States and central Russia is partially due to this effect. On the other hand, the very warm summers of continental regions are naturally more favorable to ectothermic animals than the rather cool summers of maritime regions with the same average annual temperature (Fig. 26).

Bony fishes in the ocean face a special problem in surviving the winter. As mentioned previously, their osmotic pressure is lower than that of their surroundings. The water in the open ocean freezes at a temperature of −1.7 °C, whereas a fish would freeze at only −1 °C. Here we have yet another price paid for adaptation to a new habitat. Arctic and antarctic fishes solve the problem by moving to deeper levels during the winter or by raising the osmotic pressure of the blood to approximately that of the medium. In so doing, however, they evidently enter a sort of dormant state in which they take hardly any food. Species

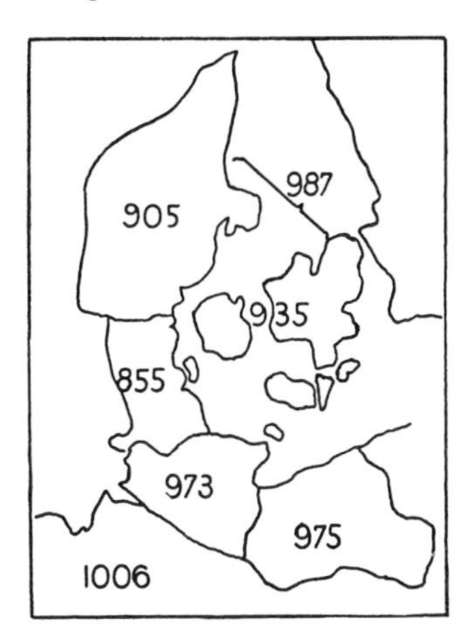

Fig. 26. The number of species of wild flowering plants in northern Germany and Denmark; it is lowest in the region with the most pronounced maritime climate (Emeis, 1950). The regions with many species have areas of continental climate, so that they also include continental species

less well adapted in some situations – for example, when ice crystals are stirred into deep water by violent storms – die in massive numbers.

Warm-blooded animals are in a quite different situation. These are organisms with a greatly elevated metabolic rate and concomitant high energy consumption and heat production. The heat is a side product of metabolism which may be given off from the body or, with suitable insulating mechanisms, can be retained. By changing the insulation properties of the body surface as environmental conditions change, such an animal can achieve a constant body temperature between 35° and 42 °C, depending on the species. When the external temperature is very high cooling mechanisms involving the evaporation of water come into play. As a result the animals are largely independent of the surrounding temperature; they can remain fully active even at low temperatures and colonize regions from which most ectothermic animals are excluded. Such regions include the antarctic continent and, to a considerable extent, the tundra of the far north (around latitude 80° N), where the temperature is so low and solar radiation so slight that ectothermic herbivores are practically negligible. The price the warm-blooded animals have paid is that they require much more food than ectothermic animals of equal size. This requirement is in fact entirely a consequence of their temperature. Extrapolating the energy consumption per unit time of a crayfish at 20°–39 °C, one obtains precisely the energy consumption of a warm-blooded organism that weighs the same. But the advantages of warm-bloodedness are easy to see. Food that is hard to utilize fully (all plant matter!) is processed a good deal more successfully in the warm digestive tract of such animals (in some cases with the aid of specially adapted microorganisms) than by any ectothermic animal. A warm-blooded herbivore thus exploits its food more thoroughly than does an ectotherm. On the other hand much of the energy obtained is lost in maintaining the necessary temperature. In effect, then, an ectothermic organism produces more animal matter than a warm-blooded animal does, for a given amount of plant matter consumed.

It is not surprising that attempts have been made to reduce this high energy requirement. The technique of insulation has been brought to perfection by some organisms; arctic animals consume hardly any more energy when the environmental temperature is low than when it is high (Fig. 27). But warm-blooded animals, like ectotherms, still face the problem of minimizing the loss of energy during the least favorable seasons. A warm-blooded animal that simply took shelter when the weather was bad would lose so much energy because of its ongoing high metabolism that it could survive only briefly. The conspicuous phenomenon of bird migration, and the comparable migrations of bats, show one way out of this dilemma. Another solution is periodically to reduce body temperature, as hibernators do. In the dormant state these animals can last out the unfavorable season without much loss of energy – but they lose the advantage of warm-bloodedness, that of being con-

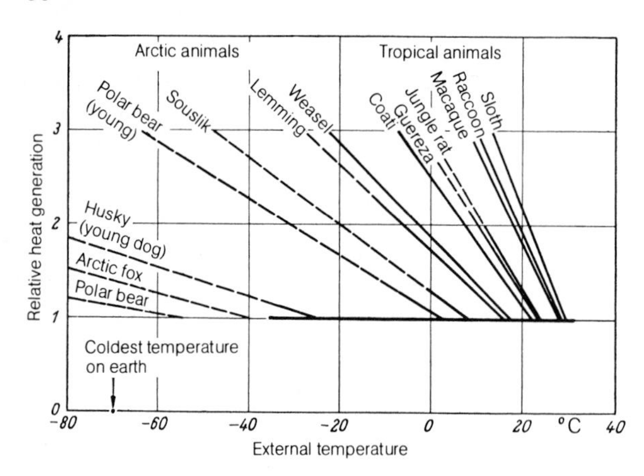

Fig. 27. The heat generated (energy consumption) by arctic and tropical animals at rest, as a function of the external temperature. The minimal turnover is set at 1 in each case. —— measured mean values; - - - extrapolated values. (Hensel in Precht-Christophersen-Hensel, 1955)

stantly prepared for activity. Hibernating mammals need special hiding places, since they cannot defend themselves against predators. Small animals with very high energy requirements (cf. the section on food requirements) cannot even afford a long rest during the night; shrews are active at regular intervals without regard to the day/night cycle. Hummingbirds can greatly alter their body temperature during a 24-h period. Those in the high Andes enter a sort of "hibernation" (torpor) every night. Many insect-eating birds can go into torpor for a few days when food is scarce. This is especially true of the young birds (swifts and swallows). Other birds adopt a less extreme form of this strategy; tits can lower their temperature slightly during the night in the arctic winter, and thus save energy. The basal metabolic rate of birds in general falls by about 20% during the resting period. Aquatic mammals are particularly vulnerable during the arctic winter. The extremities of the body (the feet of birds, and the snout, eyes, and fins of seals) are quickly cooled because of the high heat conductivity of water. In fact, these parts of the body are better described as ectothermic than as warm-blooded. Their enzyme patterns are adapted to these conditions. Tuna are comparable in this respect; the core of the body is homoiothermic. They have no swim bladder, so that their musculature is constantly in motion. The active swimming of these fish by day and night provides sufficient energy that with a certain amount of insulation the interior of the body can be kept at a constant temperature. But the gills present a problem. These thin-skinned structures must be exposed to the open water so that oxygen exchange can occur, and the blood flowing through them is cooled. At low temperatures hemoglobin takes up more oxygen than it can bind at high temperatures. Mammalian hemoglobin would take up so much oxygen that, once inside the body, the oxygen would come out of solution and cause embolism. The hemoglobin of tuna is specially adapted in a way that makes its oxygen affinity independent of temperature. Only by such means could a gill-breathing organism be effectively warm-blooded. Just as the specialized hemoglobin made possible this temperature compensation, so enzymes in the feet, fins, snout, and eyes of seals and sea birds allow these organs unimpaired function even though their temperature may be between 2° and 6 °C (Hochachka and Somero, 1973).

In view of all the different temperature effects that have been listed here, it is evident that apart from a certain large-scale temperature dependence hardly any specific predictions can be made on the basis of the available climatological data. To predict temperature relationships accurately, one would have to take account of several aspects in addition to the mean annual temperature – the difference between winter and summer temperatures, the daily amplitude of temperature fluctuation, and the elevation of the sun, which limits the degree to which organisms can be warmed by direct heat radiation. The topography of a region is also a factor. A south-facing slope in the northern hemisphere receives considerably more solar radiation than a slope with northern exposure. The south-facing slope is therefore dryer, and during the summer cooling by evaporation is less, which accentuates the difference in temperature. Finally, where diurnal temperature fluctuations are concerned, it is necessary to consider the ability of very many animals to seek out favorable positions. Because not all animals can do this, and no plants can, inferences about the species composition of a habitat based on climatic data are extraordinarily problematic. For example, the flora of the tundra in the far north (the region around latitude 80 ° N) is barely distinguishable from that of subarctic montane regions (southern Norway, 60 ° N). Because of the difference in sun elevation and the associated difference in the intensity of incident radiation, however, there are marked differences in the fauna of the two regions. These are especially apparent among the herbivores and species with special forms of behavior by which they make direct use of the sunlight. The preferred temperatures established in the laboratory give only an approximate indication of the distribution of a species in nature.

Such experiments play a central role in ecology, however, so that they deserve a brief discussion here, in the context of temperature preferences. The first question to consider is whether the preferred temperature corresponds to the optimum. This is by no means necessarily the case. Behavioral research has shown that brooding geese presented with eggs of different sizes will choose one much larger than their own. Here the preferendum is far from the optimum. And we know that similar discrepancies can occur in the realm of ecology. The Mediterranean cricket Gryllus bimaculatus has a temperature preferendum in the region of 34 °C. But the optimum for this species is much lower, between 25° and 31 °C. This range is so wide because the measured optima differ, depending on the function being analyzed – egg production, the quantity of reserve substances in the eggs, the rate of development to the imago stage, the final size of the imago, or the level of mortality. But in the life of the animal there are no distinct optima. The best indication one can give of the effective optimum temperature is a range of temperatures within which all the vital functions can operate more or less efficiently. An added complication is that in the field the temperature fluctuates. Such fluctuation has a positive effect on egg production. However, the crickets do not (as tiger beetles do) seek out locations at different temperatures at different times of day. Finally, the food supply can modify the effect of temperature (p. 90). We may well ask whether animals live within their optimum range at all. To what extent may other factors encountered under field conditions force them into a range of temperatures we would not consider optimal? We cannot answer this question here, but will return to it in the discussion of competition (p. 62f.).

3. Nutrition

Unfortunately, it is practically impossible to give a description of plant nutrition that will satisfy the physiologist. Only minute quantities of the essential nutrients are freely available in the soil and water. Even fertilization causes little change in this situation, for the added nutrients are immediately adsorbed on soil particles or react with other components of the soil and water. It requires activity of the plant (through the roots of which hydrogen ions and organic acids can be excreted) to make these nutrients accessible. Other nutrient ions are immediately incorporated by microorganisms and are thus out of reach of the plants, but when a microorganism dies – if it does so in close proximity to a root – the ions liberated become temporarily available.

The advantage of this situation to the plant is that although the supply of nutrients is small at any moment it is quite uniform in the long term. The best example is given by the primeval forests of the Amazon basin, which thrive on soil that is practically free of nutrients, with ground water containing no nutrients at all. These forests have been compared with a firm that holds no funds in reserve, but keeps its entire capital in the business (Fittkau, Beck, 1971).

The same can be said of aquatic environments. When a body of water is fertilized to increase the yield of fish, what happens to the fertilizer? It rapidly disappears from the water, sometimes in less than a week. In part this is due to the plants, which utilize it immediately, and in part to mud particles, to which the fertilizing substances adhere and from which release is slow. The latter fraction must be regarded as temporarily lost. This withdrawal of substances makes it impossible even in water to find the true relationship between the amount of minerals supplied and the resultant plant growth. Such experiments are feasible only in hydroculture, but then the results cannot readily be applied to the field situation. Only a rough idea can be obtained – for example, that diatoms thrive at mineral concentrations lower than those necessary for dinoflagellates, so that there is a strong tendency for them to take the place of dinoflagellates in nutrient-poor waters (Fraser, 1965).

The physiologist himself, on the other hand, cannot establish precisely the minimal or maximal concentrations for plant growth, to say nothing of the optima. In practice the limiting concentrations vary greatly, depending on the nature of the soil and its water supply. Furthermore, plants respond differently to variations in nutrient availability. Under nitrogen deficiency the roots of the plants are elongated and driven deeper into the soil. When nitrogen is abundant root penetration is limited to the top soil stratum, but there it is much more extensive. (As a result, fields under intensive cultivation which are heavily fertilized are extremely vulnerable to brief periods of drought. Shallow rooting is also the cause of the greater danger of erosion in intensively cultivated regions such as Icelandic pastures; cf. Ellenberg, 1969, 1971 a, b.)

In an ecological context, the uptake of nutrients involves microorganisms to a large extent. This involvement is especially clear in the case of the many land plants associated with symbiotic microorganisms or fungi that live in or near their roots. Here hardly anything can be said about the requirements of the individual species; the system fundamentally comprises two species. For all these reasons, we shall pay little attention to plant nutrition in the following discussion (for further information on production ecology see pp. 53 and 198 ff.).

In considering the nutrition of animals, we must distinguish between the qualitative and the quantitative aspects. Let us first turn to the qualitative nutrient requirements of animals.

Usually food specialization is evident in the structure of an animal. Carnivores all have shorter digestive tracts than herbivores, even when they are members of the same species. The intestine of a plant-eating tadpole is very long, whereas that of the adult, carnivorous frog is very short. The modification of the forelimbs of arthropods to form a great variety of mouthparts – biting, licking, piercing-sucking – is a well-known phenomenon. Specialization can lead to extreme selectivity; the snake Dasypeltis scaber subsists entirely on the eggs of birds, and some nematodes eat only diatoms (Fig. 28). There are a particularly large number of specialists among the insects, some of which are restricted to one or a few species of plant. Such obligate feeding is based predominantly on secondary substances in the plants. But it has been established that only a few animals

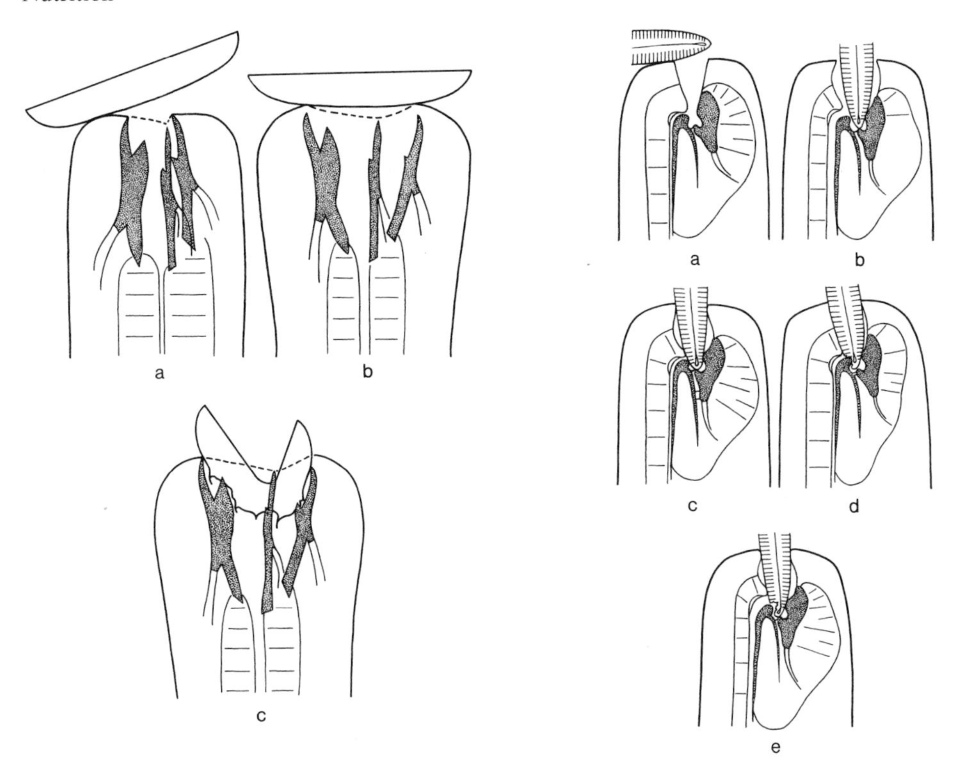

Fig. 28. Food specialization among nematodes of the marine sand fauna. *Left:* Adoncholaimos thalassophygas. When a prey animal is touched the worm adheres to it by suction, sucks it in, and tears it to bits. *Right:* Hypodontolaimus balticus. A diatom is encountered and received into the oral cavity, the cavity is closed, the diatom is bitten open, and its contents are sucked out. (von Thun, 1968)

really require these secondary plant substances. In actuality the alkaloids, terpenes, and phenols were developed in evolution as defense mechanisms against grazing. Through coevolution specific animals developed resistance to these defense substances, and at the same time they came to use the substances as a means of recognizing their specific plant. If the sense organs on the mouthparts of the tobacco-moth caterpillar are destroyed it no longer feeds exclusively on tobacco, as before, but will eat and thrive on a great variety of other plants. These specialists have "made a virtue of necessity". Not only do they tolerate the poisonous secondary plant substances; most can store them in their bodies and as a result become unpalatable, inedible or even toxic to predators. In fact, not only herbivores protect themselves in this way. Marine slugs that feed on coelenterates with nematocysts not only are unharmed by these stinging organs but carry them in their own body walls, in firing condition; these "kleptocnids" make life difficult for the slugs' predators (cf. p. 72).

The animals that have coevolved to tolerate the defense substances of plants are not physiologically dependent on them, although they benefit from a food supply inaccessible to other herbivores under the competitive conditions in the field. But not all secondary plant substances are dispensable. It is a familiar fact that animals have lost the ability to synthesize a great many biologically important materials, which they must then obtain in their food. A few remarkable cases, of interest in the ecological context, deserve mention. Barnacles (Balanus balanoides) can use a great variety of living or dead foods, which they filter out of the water, for growth. But to reach sexual maturity they need the planktonic diatom Sceletonema costatum. Without this food they can attain a normal size but cannot reproduce. Similarly, the marine isopod Idotea requires green algae plus the attached diatoms to get through the first stages of growth, whereas later the green algae alone suffice (Jansson, 1967). The butterfly Iphiclides podalirius enters diapause in the fall, when the days become shorter. The mechanism that elicits diapause functions better when decreasing day length is accompanied by the availability of autumn leaves as food – which ordinarily occurs in nature. But autumn leaves can also have a diapause-triggering effect if provided when the days are growing longer. The males of many danaid species were found to be utterly unattractive to the females under laboratory conditions. It turned out that the male imagines are incapable of synthesizing the pheromones that stimulate the females unless given the opportunity to suck on dried Boraginaceae (Heliotropium), from which they obtain the precursors of the pheromones (Schneider, 1975). Normally the food consumed by an adult butterfly or moth is considered a negligible element among the insect's ecological requirements. But it is quite significant apart from its energy contribution. This example demonstrates that a plant that is not food for the larvae and is not alive during the flight phase of the imagines (the males suck on the dried stems) can be crucial to the existence of the species.

The subtle requirements of animals for particular food constituents imply adaptations of sense organs and nervous system that are now receiving greater attention from neurobiologists (cf. the summaries by Roeder, 1968; Ewert, 1976). Toads, for instance, when presented with a horizontal stripe moving along its long axis (a sight resembling an earthworm), respond with prey-capture behavior; they exhibit no interest when the stripe is moving in the perpendicular direction. Grasshoppers have a specific grass receptor (Boeckh, 1967a); sexton beetles (Necrophorus and Thanatophilus) have specific carrion receptors (Boeckh, 1967b). The caterpillars of the oleander hawk-moth (Daphnis nerii) flourish with privet leaves as food if such leaves have once, at the outset, been sprayed with an extract of oleander leaves. Evidently the caterpillars recognize their food plant by a chemical substance in the leaves which they do not necessarily require for growth and health (Koch and Heinig, 1977). The complicated ultrasonic echolocation system of the bats and the coevolution of these animals with the moths they hunt is a further example of this sort (cf. p. 168), showing that to understand ecological questions one must take into account not only the aspects of metabolism and physiology (what is eaten and how much?) but also that of neurobiology (how is it recognized?). The latter has been given short shrift by both ecologists and sensory physiologists (cf. p. 75; for a summary see Chapman and Bernays, 1978).

So far, the amount of data accumulated in this area is extraordinarily modest. Meticulous field observations will probably never be adequate to solve such questions. Experiments are worth doing only if they lead to culture of the animals over several generations.

Even with respect to the basic nutrients – carbohydrate, fat and protein (in the ocean waxes play a large role; Benson and Lee, 1975) – animals make qualitative demands on their food. These substances must be available in a form the animal can utilize and in the right proportions, and they must contain the right amounts of the essential amino acids and fatty acids. Roe deer fed with the highest quality meadow hay, on which red deer and cows thrive, lose weight and eventually die. The roe-deer rumen is very small, and cannot break up the cellulose of the cell walls fast enough. Roe deer need food that is much more easily digestible, containing little cellulose or lignin. In the field, therefore, they subsist almost entirely on the buds of leaves and flowers. It is because of this diet that they have such an impact on the ecosystem – an effect far greater than would be expected from the number of animals and the amount of energy each requires (Eisfeld, 1975).

Other animal species have been less thoroughly studied. But we can take it as established that the specialization of the different ungulates on the African steppe to different plants or plant parts has similar causes, at least to some extent. And the situation is no doubt similar with nonmammalian animals – birds and ectotherms. In the case of ectotherms, the degree to which plant food can be utilized is also strongly temperature-dependent (cf. p. 90). The lignin and cellulose ("fiber"; cf. Helfferich and Gütte, 1972) that make old leaves harder to digest than fresh buds (which contain very little) cause great difficulty to the ruminant digestive system. It is not a matter of breaking the substances down, but of doing so as rapidly as possible; otherwise more energy is used up than is gained. When an organism selects its food on the basis of digestibility it uses much energy in looking for the food but does not need any to digest it, and very little ballast material is consumed. The alga-eating chironomids are less selective, and consume far more than they utilize; many algal cells leave the gut of such an insect undamaged and fully capable of further growth.

Special significance attaches to the composition of food. If this is not appropriate, animals (e.g., some that suck plant juices) may be assisted by symbionts; in other cases, considerably more food is eaten than the energy requirements of the organism demand. The surplus components in the food are absorbed in the intestine but then eliminated. This phenomenon has become famous in the case of the aphids, which excrete "feces" in which sugar is highly concentrated and can be used by other animals such as bees. Aphids are forced to excrete so much sugar because their food contains such great amounts in comparison with the low content of protein or amino acids. Similar relationships can be found with regard to many nutrients. Herbivorous mammals living in regions with a long winter or a long dry season must make do for extensive periods with food of very low quality, consisting almost entirely of lignin and cellulose. In the animals' very warm intestinal canals this food is broken down with the help of symbionts, so that it provides sufficient energy but not enough protein or amino acids. When mammals with several-chambered stomachs (kangaroos, camels and llamas, ruminants) are restricted to such a diet, they can recycle nitrogen. The urea formed in the liver is transported to the stomach, taken into the rumen, and there converted by microorganisms to amino acids and protein that can be used by the animal. As a result, the urine contains hardly any nitrogen. This sort of nitrogen processing has been shown to occur in kangaroos, camels, and most ruminant groups. The animals also conserve water in this way; it is not necessary to produce urine in order to get rid of nitrogenous products of metabolism. Naturally this procedure is not used on a permanent basis, for it does not permit growth. But it does permit an animal to maintain the status quo for fairly long periods of time (Fig. 29).

Nagy and Milton (1979b) showed that the howler monkey (Alouatta palliata) is adequately supplied with minerals only if its food is very diverse. The figs and young leaves of Ficus insipida that constitute the most important food of these animals during the dry season do not contain enough copper, sodium, or phosphorus; these minerals must be obtained from other food. Many tropical animals leave the crowns of the trees to supplement their diet with minerals from the soil. Howler monkeys do not descend to the ground; evidently they can find the appropriate nutrients in the treetops.

The alternative approach to specific nutrient scarcity – eating enough to supply the scarcest substance and eliminating the excess of the others, as the aphids do – does not seem to be realized in other animals.

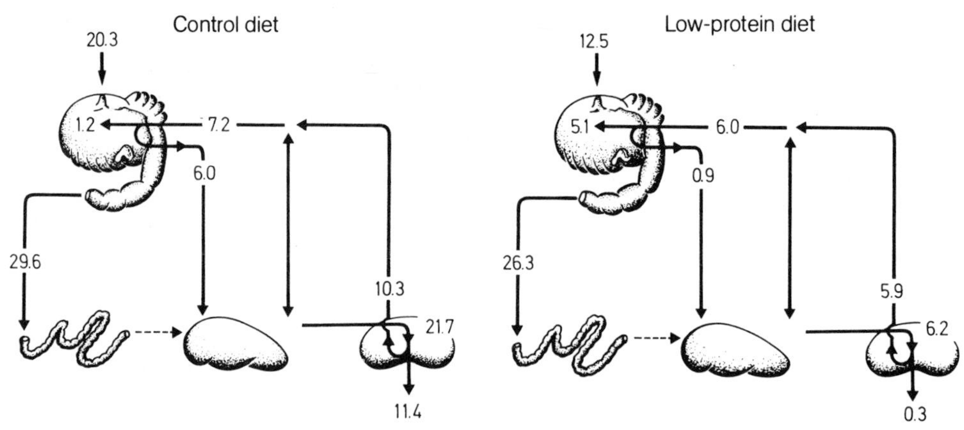

Fig. 29. Nitrogen uptake, urea exchange, and protein flow out of the anterior stomachs (rumen and reticulum) of the llama, when fed on a control and a protein-depleted diet. All data in g nitrogen per 24 h. When the diet is low in protein urea is returned to the digestive stomach, and the permeability of the walls of the anterior stomachs to urea is increased. Despite the reduced excretion of urea through the kidneys, the amount of urea in the body was thus reduced. The urea-nitrogen not used up is resorbed as ammonia and resynthesized to urea in the liver (Ali, 1977). (Because of the internal fluxes and additional measurement problems, it is not possible to give an overall numerical analysis.)

The regulation of amount eaten appears to be based entirely on energy requirement. Naturally animals choose the qualitatively best food available, but if this contains too little protein either recycling occurs or the animals show deficiency symptoms. Hyperphagy in insects has not yet been completely explained. When given food deficient in protein crickets eat distinctly more than they do of better-balanced food. But this hyperphagy cannot compensate for the protein deficiency, because it begins only when the protein content of the food is extremely low. However, some leaf beetles are said to be able to compensate for the low nitrogen content of leaves by hyperphagy. This complex of questions deserves considerably more attention in the future, because of the abundance of trace elements present in widely varying amounts in potential food sources. The ecological effect of animals in a system could, under certain circumstances, be increased many times by hyperphagy – but no more than this can be said with the limited data now available.

Experiments along these lines can be laborious and frustrating. Hahn and Aehnelt (1972) were able to demonstrate that the fertility of male domestic rabbits and cattle was greatly reduced when the animals were fed on hay from a heavily fertilized meadow in which only one grass species was growing. Animals fed on varied sorts of hay from unfertilized meadows had considerably more reproductive success. The search for ions responsible for this difference in the various foods gave consistently negative results. This is the problem that confronts us everywhere: minute differences can have marked effects. One can speculate further, that a massive increase in the size of the herbivore population leads to increasing uniformity of the plant community. Is this uniformity in food one of the factors responsible for the collapse of a herbivore population (cf. p. 130 and pp. 144, 227f.)?

Recently we have begun to appreciate the significance of the discovery that many marine animals can nourish themselves directly on dissolved substances (amino acids, carbohydrates). Such a capacity seems obvious in the relatively large pogonophorans, which lack a gut, but it also plays a considerable (though not decisive) role in many other animals. Sea anemones (Actiniaria, Cnidaria), for example, bear

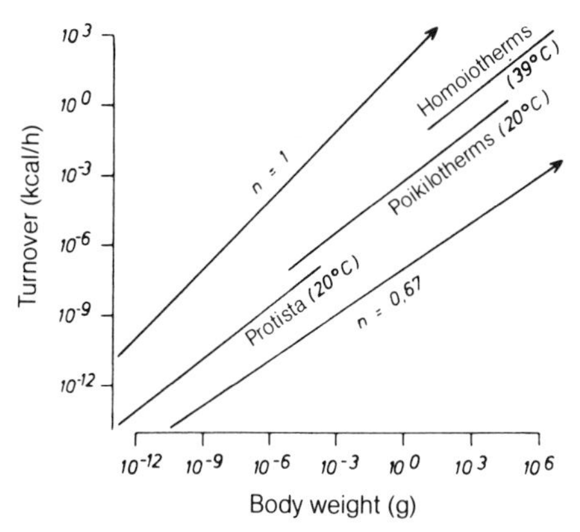

Fig. 30. Turnover in three groups of organisms, as a function of body weight; *n* is the exponent in the basic equation. The data for ectothermic and unicellular organisms have been corrected to 20 °C. (Aschoff et al., 1971)

on the body surface an absorbing epithelium with microvilli, and molluscs and polychaetes take up dissolved substances from their surroundings. Calculations by Schlichter (1975) have shown that the quantity of such dissolved substances is quite sufficient to supply the entire nutrient requirement of these animals. In fresh water and on land, this principle is probably of no importance. One reason is that such direct absorption is more difficult in view of the osmotic gradients involved; furthermore, the numerous bacteria in fresh water remove from it essentially all of the free amino acids and carbohydrates (most of which are derived from the metabolism of planktonic algae and lower animals). The fresh-water clam Pisidium does take up glucose even when it is present only at the concentrations found in the natural environment. But the amounts so accumulated meet less than 0.05% of the animal's energy requirement – in marked contrast to the marine situation, in which 100% of the required energy can be obtained from dissolved substances. Nor can the fresh-water sponge Ephydatia meet its requirements; it takes up dissolved proteins by micropinocytosis, a method quite different from the enzymatic uptake by marine animals (Efford and Tsumura, 1973; Weissenfels, 1976).

It is difficult to determine the quantities of food that animals need, particularly in the field. Repeated attempts have been made to arrive at generalizations that would at least permit certain predictions. The formula for allometric growth can be applied to the resting metabolism of animals (Fig. 30):

Resting metabolism per unit time (V) $= \mathrm{N} \times (\text{body weight})^{b}$,

where the exponent b is a function of the surface area and is thus in the region of 0.75, with very few exceptions. The factor N depends on the temperature (in ectotherms) and on the units of metabolism and weight chosen. In mammals, according to Kleiber (1967),

Resting metabolism (calories per day) $= 70 \times (\text{g body wt.})^{0.75}$.

Once the resting metabolism is known the minimal food requirement can be calculated. It appears possible to make rough estimates even for animal groups that have not yet been studied. If one can obtain additional information by collecting the daily excreta of the animal, its energy content can be added to the computed calories for metabolism to obtain the caloric value of the food consumed per day. But even this approximation is subject to fairly large errors, because in mammals frequently much of the energy in the food is given off in gaseous form (as methane, especially, by ruminants). Nevertheless, this method sets an approximate lower limit on the food requirement of a species.

The general implication of the formula for basal energy turnover is that small organisms require relatively more energy, and thus food, than large ones. A good illustration has been given by Kleiber (1967) in his comparison of cattle and rabbits (Fig. 31). We can see from the figure that given two habitats with the same amount of available food, we cannot expect to find the same consumer biomass if the consumers in one habitat are very small and in the other, very large. Where large animals

Animals	1 cow	300 rabbits
Total body weight	600 kg	600 kg
Daily food consumption	7,5 kg hay	30 kg hay
Daily heat loss	20000 kcal	80000 kcal
Daily weight increase	0,9 kg	3,6 kg
Weight increase per t hay	108 kg	108 kg
Meadow with 3 t hay suffices in theory for	1 year	90 days

Fig. 31. A cow weighs as much as 300 rabbits. The ecological effects of the two, however, are quite different. (Kleiber, 1967)

are concerned, the possible biomass per unit area is much greater. To continue with the example in the figure: a meadow that produces 7.5 kg of hay per day (2,738 kg over the entire year) can theoretically feed a 600-kg cow for an entire year. But only 75 rabbits, totalling 150 kg, can be fed for a year on the same area (cf. p. 205).

Statements about the biomass of animals in a habitat, then, are not in themselves informative. At least the mean body size of the species must be given (Fig. 32). The relatively higher food consumption and the relatively higher metabolic rate of small organisms naturally finds expression in other fundamental ecological parameters. Small organisms, for example, grow more rapidly than larger animals (this is also indicated in Kleiber's table); the productivity per unit time is greater and the rate of reproduction is higher. The consequence of all this is that small organisms are more likely to starve, and do so more quickly, than large ones. Shrews must eat at intervals of at most two or three hours; they cannot follow a light/dark cycle of rest and activity. Field mice, too, must leave their burrows at least once every 2–3 h, although their chief activity period is at night. An ecological consequence of this distributed activity is that both nocturnal predators (owls) and those that hunt by day (buzzards) can feed on the same prey. Small animals are also better suited than large animals to take advantage of the sudden occurrence of favorable conditions by reproducing very rapidly. Large animals are not so flexible, but they are better able to survive unfavorable periods without serious losses. We shall return to these questions in considering predator-prey relationships (p. 133f.).

The next step is a detailed analysis of the food requirements of an animal in the lab-

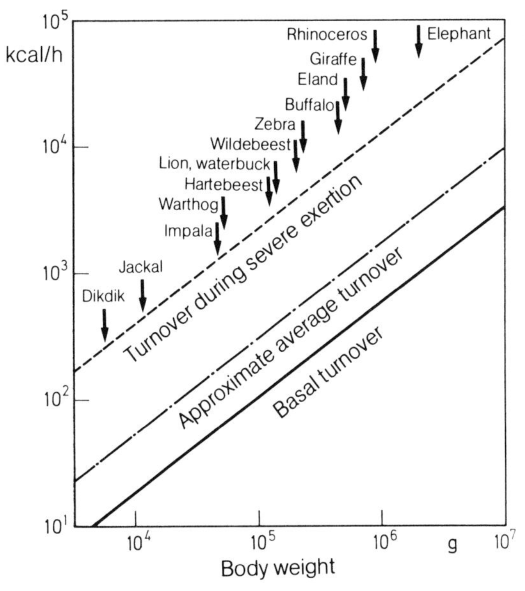

Fig. 32. Turnover in large mammals of eastern Africa. The energy requirement must fall somewhere between the basal turnover and the line for hard work. This presentation is also unsatisfactory; there is too large a probability of error. The line assigned to average metabolism is based on pure conjecture. (Lamprey, 1964)

oratory or in captivity. First, the validity of the general formula can be tested; it can be expanded by determination of the additional food required when the animal is regularly in motion. With small animals (for example, insects or clams) the quantitative food requirement estimated in the field can be made relatively precise. But correction factors must be introduced to allow for the complicated effects of temperature; at higher temperatures food is more readily utilized, metabolic rate rises, and the amount of food necessary increases (development is more rapid). Such effects present no problem in the case of warm-blooded animals; on the other hand, with warm-blooded animals it is more difficult than with most ectotherms to transfer the data obtained in captivity to field conditions. For many animals – from filter-feeding clams to fish, frogs, birds, and mammals – there is great uncertainty as to how the two situations compare.

Some plant-eating insects have been usefully studied in the field, with analysis of both the plant substance utilized and the excreta produced. Under favorable circumstances feces production by mammals and birds can be determined in the field, and such data can be used to infer consumption if the digestibility of the food is known. This approach has been moderately successful with wild geese in Iceland and reindeer on Spitsbergen.

A promising new method of estimating energy requirement in the field has been developed by Nagy and Milton (1979, a, b). They captured howler monkeys (Alouatta palliata) and injected them with weakly radioactive (tritium-labeled) water. The blood thus acquired a certain initial level of radioactivity. Because water is continually produced by metabolism and then eliminated, the radioactivity in the blood steadily declines, providing a measure of metabolic rate. The results implied that the active howler monkey under field conditions uses up about twice as much energy as when at rest.

But there are still many gaps in our knowledge. Consider, for example, the uncertainty that prevails in the current estimates of consumption by mammals. A widely used procedure is simply to multiply the basal metabolism by 3 and take that as the food consumption of herbivorous mammals. Recently, with no particular justification, people have begun to regard a factor of 1.5 as sufficient. In the case of roe deer there are indications that a factor of 4 is more realistic. But probably there is no generally applicable factor. Rate of locomotion need by no means be linearly related to energy consumption. Birds have an optimal flight speed well above the minimum; slower flight costs more energy. The relatively quick "five-legged" shuffle of a kangaroo looking for food on the ground costs considerably more energy than the smooth forward leaps it uses for long-distance travel.

It will probably be impossible to find any general solution to these problems. The

Table 3. Comparison of the food intake of free-living snowy owls with that of snowy owls in the Nürnberg Zoo

	Alaska	Zoo
Food intake of 1 adult per year, in kg live weight	600–1,600 lemmings ∅ 80 g per lemming ≙ 55–130 kg	2,760 mice ∅ 25 g per mouse ≙ 69 kg
Food intake of 1 adult per day, in g live weight	Estimated 150–350 g	Average 219 g
Food requirement of the young until fledging, kg live weight	1,300 lemmings ∅ 80 g per lemming ≙ 100 kg (9 young)	2,360 mice ∅ 25 g per mouse ≙ 59 kg (5 young)
Food intake of 1 young bird per day until fledging, g live weight	160 g ≙ 2 lemmings ∅ 80 g	132 g ≙ 5.6 mice ∅ 25 g

particular feeding strategy of a species has a marked effect on its energy consumption in its habitat. A "sit-and-wait" animal needs little more than its basal turnover, as experiments of wolf spiders have indicated (Ford, 1977 a, b). And the extra energy required by a web-building spider is only that needed for the construction of the web. Studying a pair of snowy owls with young in a zoo, Ceska (1974) found the same food requirement as measured for snowy owls with an equal number of young in Alaska (Table 3). When the lemming population is in the exploding phase, at least, the owls do hardly any physical work when hunting. Things are quite different for the roe deer, which accepts only the most fresh and delicate buds and must therefore hunt for each separate bite. An animal feeding by this method is in constant motion. The fact that metabolism is regulated by size also implies that no simple factor is applicable. Movement increases the metabolic rate of a small animal much more than that of a large animal.

Finally, short-term analyses of an animal's food requirements are extremely uninformative. Even adult animals undergo great changes in food requirements during the year, although the supply may be constant. This is true of both birds and mammals (Fig. 33). The situation is doubly complicated in that the quality of the food in the natural habitat changes during the year. And the feeding strategy and energy consumption of a given species is not always the same; in a dry summer the grasshopper

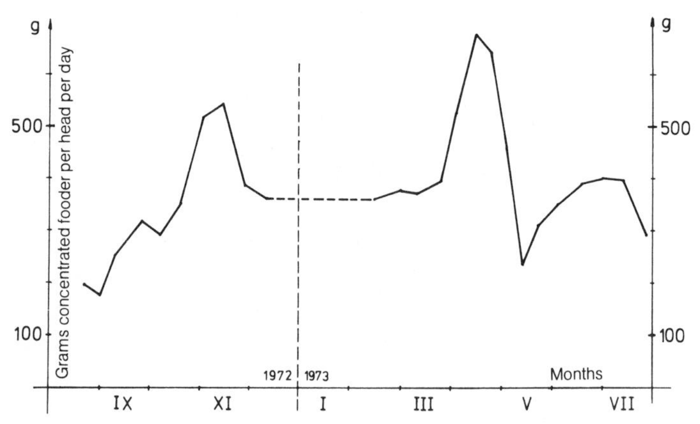

Fig. 33. Food consumption by a herd of roe deer in the Stammham enclosure, over the course of a year. (Ellenberg, 1974b)

studied by von Gyllenberg (Chorthippus parallelus) had to travel further than before to find favorable food, so that Gyllenberg (1969) had to expand his population model by adding an "activity block." The activity of roe bucks and does during the mating season expends considerable effort and they lose weight correspondingly; later this loss is made up by increased food consumption.

Of course, the nutrient content of food – whether its chemical composition or its caloric value – is also subject to change. When a wild mouse is caught and fed in the laboratory, the energy content of its body rises distinctly after only a few days because of the accumulation of fat; laboratory mice have a higher caloric value (per unit weight) than wild mice. The composition and caloric value of plant parts change

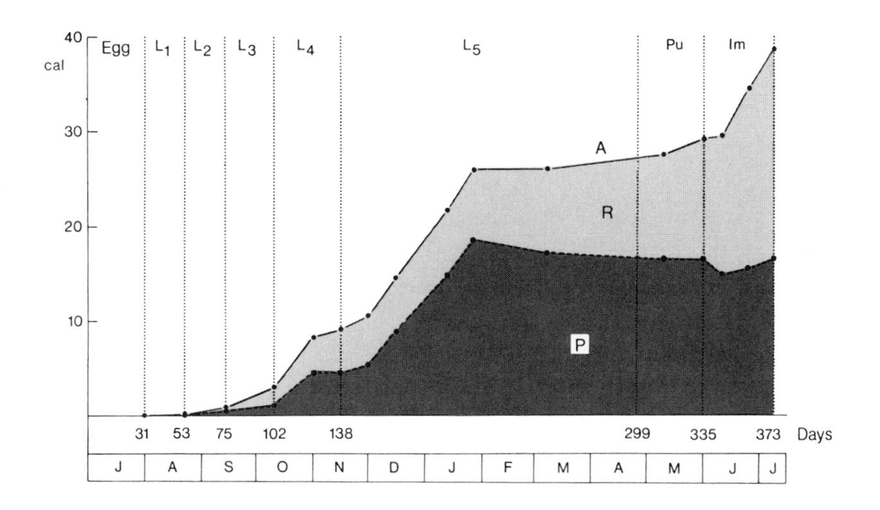

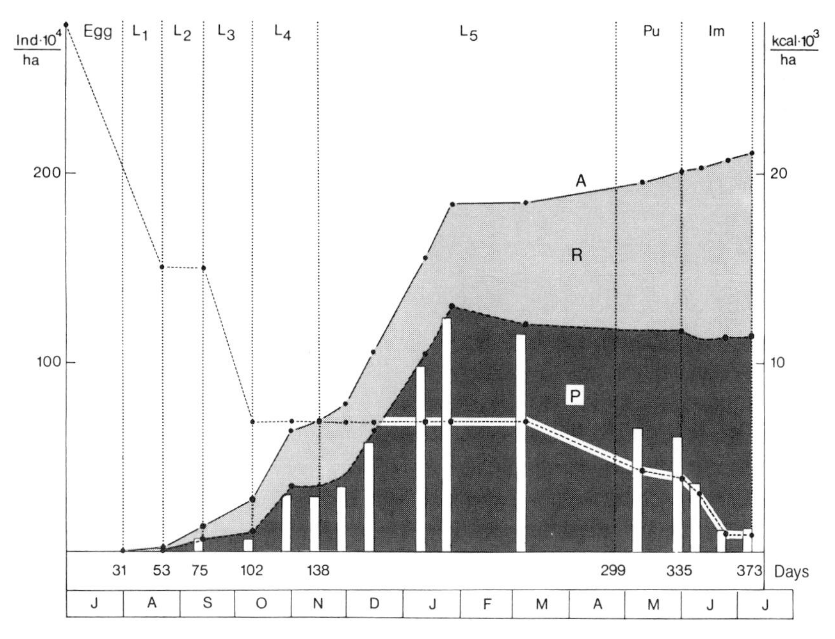

Fig. 34. Cumulative energy balance of an average individual *(above)* and of a population *(below)* of the weevil Phyllobius argentatus in 1969/1970. *White columns:* biomass; *dashed line:* survival curve. *A* assimilation; *R* respiration; *P* production. (Schauermann, 1973)

in the course of a year and even during a day.

As important as the problem of quantitative animal nutrition is to ecology, then, it is still far from a solution. Very rough estimates can be made on the basis of approximate calculations of basal metabolism, if the feeding strategy of the animal is known so that the required locomotor activity can be taken into account. But for any more precise estimate one would have to know the energy budget of each individual from birth to death, and this information is

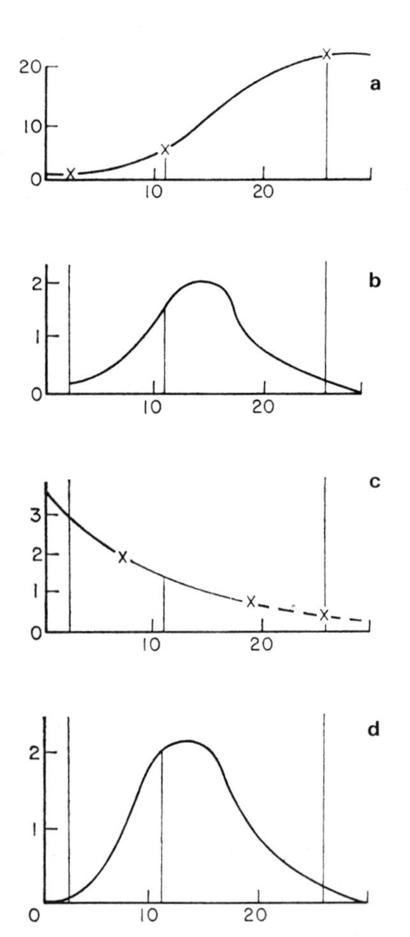

Fig. 35 a–d. Calculation of production by organisms with distinct generations or distinct stages (Winberg, from Petrusewicz and Macfadyen, 1970). *Abscissa:* time in days; the vertical lines demarcate the different stages. **a** Growth curve of an individual (fresh weight), **b** weight increase per unit time (calculated from **a**), **c** survival curve, **d** production per unit time (calculated from **b** and **c**)

available only for certain plant-eating insects. In the International Biological Program studies of this sort have been done and a number of appropriate formulas and units established; on this basis, an ecosystem analysis can be undertaken (Figs. 34 and 35).

There are hardly any detailed analyses of the degree to which animals satisfy their needs. The few studies of wild mammals, birds, springtails and crickets, though, indicate that "the normal animal is a hungry animal." Times of abundance probably occur everywhere, but only rarely. The normal free-living wolf is undernourished, just like the normal springtail.

The choice of suitable food for larvae is determined not only by quality, but by quantity as well. Parasitic wasps are famous in this regard, for in general they check very closely whether their victim has already been invaded by parasites or not. When Trichogramma finds a host egg already parasitized, it adds no eggs of its own. Most other insects behave in the same way. The number of eggs laid by the beetle Oryzaephilus depends on the mass of food on which they are deposited and which the larvae will eat; only about $^1/_{16}$ as many eggs are laid as the mass could feed. Flies and gnats with larvae that live in a rotting substrate (Drosophila, Limosina, Pseudosmittia) also measure out their eggs with relative precision. If there are already very many eggs in the substrate, no more are laid.

The eventual disposition of the energy acquired by feeding is of fundamental ecological interest. Some of the food is digested and incorporated into the body, some is used up in respiration, and some is excreted as feces. In simplified terms, one can define assimilation as production plus respiration: A = P + R. In times of a world nutritional crisis it is of course not a matter of indifference what capacities animals have for assimilation. Simplifying the situation greatly, we can say that warm-blooded animals assimilate 80%–90% of the energy in the food they eat. Similar values are reached by ectothermic predators, which feed on easily digestible, high-protein flesh. Ectothermic herbivores, by contrast, in general can assimilate only 20%–40% of the energy they take in. Departures from this rule occur, for example, when warm-blooded herbivores receive food extremely hard to utilize; then their assimilation efficiency can fall as low as 30%. Of the assimilated energy, some must be used to maintain the normal vital processes, and some of the remainder goes into production of new matter (growth and reproduction). It costs a great deal of energy to maintain a high body temperature, so that the ratio of consumption to production in warm-blooded animals is incomparably worse than that in ectotherms. For this reason it has occasionally been suggested that instead of raising warm-blooded domestic animals for meat, ectothermic herbivores should be provided with favorable food and used as a source of protein. But this suggestion is not realistic. The composition of the protein in ectothermic animals is not as suitable for our needs as that of warm-blooded animals. If we relied on ectothermic animals for protein we would need more than we would if the protein came from a warm-blooded animal. Instead of the effect we were hoping for, we would achieve the reverse (Table 4, Block 1).

Data on ecological efficiencies should never be taken too literally. When food is abundant, owls dissect their prey and eat only certain parts; the American badger (Taxidea taxus) has a digestion efficiency of 85.6% when food is scarce and only 72.2% in times of plenty (Lampe, 1977).

A general rule of thumb is that an animal uses 1%–10% of the food eaten to synthesize new matter for its own body. Up to 90%–99% of the food assimilated is burned in metabolism. The end products water and carbon dioxide are eliminated. Other components of the food (protein contains nitrogen) are given off as excreta. It is on the picture of energy flow through an organism so obtained (cf. Fig. 34) that subsequent calculations and discussions in the realm of ecosystem research are based.

Materials that can neither be burned nor be eliminated accumulate in the body. Examples of such substances are DDT, PCBs, lead, and mercury. When one animal is eaten by another and that by a third, there is a progressive accumulation of these substances in the food chain. This is the physiological basis of the danger inherent in modern environmental poisons (cf. Ehrlich et al., 1975). There is an additional consideration. Because predators are always fewer in number than their prey, the predator population is more likely to be affected by any biocide application than the prey it controls. The consequence of any generalized poisoning will thus, from the outset, probably be a mass multiplication of the pest, for the predators that had previously kept its numbers down will have been destroyed. Finally, the much higher number of prey individuals has yet another effect. The probability that offshoots resistant to our poison will appear rises, the greater the number of individuals. Such poisoning, then, brings the danger of eradicating the predator and making the prey immune.

The spatial distribution of food is highly significant. A dense monoculture – whether animal or plant – is easier for a predator or grazer to exploit than a very irregular distribution, which requires an animal to spend considerable time and energy in finding a suitable food source (cf. p. 147). There is a certain minimum food density below which the effort per unit time to obtain enough to eat is prohibitive. Here, again, we have the problem of cost vs benefit. If the water flea Daphnia pulex is offered the diatom Scenedesmus acutus as food, it filters the diatom from the water and eats it. A 1-mm-long water flea in water at 10 °C can survive if 0.04 mg carbon is available per liter of water (1 mg carbon corresponds to about 20 mg animal matter). Water fleas 3 mm long at 25 °C need much more to maintain themselves – 1.9 mg carbon per liter. Copepods filter-

Table 4. Efficiency of food utilization (in %) by animals of various species. Warm-blooded animals have a high digestion efficiency (Assimilation/Consumption), and poikilothermic animals have a high ecological efficiency (Production/Consumption). Cf. Block I. (Data from Schwerdtfeger)

Species	A/C	P/C	R/C	P/A	R/A
Rotatoria					
Brachionus plicatilis	19	11	8	57	43
Annelida					
Polychaete Nereis virens	85	44	41	52	48
Enchytraeidae	54	36	18	67	33
Arachnida					
Harvestman Mitopus morio	46	20	26	55	45
Orb-web spider Araneus quadratus	85	57	28	67	33
Wolf spider Pardosa lugubris	82	24	58	30	70
Crustacea					
Water flea Gammarus pulex	37	24	13	65	35
Crab Menippe mercenaria	96	68	28	71	29
Isopods, several species	25	4	21	16	84
Sowbug Oniscus asellus	29	6	23	21	79
Insecta					
Mayfly Stenonema pulchellum	53	15	38	28	72
Stonefly Preronarcys scotti	11	4	7	43	57
Damselfly Pyrrhosoma nymphula	90	53	37	58	42
Grasshopper Orchelimum fificinium	28	10	18	37	63
Grasshopper Melanoplus sp.	33	5	28	16	84
Grasshopper Chorthippus parallelus	40	17	23	41	59
Lepidoptera, Chimabacche fagella	24	11	13	47	53
Lepidoptera, Ennomos quercinaria	32	20	12	60	40
Lepidoptera, Hyphantria cunea	29	17	12	57	43
Bug Leptopterna dolabrata	33	18	15	55	45
Cicada Neophilaenus lineatus	33	18	15	52	48
Mollusca					
Clam Scrobicularia plana	61	13	48	22	78
Snail Littorina irrorata	45	7	38	14	86
Pisces					
Gudgeon Gobio gobio	80	7	73	8	92
Roach Rutilus rutilus	80	7	73	9	91
Bleak Alburnus alburnus	80	7	73	7	93
Perch Perca fluviatilis	76	20	56	27	73
Aves					
Bunting Passerculus sandwichensis	90.0	1.0	89.0	1.1	98.9
Mammalia					
Waterbuck Adenota kob thomasi	84.4	1.1	83.3	1.3	98.7
Elephant Loxodonta africana	32.6	0.5	32.1	1.5	98.5
Ground squirrel Citellus sp.	68.0	2.0	66.0	2.9	97.1
Mouse Peromyscus polionotus	90.5	1.6	88.9	1.8	98.2

feed more effectively and can live even in nutrient-poor lakes, in the winter and at depth; in this regard they are superior to water fleas (Lampert, 1976). Quantitative data of this kind are scarce; there are only a few other examples. In some places redshanks feed primarily on the crab Corophium, which is found in enormous numbers in mud flats. If the density of the crabs is about 200 animals/m² or less, the redshank switches to polychaete worms (Nereis, Nephtys), which provide more food for a given amount of time spent searching (Goss-Custard, 1977). The sand-dwelling shrimp Crangon feeds chiefly on polychaetes (Nereis, Nephthys) and crus-

Block 1. Symbols and abbreviations used in production ecology, as recommended by the International Biological Program. Diagram of the flow of matter and energy through an ecological unit (organism or population)

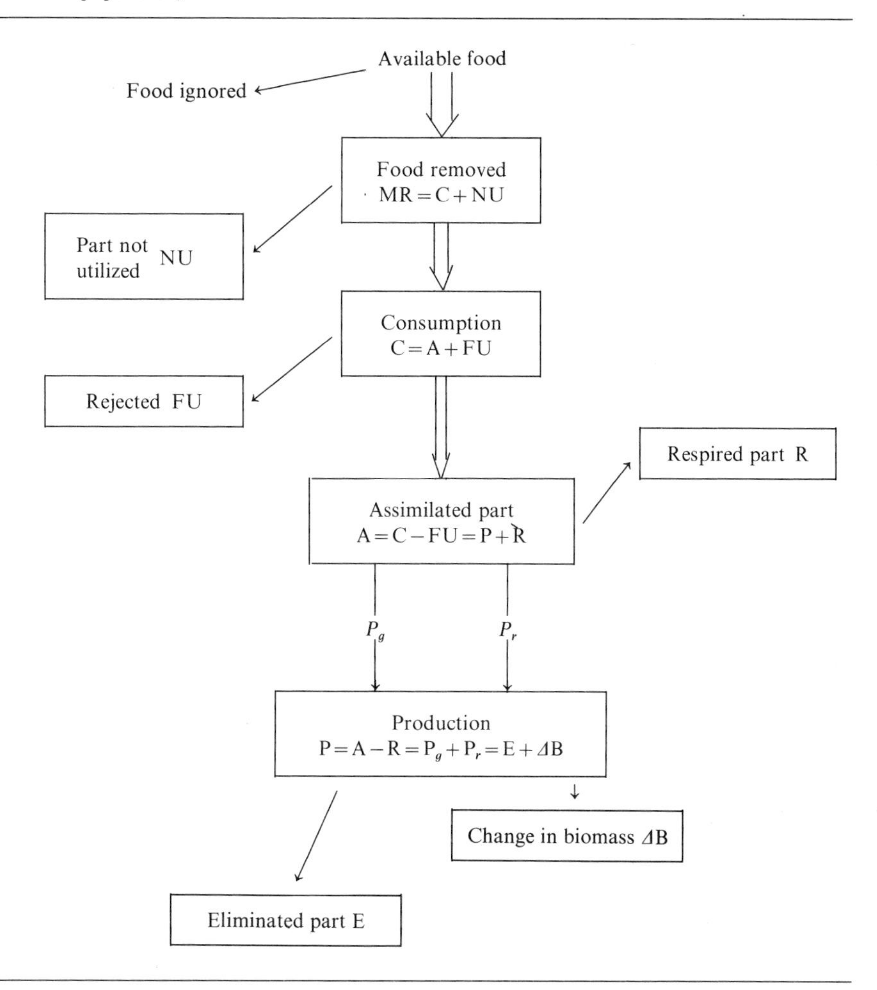

MR: Food removed from the system
NU: Unutilized part of the food removed
C: Consumed part of the food removed
F: Feces
U: Urine and other excreta
A: Assimilated part of the food
D: Digested part of the food
P: Production
R: Respired part of the food
E: Eliminated part of the food (molted, shed)
P_g: Production associated with growth
P_r: Production associated with reproduction
ΔB: Change in biomass

These are terms in a number of equations, for example:

$MR = NU + C$ $\quad C = P + R + FU$ $\quad A = P + R$ $\quad P = A - R = C - (R + FU)$

Ratios (usually given in %) indicate the ecological efficiency; for example:

MR/P C/P C/A C/R R/P etc.

These conventions should be adopted in publications on animal ecology, so that the data are comparable; when ecological efficiency is concerned it should always be made clear which of the possible efficiencies is meant (see Table 4). (Petrusewicz and Macfadyen, 1970).

taceans (Corophium, Mysidaceae). Crangon is also capable of seeking out 1.5-mm-long free-living nematodes in the substrate, seizing and eating them. But experiments by Gerlach (1969) showed that this laborious hunting strategy costs more energy than it brings in, so that the weight of the shrimp slowly declines. Badgers (Meles) in southern England feed almost exclusively on earthworms (Lumbricus), catching them on the surface of the ground by night. There are vast numbers of these worms everywhere in the soil, but they do not come to the surface everywhere, and not enough are available to the badgers to satisfy their needs. Therefore no relationship can be established between the density of the earthworm population and that of the badgers. The sole decisive factor is the availability of the food, which depends on specific geomorphological features (Kruuk, 1978). In the case of colony-breeding sea birds, a relationship has been found between reproductive biology and the distance that normally must be travelled to find food. Terns usually lay three eggs and feed the young until they can fly. They can manage this only if there is a good feeding site within 5 km of the nest, where they can fish without losing time. Among the auk species one can find a progressive series of brood-feeding behavior. The black guillemot is the only bird in this family to lay two eggs; it breeds individually near the coast and usually finds its food by diving in the water just off the coast. The fish it catches are brought directly to the young, which very soon leave the nest and go along on the fishing expeditions. The other species breed in very large colonies and each bird lays only one egg. The parents usually fish relatively far from the coast; they feed their young until they have grown to about one-quarter of the adult size. Then the young leave the nest and follow their parents out to sea. The puffin flies out to the high seas in its search for food. Its young live in holes in the ground and are thus particularly well protected; they are fed for a very long time. Eventually they begin to leave their nest-holes at night to venture over the water; once the young bird has left its hole the relationship between it and its parents is destroyed. This sort of behavior is taken to the extreme by the tube-nosed albatrosses and their relatives. These birds also lay only one egg; they hunt for food on the high seas, out of sight of the coast. Therefore the young can be fed at most once a day, and those of many species are fed only every second day. The adult birds fly at great speed; their feeding grounds in general lie between 30 and 200 km from the nest, and it is quite possible for them to extend to even greater distances. Among the larger forms, then, the period during which the young are fed lasts more than three months. By that time the young bird has grown to a weight greater than that of the adult. The young of most species at this stage leave the brood-hole at night and wander independently out to sea, where they develop their flying skills. Some other species (for example, the arctic fulmar, Fulmaris glacialis) are fledged while still in the nest, with marked loss of weight, for the adults have already stopped coming to the nest rocks and feeding them. The young bird is fully fledged when it leaves the nesting site. The large procellariids and albatrosses can breed only every second year; the breeding ground is occupied each year, however, by different pairs of birds in alternation. The price these birds pay for the opportunity to exploit the food offered by the high seas is a reduction in rate of reproduction (only one offspring per reproductive season) and an extremely long breeding period. They can afford this cost because their life span is so long.

4. Light

Light is the factor that permits life to exist on earth at all. But it is difficult to distinguish among various habitats on this basis, for there is enough light to support photosynthesis over the earth's entire surface. Very few habitats – the ocean depths, the deep zones of some inland waters, and caves – do not receive enough light for plants to be productive. Everywhere else in the biosphere there is light in adequate amounts and with the correct spectral composition. It is only natural, then, that the quantitatively significant production of organic matter everywhere is in the final analysis based on the same mechanism – photosynthesis by means of chlorophyll. The different kinds of chlorophyll, as far as we know, hardly differ at all in productivity. We can consider them as a single entity.

As an initial consideration, we can take it that light energy is a factor in excess. Plants make use of but a small fraction of the incident radiation; the efficiency of light utilization is always less than 5% and usually 1% of the available energy (Block 2). It follows that productivity within a habitat is usually not dependent on the amount of light – there is always enough. The level of production is dictated by temperature and water supply, as well as by the presence of minerals. Therefore it must be possible to calculate the level of production possible (Table 5). Under the same conditions of soil, nutrient availability, and climate all plant communities must in theory exhibit roughly the same production, and observations have shown that in fact they do. Studies carried out in the Solling Project (supported by the German Research Foundation as part of the International Biological Program) showed that the production of organic matter in a meadow, an area of coniferous forest, and an area of beech forest was quantitatively comparable.

This fact is hard to accept. We know, after all, that there are shade plants and sun plants. We know of plants that carry on photosynthesis during only part of the year – spring flowers, for example, which later die back into the ground. How can we reconcile such things with the above claims regarding uniform productivity and a single basic biochemical process?

The only difference between shade leaves and sun leaves is that the former contain more chlorophyll; the same is true of shade plants as compared with sun plants. The greater amount of chlorophyll compensates for the lower relative light intensity. Greater amounts of chlorophyll are obtained at a price; there is room for them in the leaf only if the thick epidermis and cuticle are eliminated. Therefore shade plants are very vulnerable to drought. In their normal habitat this is irrelevant, because shade is practically always associated with ample moisture. The essential point here is that the mechanism of photo-

Block 2. Production by a field of maize, as an example of the calculation of primary production (Tischler, 1965)

Total dry weight of the maize from 0.4 ha (=1 acre) (10,000 maize plants)		6,000 kg
Ash (inorganic components) subtracted		– 300 kg
Total weight of organic components		5,700 kg
Their equivalent in glucose		6,700 kg
Plus organic matter lost by transpiration appropriate to the season (expressed as the glucose equivalent)		2,000 kg
Total weight of the glucose formed by 0.4 ha maize		8,700 kg
Energy required for synthesis of 1 kg glucose		3,800 kcal
Energy required for synthesis of 8,700 kg glucose	ca.	33×10^6 kcal
Total solar energy available to 0.4 ha		$2{,}040 \times 10^6$ kcal

$$\% \text{ utilization of the available energy} = \frac{33{,}000{,}000 \times 100}{2{,}040{,}000{,}000} = 1.6\%$$

Table 5. Theoretical maximum yields under the geographical conditions in various habitats, and the theoretically calculable utilization by a warm-blooded animal. (After de Witt in Remmert, 1973)

Theoretically possible annual maximum yield (kg dry matter/ha, from de Witt as cited by Baeumer, 1971)	Location	Number of sheep theoretically supportable by this amount of energy (per ha per year)
25,000	Stockholm	68
30,000	Berlin	80
51,000	Puerto Rico	140
57,000	Tropical Australia	156

synthesis is identical to that of the sun plants. The assertion that productivity of different plant communities is similar under similar conditions does not refer to single species within the communities; no one could maintain that productivity is the same among different species. Differences at this level are inevitable in view of the variations in length of growing season, which is controlled by other factors. The comparison here is between one plant community composed of many species and another of equally complex composition. The overall production in such a community results from the activity of different plants occurring at different times; but because the underlying biochemical system is the same in all cases, the total production of the community per unit area and per unit time is the same. To recapitulate: the uniformity in the productivity of different plant communities under the same conditions is ultimately based on the fact that the amount of chlorophyll exposed to the light is independent of the species composition of the community.

The modern high-yield varieties developed for agricultural purposes produce no more organic matter than the wild varieties, but what is produced is differently distributed. Instead of an extensive root system and strong stems more is produced of the part usable by humans – the kernels of grain, for example (cf. p. 38). There is one exception in principle to the uniform-productivity rule – one which leads to a slightly but distinctly increased productivity. Whereas the first product of normal photosynthesis is a triose, a sugar with three carbon atoms (hence the term "C_3 plants" for those with this form), in a number of plants the CO_2 taken up is first coupled to phosphoenolpyruvate — a C_3 molecule — to form oxalacetate, a C_4 molecule. Since the first product of photosynthesis here is a C_4 molecule, these are called C_4 plants. Later the oxalacetate is split up again; the carbon dioxide becomes available to the normal photosynthetic apparatus and is further processed in the normal pathway of photosynthesis, described previously. This procedure uses up more energy than normal photosynthesis, and it is probably for this reason that C_4 plants are restricted to regions where solar radiation is very intense. But it conveys advantages. The limiting factor for the plants is carbon dioxide. Phosphoenolpyruvate (PEP) carboxylase has a higher affinity for carbon dioxide than the ribulose-diphosphate (RuDP) carboxylase that acts as the acceptor for the CO_2 molecule in normal photosynthesis. As a result, more carbon dioxide can be fixed per unit time and productivity can be greater than that possible with normal photosynthesis (Fig. 36 and 37). Examples of plants with this kind of photosynthesis can be found among the grasses and dicotyledons (e. g., Atriplex) in very sunny regions which (usually) have an irregular water supply. The familar crops maize and sugar cane are in this group, as well as aggressive, wide-spreading tropical

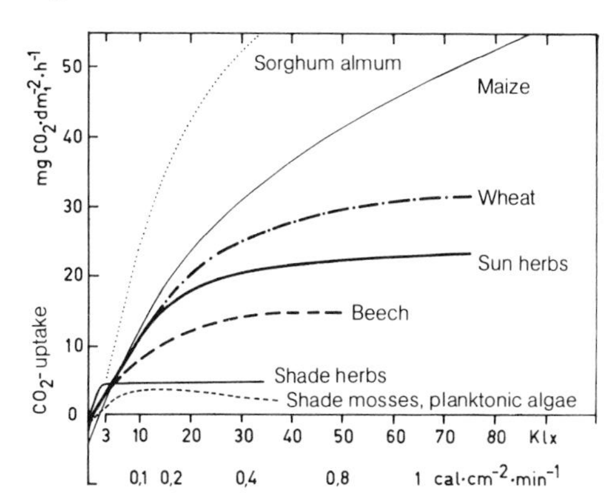

Fig. 36. Light-dependence of net photosynthesis in different plants with the natural amount of CO_2 available. The C_4 plants sorghum and maize are more productive than the C_3 plants. (Larcher, 1973)

weeds. Because the difference between the two photosynthesis types is not clear-cut but gradual (all plants can bind slight amounts of carbon dioxide to PEP), a number of attempts are currently being made to find and cultivate varieties of other crop plants with C_4 photosynthesis. The superiority of C_4 plants over C_3 plants is based on yet another principle: C_3 plants have so-called photorespiration. That is, they lose considerable quantities of organic matter by respiration even during the day – in fact, the daytime respiration level is about 5 times as high as that at night. The enzymes involved are not located in the mitochondria, but in the very small peroxysomes. For a long time there has been speculation about the significance of this photorespiration, which cannot be found in C_4 plants. Now it has turned out that RuDP carboxylase, which couples carbon dioxide to the RuDP molecule, often "confuses" oxygen with carbon dioxide, especially at high oxygen concentrations. When this happens the ribulose diphosphate is oxidized to form phosphoglycerate and phosphoglycolate. The latter is oxidized to glyoxylic acid; the hydrogen peroxide produced in this reaction is converted to water by peroxidase. During photorespiration – the oxidation of glycolate – the plant obtains no energy; the process represents a pure waste. The entire pathway is probably explicable only as a "historical relict." It originated at a time when the oxygen concentration was still significantly lower and that of carbon dioxide higher than today (see p. 1). The C_4 plants, with their higher specific affinity for carbon dioxide, have solved the problem of CO_2 deficiency, but the C_3 plants have not. On the other hand, the C_4 plants require greater amounts of energy for their actual photosynthesis – a higher level of solar radiation per unit time. For this reason they cannot spread into regions too far from the equator (Ehleringer, 1978).

A very similar modification, but complicated by introduction of a timing factor, is found in many succulent desert plants (Fig. 38). Their stomata are opened only at

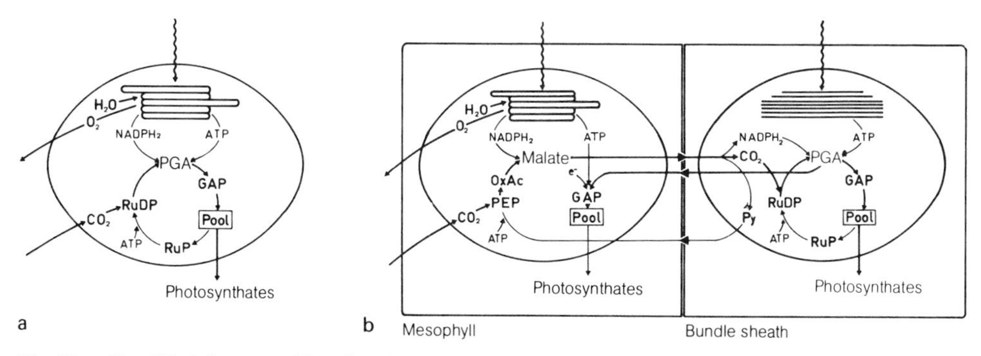

Fig. 37. a Simplified diagram of CO_2 fixation and the production of photosynthates by C_3 plants. **b** C_4 plants. *PEP* phosphoenolpyruvate; *RuDP* ribulose diphosphate; *PGA* 3-phosphoglyceric acid; *GAP* 3-phosphoglyceraldehyde; *Pool* intermediate C_3 to C_7 compounds; *RuP* ribulose-5-phosphate; *OxAc* oxaloacetate; *Py* pyruvate. The regeneration of *PEP* from *PGA*, in which water is given off, is not shown. (Larcher, 1973)

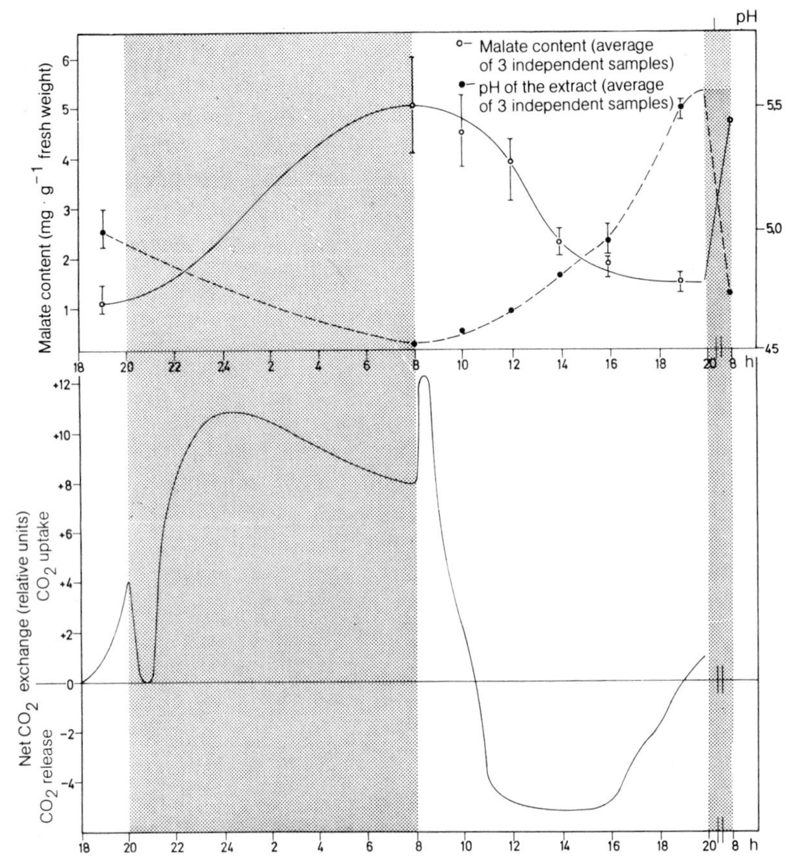

Fig. 38. Changes in the malate content and pH within the vacuole of Tillandsia usneoides in the course of a day. (Kluge et al., 1973)

night; at this time they bind carbon dioxide to PEP as the C_4 plants do. Most of the oxalacetate is converted to malate and stored in the vacuoles. As it accumulates during the night, the pH of the vacuole contents falls dramatically with the increasing concentration of organic acids. In the morning the stomata close, and carbon dioxide is split off from the malate and enters the normal photosynthetic pathway. The pH of the cell sap again approaches the alkaline range (diurnal acid rhythm). By keeping the stomata closed in the daytime the plant loses no water vapor during the period of greatest heat and low relative humidity – and for desert plants, water conservation is important.

It may be that the C_4 pathway of photosynthesis gives still another selective advantage. In the high-performance plants conversion of the C_4 acids to carbon dioxide and pyruvate occurs in specific bundle-sheath cells that are indigestible by ectothermic herbivores. Grasshoppers that eat such grasses excrete undigested bundle-sheath cells. These plants are therefore a relatively unprofitable source of food for such animals (Caswell et al., 1973, 1977; warmblooded herbivores have no difficulty in digesting the bundle-sheath cells). The pronounced daily rhythm of acidity in the succulents may also provide protection from plant-eating animals, but this possibility has not yet been documented.

As far as ecology is concerned, a crucial aspect of productivity is the site of incorporation of the organic matter produced by a vascular plant – does the new substance take the form of nectar, seeds, leaves, roots or wood? At present there is no satisfactory answer to this fundamental question. Evidently the different plants vary in this regard; annuals differ from perennials. In addition, there are endogenous control mechanisms, most conspicuous and self-explanatory in biennial plants. And within a given species "resource allocation" depends very much on external factors. A well-fertilized and well-watered plant develops only a small root system, whereas under water stress or conditions of mineral scarcity a larger, and usually deeper, root system is elaborated. Flowers and seeds account for a smaller proportion of the total growth of a plant when water is abundant than when it is in short supply. This may be an absolute rule for some species; these form flowers and seeds only under conditions of water stress. Finally, the growth of a plant is critically affected by the amount of light received (photomorphogenesis, the regulatory hormone phytochrome).

These considerations lead us to the effects of light on animals. For animals, light is almost never a source of energy; it is most significant with respect to orientation of an animal in space (via the eyes) and in time (the alternation between light and darkness acts as a timing signal for the daily and annual rhythms of animals). Light thus appears not to be a factor necessary for animal metabolism, so that one would expect there to be no problem in raising animals in permanent darkness if other conditions were favorable. Such experiments have been done with only a few species, and most of them have failed. Drosophila, terrestrial chironomids and cicadas, for example, have very high mortality rates in maintained darkness. No one knows why.

As a model of the ecological effect of light in the context of the physiology of metabolism, consider Vitamin D. This is taken up as 7-D-hydrocholesterol; in the skin, by the action of short-wavelength sunlight, it is converted into cholecalciferol. This conversion does not occur if insufficient light reaches the skin. When the supply of Vitamin D is inadequate the skeleton is incompletely ossified, leading to the typical symptoms of rickets in children. Rickets is found predominantly at northern latitudes, where the winters are long und dark and UV radiation is weak even in summer. In tropical regions with intense sunlight the illness is essentially unknown. Dark-skinned humans living in the north are particularly subject to rickets, because most of the limited radiation available is absorbed by their skin pigment and cannot be used for conversion of the precursors into the vitamin. Conversely, people with light skins exposed to the strong tropical sunlight are in danger of excessive ossification. The problem of insufficient irradiation can be simply solved by taking vitamin pills, but so far no means have been found by which overproduction of Vitamin D can easily be prevented.

This example suggests that the distribution of animals may be closely related to light, but there is as yet practically no confirming evidence. It also appears plausible that animals of valleys and the high mountains might differ in their radiation requirements and tolerance. Such a possibility has often been proposed, but again evidence is almost entirely lacking. Glück (1979), however, has shown that there is a clear correlation between the locations where birds breed successfully and the amount of light falling on their nests. He found that the nests of chaffinches, goldfinches, and hawfinches were located where the total irradiation was relatively high, whereas greenfinches, serins, and linnets choose nest sites with a distinctly smaller amount of light.

Some animals – unlike plants – are active at night. The cause of nocturnal activity is frequently the higher relative humidity that prevails at night. In this case, then, light acts solely as a timing signal.

5. Oxygen Supply

Organisms need oxygen to break down organic matter and thus obtain energy. When no oxygen is available other, less efficient metabolic pathways can be used to provide energy. A lack of oxygen can occur temporarily when there is a sudden marked increase in energy consumption (this probably happens only in animals), or it can be a maintained condition; there are organisms capable of living in a milieu extremely low in oxygen or permanently lacking it. Such a situation arises only underwater or in the soil; actual terrestrial organisms never encounter it.

On the land – that is, in the air – there is always an adequate supply of oxygen. It is limited to a certain extent in the high mountains. Here birds and mammals find it very difficult to achieve peak performance. Land animals seem to have had hardly any success in evolving hemoglobin with greater oxygen affinity. This fact is remarkable, in that aquatic animals have been able to take this evolutionary route; modifications utilizing the Bohr effect and the Root effect (cf. Hochachka and Somero, 1973) have made it possible for the oxygen to be released to their organs as required despite the high oxygen affinity of their hemoglobin. Adaptation to the high mountains mainly involves an increase in the amount of hemoglobin in the blood, so that more oxygen is transported. To some extent this increase in blood hemoglobin content is a simple acclimatization acquired after a prolonged stay at altitude, but in some cases genetic factors play a role. At the moment it is impossible to decide whether these measures – a slightly increased oxygen affinity of the hemoglobin and a greatly increased hemoglobin concentration – really represent the only response of animals to the low oxygen supply at great heights. Relevant findings are few, although an intensive program of physiological research is currently underway. Data are particularly scarce for all ectothermic animals and for plants.

To permit brief bursts of high-performance activity, with short-term high oxygen consumption, many animals store oxygen in their musculature. Oxygen is transferred from the hemoglobin in the blood to the myoglobin in the muscles, giving the muscle tissue of these animals a red color, and enabling the animals to run rapidly or fly over long distances. The breast musculature of pigeons, falcons, curlews and gulls is, because of its myoglobin, as red as the muscles of hares, deer and antilopes. By contrast, animals that move at top speed only briefly, and then hide, do not store oxygen in their musculature; such muscles (those of rabbits, for example, and the pectoral muscles of most fowl) are white. They consist entirely of contractile fibers, and are capable of maximal activity only for a short time. Oxygen is also stored in the myoglobin of whale muscles, which are a deep red; whales dive without first filling their lungs with air and so must rely on the oxygen in the myoglobin while they are under water.

There are additional ways of supplying energy where it is needed during such "active oxygen deficiency," via metabolic pathways other than the normal glycolysis followed by citric-acid cycle and respiratory chain. These alternative pathways are present in bacteria and thus do not represent recent evolutionary acquisitions. We must assume that they have been carried along during evolution as a "genetic load," until it suddenly again became selectively advantageous to use them (Thauer, 1977). Chief among them in warm-blooded animals is lactate fermentation; here glycolysis does not feed into the citric-acid cycle but branches off with the conversion of pyruvate to lactate and ends there.

The yield of energy is only 2 ATP per mole of glucose, as compared with 38 ATP for the classical pathway (Fig. 39). This alternative may be more widespread than was previously thought. It has been demonstrated, for example, in marine worms. During the escape movements of cephalopods and molluscs (Figs. 39 and 109) octopine fermentation occurs, another strictly temporary, low-yield alternative. Octopine, like lactate, is subsequently reconverted under normal conditions and returned to the metabolic pool.

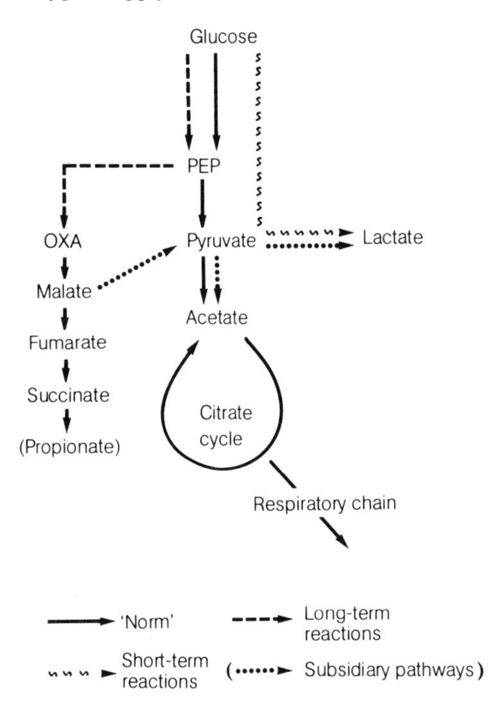

Fig. 39. Biochemical pathways by which animals obtain energy. The normal (aerobic) pathway involves glycolysis, the citrate cycle, and the respiratory chain; the (theoretical) energy yield is 38 ATP per mole glucose. "active anaerobic" pathway during short-term maximal muscular work, energy yield 2 ATP per mole glucose, end product lactate. long-term anaerobic pathway, used in oxygen-deficient media; energy yield 8 ATP per mole glucose, end product propionate. *OXA*, oxalacetate; *PEP*, phosphoenolpyruvate. (Zebe, 1977)

If an oxygen deficiency should arise in the surrounding medium many organisms can switch, for prolonged periods or even permanently, to other means of obtaining energy. Familiar examples are the fermentation of acetic acid or alcohol in microorganisms and that of succinate by worms parasitic in the intestinal canal (Fig. 39). These forms appear to have discarded the "normal" (and phylogenetically younger) pathway by which energy is derived. But recently it has become increasingly evident that many aquatic animals are very flexible; whereas they ordinarily operate with the citric-acid cycle and the respiratory chain, they can use fermentation pathways under short- or long-term environmental oxygen deficiency. This is true of Arenicola, many marine and fresh-water clams, Tubificidae and earthworms – and no doubt of many other organisms. The alternative pathway as a rule involves conversion of succinate to propionate; a small fraction of the malate is converted via pyruvate to lactate or to acetate. The energy yield of succinate fermentation is probably 8 ATP, higher than that of lactate fermentation though not as high as is obtainable with the citric-acid cycle and respiratory chain. It suffices to guarantee the existence of many aquatic animals for relatively long periods in a medium lacking oxygen.

One must bear in mind, however, that in all these cases there is not only a reduced energy yield but a very high food requirement. The increased food supply cannot be actively provided, for any extra activity would use up more energy. The amount of food immediately available must therefore be extraordinarily high if the alternative pathways described are to mean more to the animals than just survival with loss of energy (for a summary see Zebe, 1977).

In the ocean, oxygen deficiency occurs primarily in the tidal zone during ebb tide; clams and mussels close their shells, thus shutting off the flow of fresh oxygenated water past their gills as effectively as it is shut off from the tubes occupied by crustaceans and polychaetes. In fresh water and in enclosed bodies of sea water (the Baltic and Black Seas) the conditions bringing about oxygen deficiency are of particular current importance to humanity.

The oxygen in the water is derived from two sources. First, it can diffuse into the water from the air. This is a slow process, and it supplies oxygen to underlying layers only if the surface water moves into the depths. In lakes such circulation is brought about only by high winds; because such strong wind is rare during the warm summer months, there is hardly any stirring of the water at the time when the temperature is high – just when the metabolic rates of all the organisms, and thus their oxygen consumption, are especially high. The second supply of oxygen comes from aquatic plants, especially algae floating in open water, in the plankton. These are of course restricted to the upper, illuminated water levels. If the body of water contains abundant nutrients for the plants, a particularly large mass of planktonic algae develops. The consequence is increased turbidity of the water, so that light penetrates less deeply and photosynthesis is concentrated in the uppermost layers. When the water is heavily fertilized, then, it is precisely the most endangered levels that are deprived of oxygen. The danger is increased in that during such a "bloom" dying planktonic algae sink to the depths in vast numbers. There they are decomposed by bacteria – a process that naturally requires oxygen. The oxygen at the bottom of the river or lake can be used up completely, so that all the animals living there die. This is the reason why eutrophication of inland waters is such a severe threat; fertilization here has effects quite different from those on land. In a lake poor in nutrients light penetrates deeply; planktonic algae can carry on photosynthesis and produce oxygen at depths of more than 50 m. The end result is that fish production in a nutrient-poor lake is higher than in a lake rich in nutrients. Of course, there are lakes that have a high nutrient concentration even without human intervention. Reichholf (1975) showed that flocks of ducks overwintering on nutrient-rich reservoirs along the Inn River graze off all the water plants, removing from the water organic material which otherwise would sink to the bottom and rot there – a process that would use up oxygen. The ducks themselves, being air breathers, withdraw no oxygen from the lake. The important point is that the ducks must not be disturbed, so that they will remain in large numbers. They will then prevent loss of oxygen from the water and thus postpone the reversion of the lake to land (Reicholf, 1975b, 1977).

6. Fire

Fire is a regularly recurring factor in many natural ecosystems. Spontaneous combustion and lightning are its most common causes. Fires at regular intervals are a feature of the dryer parts of the tundra, throughout the taiga, in all savanna and steppe regions, and in all Mediterranean plant communities in the broadest sense – including the chapparal of California and the pine forests of Florida as well as the sclerophyll woodland around the Mediterranean Sea. Some eucalyptus forests in Australia are also exposed routinely to fires of natural origin. The detailed investigation Zackrisson (1977) made of the burn scars on very old trees showed that in the northern European taiga, before humans intervened to protect it, the forests were swept by about two fires per century. Similar figures have been obtained for the Canadian tundra and the tundra of Alaska. Fires are still more frequent in savannas, steppes, and the Mediterranean regions. To simplify matters, one can say that in general pines (the genus Pinus), oaks (Quercus) and all the Ericaceae are typical "fire plants." Because fire is a regular event in their natural habitats, they have become highly modified in adaptation to such conditions.

Their thick bark gives excellent protection against fire, allowing them to survive without difficulty. If large regions of bark should be destroyed by the fire, regeneration hardly ever presents a problem. The seeds of pines and of the heather Calluna germinate particularly well after being subjected to heat stress. Indeed, the cones of many pine species release the seeds only after they have been warmed to 70°–80 °C. Germination thus occurs after competitors have been eliminated; the seedlings of pines and of Calluna are very sensitive to competition. The lichens Lecidea anthracophila and Lecidea friesii grow only on charcoal and are therefore strictly dependent on forest fires for their existence. Similar adaptations to fire are exhibited by the animals inhabiting these regions. Beetles of a number of species seek out very warm wood in which to lay their eggs. One metallic wood borer has actually been found to have infrared sensors that enable it to find freshly burned-over areas.

The heath that germinates and grows rapidly after a fire provides very many birds and mammals (hares, especially) with far better food than old or cut heath. For centuries the heath in Scotland has been burned regularly to keep the population of willow grouse and alpine hare large; the effect on black grouse populations is similar. The new growth of heath after a fire contains more nutrients and appears to be more palatable to animals than either old heath or the new growth following mowing. In the pine savannas of the southeastern United States the parts of the plants that grew out after either fire or cutting were found to contain distinctly more N, P, K, Ca, and Mg, and the amounts of N, Ca, and Mg were greater after fire than after cutting. Not until 4–6 months had elapsed did the mineral content of these plant parts return to normal (Christensen, 1977).

Other species are extraordinarily sensitive to fire – spruce, for example, and most deciduous trees such as beech and linden. Pure spruce forests or beech forests catch fire only after strong winds have blown sufficient combustible material to the ground. It is generally valid to say that fire is not a natural event in most deciduous forests (apart from oak woods and Mediterranean sclerophyll regions).

Fire plays a leading role in determining the composition of plant communities. The northern European-Siberian taiga belt, in which spruce and pine predominate in different proportions, is a product of fire. Without repeated burning the forests here, where it is not too dry, would consist exclusively of spruce. The recurrent fires do severe damage to the spruce trees, and it is only for this reason that the pines can maintain themselves. Fire resistance could be regarded as a direct mechanism for competition; pines shed a great deal of burnable material onto the ground, where it accumulates in loose piles. The fire sweeps rapidly through this fuel, injuring the spruce as it goes. By contrast, a dense stand of spruce is practically invulnerable to fire. The short needles become tightly packed in the ground and so are almost impossible to set alight. Once it covers a certain minimal area, then, a spruce forest is to a great extent protected from burning. The situation is similar in the oak savanna of North America, where the oaks (like the cork oaks in the Mediterranean region) develop specific kinds of bark that enable them to survive even intense fire. When the acorns germinate, they are free of competition. The well-known oscillation of hare (and thus of lynx) populations may perhaps be associated with fire cycles.

Man has fought fire wherever he could. As a result, the frequency of fires in all regions inhabited by humans soon declined sharply. The result was a drastic change in the composition of the local plant communities. In northern Europe the range of spruce distribution expanded, and in the North American oak savanna brush displaced the original oaks. Bushes also invaded the savannas of Africa. Specifically fire-adapted animals became rare.

To offset such changes, a system of controlled burning was introduced some time ago in North and South America, Africa, and recently in Europe. Because of this conservation measure the oak savannas of North America have regenerated in many places. Fire has also been used to keep the spruce population down in parts of northern Europe and in the national parks of North America, where the growing spruce were slowly inhibiting germination and growth of the giant sequoias the parks were established to protect. In the Florida Everglades, too, fire has been and is being widely employed to maintain the natural vegetation [cf. the papers of Riess (1975, 1976) and that of George (1972)]; without

its aid, the natural stands of pine would disappear.
It was a laborious process to learn how to handle and live with such intentional fires. Today we know how to light very cool fires, with an effect like that of mowing, under certain moisture conditions; we can also lay very hot fires that burn off part of the raw-humus layer and speed up the breakdown to real humus. We know the difference between hot fires that go with the wind and the relatively cool fires against the wind. The insect world suffers to varying degrees from the different kinds of fire, but in general the effects are less severe than was expected.

7. Interspecific Competition

In many cases competition between species has been considered responsible for the spatial distribution of organisms. It is particularly likely to be invoked when no ecological factors can be found to account for the absence of a species from a certain place. But often those who fall back on this explanation are making things too easy for themselves. Without real evidence it is not justified to speak of competition as an ecological factor.
In each case a distinction must be made between historical and contemporary competition. There can be no doubt that the adaptations of organisms to different conditions of life arose by competition. The question is whether today, now that the organisms have become adapted to different environments, competition between them continues. The question must be investigated separately for animals, plants, and microorganisms. Because of their mobility and their sense organs animals are capable of choosing extensively differentiated biotopes. This choice of biotope affects not only the stage that makes it, but the next generation as well – whether the young are actually cared for or simply provided for. The majority of land animals make at least some provision for their offspring, by laying their eggs in places that offer favorable conditions for development of the next generation. In flies of the genus Limosina, with larvae that live in a variety of rotting substrates that includes dung heaps, the litter in stands of plants, and the debris left on the beach by the tide, selection of a site for egg-laying is highly specific, depending on the species of the rotting plants and the salinity of the fluid between the plants. The dragonfly Leucorrhinia dubia lays its eggs only in acid water; although the larvae would be quite capable of surviving in alkaline water, they are restricted by the action of the mother to water of low pH. The extremely selective choice of a host by parasitic wasps has been studied by many researchers. Not only do these wasps seek out a particular host species; when a suitable host is found the wasp inspects it to see whether it is already infested by a parasite that might offer competition to the egg that is about to be laid. When such active distinctions are made during the choice of a habitat competition is kept to a minimum. How realistic, then, are experiments in which animals of two species are brought together in an artificial habitat? All such experiments result in the dying out of one or the other species in a more or less short time. Under natural environmental conditions the two species would probably have chosen to occupy different places. Such exercise of choice can also be demonstrated experimentally. If rice beetles of the genera Tribolium and Oryzaephilus are put together in flour, Oryzaephilus dies out relatively soon. If small glass tubes are added to the medium, the larvae of Oryzaephilus hide in them. The system now actually consists not of a single habitat, but of two. Under such conditions the two beetles can thrive side by side for many generations (Fig. 40).
All this evidence implies that under present-day conditions competition between different species in the field does not occur, because of the animals' choice behavior. These experiments underline a principle that has always been given special consideration in ecology — the Exclusion Prin-

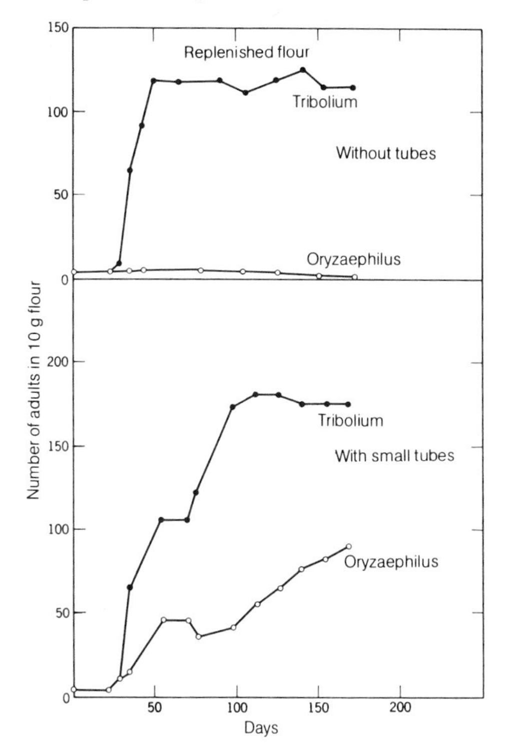

Fig. 40. When two beetles are living in flour that is continually replenished, Tribolium eliminates Oryzaephilus. But if small glass tubes are put into the medium, the two species can coexist. (Crombie in MacArthur and Connell, 1970)

ciple, to give it its descriptive name, though it is also called Monard's Principle or Gause's Principle after its discoverers. According to this principle, two animal species found together in a particular habitat must nevertheless differ ecologically. Because the biology of very closely related species – those in the same genus, for example — is in general quite similar, the principle would require that very closely related species either do not occupy the same habitat simultaneously or, if they do, their requirements within it are very different (as in the example just cited).

Excellent examples of the ecological separation of very closely related animal forms are found among the parasites. Various species of bird lice and mites take up highly specific positions on the bodies of their vertebrate hosts (Fig. 5). Work by systematists has laid a good foundation for understanding this phenomenon, but there have been few experimental studies of the problem. Site selectivity by parasites offers a particularly good approach for experimental research on the ways resources are utilized in a system. It is to be hoped that experimenters will soon begin to build on the existing foundation, for the area is one of great importance to ecology.

A way in which ecological boundary-lines can be drawn between closely related species that evidently live on the same food within a given habitat has been demonstrated by Hylleberg (1976), who studied three barely distinguishable species of Hydrobia (Fig. 41). These three species are to be found living together in saline meadows. Even a specialist can hardly tell the difference between them. But they differ distinctly in their development. Hydrobia ventrosa undergoes direct development, whereas Hydrobia ulvae has planktonic larvae. Moreover, the feeding habits of the adults are different; although this is very hard to observe in the field, it can be de-

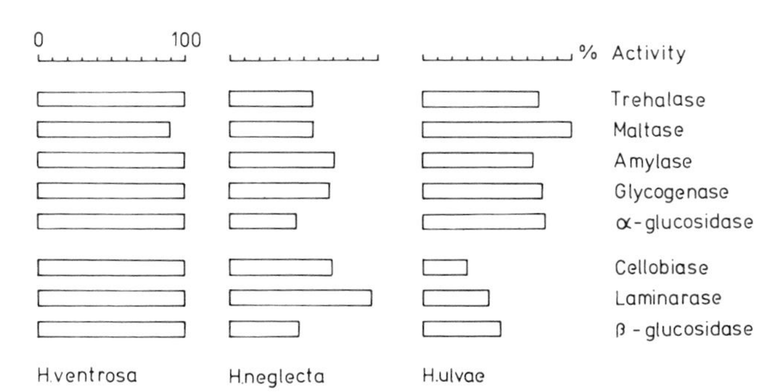

Fig. 41. Differences in the activity of basic enzymes in three coexisting species of Hydrobia. (Hylleberg, 1976)

duced from the differences in activity of their digestive enzymes.

Recently another sort of situation has attracted interest – that in which there is obviously no possibility of making an ecological distinction between species living side by side. Experimental analysis and the subsequent calculations have made plausible the hypothesis that these species are prevented by other factors (predators, parasites) from attaining population densities anywhere near those that could be supported under the existing conditions of space and food availability. In such circumstances, it is of course possible for closely related, ecologically indistinguishable species to live together. Competition arises only when the number of individuals strains the capacity of the habitat, which is not the case here.

The quantitative description of the requirements of a species in such a system and of the delimitations between these species has lately been denoted by the word "niche"; during evolution, organisms become adapted to a particular niche. Even before this term became popular, exactly the same sort of analysis of the occurrence of a species in the context of its ecological requirements was regularly being done – by Strenzke (1951) and Zahner (1960), for example. The original niche concept of Elton has an entirely different significance. In Elton's usage, "niche" denotes the "profession" of an organism in the system, its ecological function (Figs 1, 42, and 43). According to Elton, penguins in the southern hemisphere and auks in the northern hemisphere would occupy the same ecological niche. Moreover, other authors have used the term in still different senses. To avoid confusion, the word "niche" should be qualified whenever it is used – for example, the "spatial or habitat niche," the "trophic niche" (Elton's connotation), or the "multidimensional niche" (to describe position in gradients of environmental factors).

Competition can become very vigorous between animals simultaneously attempting to colonize a newly formed habitat. It is largely a matter of chance which is the first colonizer of a pool or puddle left by a rainstorm; very many species are capable of entering such a new environment. But the one that arrives first occupies the space almost entirely, so that most later arrivals are excluded. The initial formation of a habitat is thus by no means the same everywhere (this also holds for plants, q. v.; cf. also p. 113f., K and r selection).

This element of chance can have long-term implications for the colonization of shallow seas by animals. During a very harsh winter, a large fraction of the fauna on the bottom of the North Sea dies, leaving this habitat practically unpopulated. But the plankton includes larvae of a great variety of bottom-dwelling animals – long clams (Mya arenaria), brittle stars (Ophiura albicans), cockles (Cardium edule), polychaetes (Arenicola marina), and many others. The individual larvae are not mixed up at random, but float in swarms comprising a single species. As soon as such a swarm reaches the vacant bottom, it decides the future of the colonization process. A swarm of long-clam larvae results in a dense population of long clams. Larva swarms that arrive afterward have no chance, for they are swept into the feeding currents of the clams, filtered out of the water and eaten. On the ocean floor, then, there is considerable competition between different species; the first to come excludes all the others, until it is exterminated by the next severe winter.

Moreover, the various facets of habitat and animal requirements must be distinguished. The resources of the African steppe are precisely distributed among the inhabitants of this steppe. There is very little competition between the individual ungulates. Giraffes graze the crowns of trees at the very top, while gerenuks eat the leaves lower down. The black rhinoceros strips leaves from the bushes; zebras graze on the tall grass and are followed by gnus and finally Thompson's gazelles, which can eat the short grass that remains. Food, then,

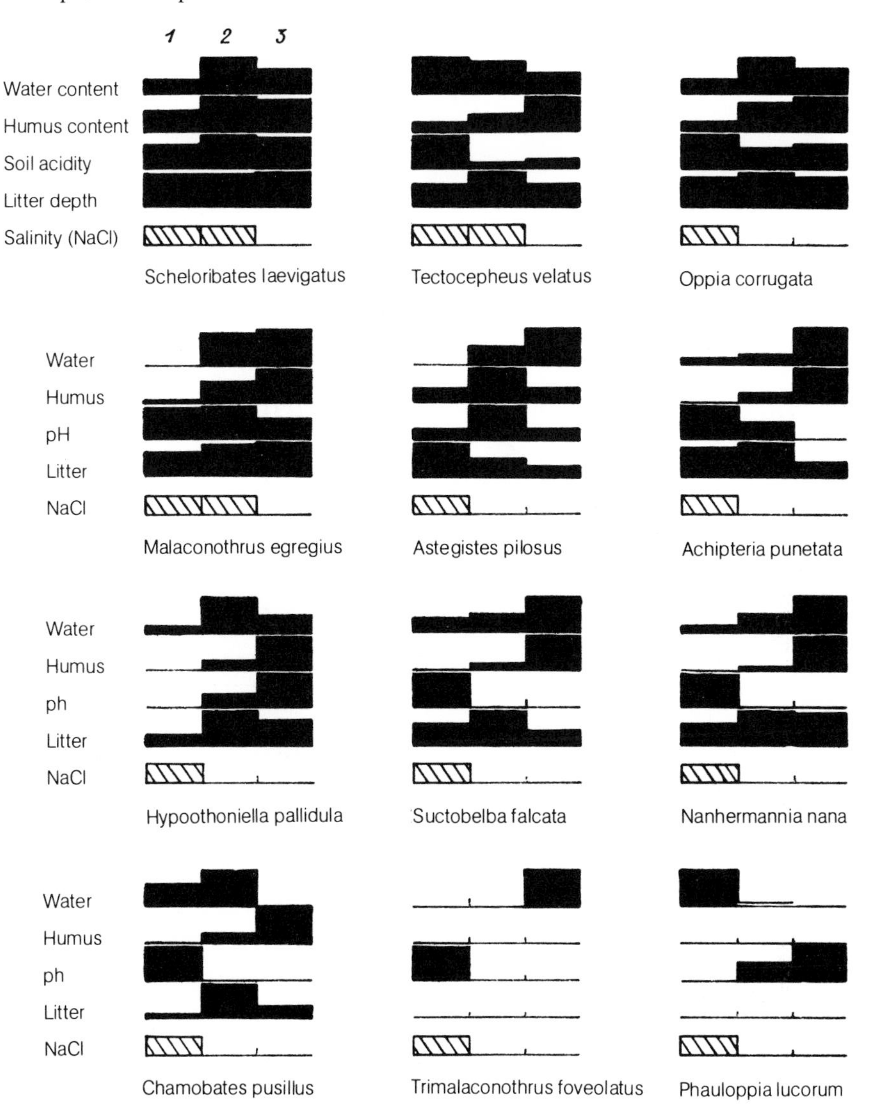

Fig. 42. Quantitative description of the requirements of 12 species of mite (Oribatidae), with respect to 5 edaphic factors. Columns 1, 2, and 3 represent increasing levels of the indicated factors. The height of the black bar in each column represents the percentage of all the samples for each condition in which the species was found to be present. To facilitate comparison these are normalized, so that the highest percentage found for each species and factor is represented by the maximal column height (0.5 cm). For example, Astegistes pilosus prefers very moist soil covered by a thin layer of litter, with intermediate amounts of humus, a pH near 7, and low salinity. Only the data for NaCl are estimated. This quantitative description of the distribution of an animal in terms of habitat characteristics, as done by Strenzke, is often referred to by the term "niche"; this current use of the term is fundamentally different from that of Elton (cf. Fig. 1). (Strenzke, 1951)

can be excluded as an object of competition between species.

But all these animals would vanish if there were not enough water holes distributed over the countryside. The animals visit these at fairly long intervals to drink; in certain circumstances, toward the end of the dry season, there can be fierce competition for water among the species mentioned. Interspecific rank orders become established, similar to the familiar pecking orders within a species. In general the

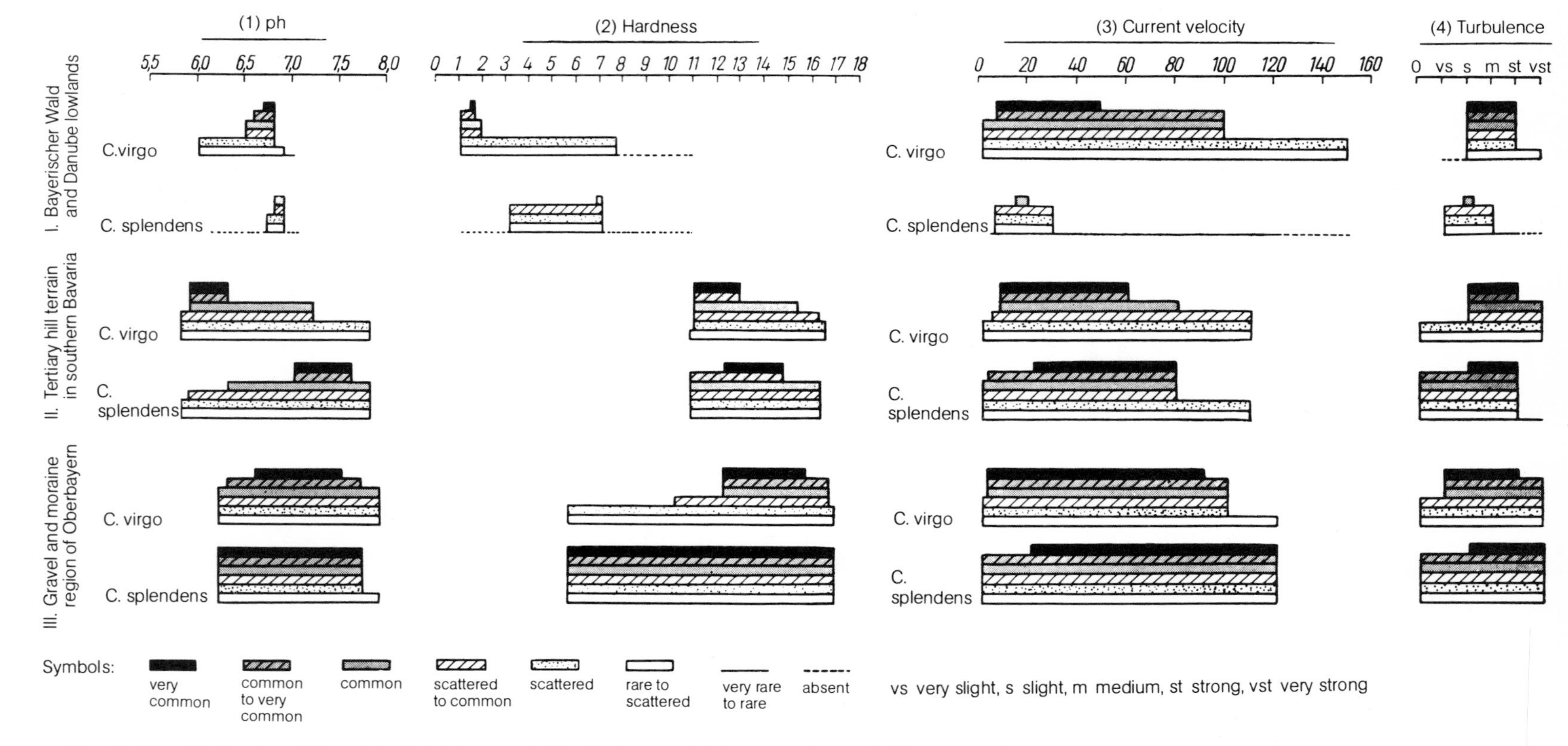

Fig. 43. Quantitative description of the habitats occupied by the larvae of Calopteryx virgo and Calopteryx splendens (Zahner, 1959). The bodies of water sampled were in the regions of Bavaria indicated at the left. Hardness is given in German degrees, °d (1°d = 17.8 ppm $CaCO_3$)

Fig. 44. A sable antilope driving zebras away from a water hole. (Klingel in Leuthold, 1977)

elephants are dominant, followed by rhinoceros and hippopotamus, and finally zebras and antilopes, in that order. But it is characteristic of these rankings that they can be rearranged. Sable antilopes can drive away zebras, and under some conditions even elephants (Fig. 44). Interspecific competition is thus a contemporary phenomenon in the case of vital resources which are called upon only briefly at long intervals, and thus can be sparsely distributed over the terrain; in fact, such competition is of widespread occurrence. Consider the following examples.

In North America sapsuckers (Sphyrapicus) drill particular patterns of holes through tree bark. The sap that emerges is licked up by the birds; moreover, they return at regular intervals and catch the insects that have been attracted by the sap flow. However, both sap and insects are eaten by a number of other vertebrates. There develops a rank order at sapsucker trees, with squirrels at the top. The male and female sapsuckers are relegated to second place, followed by downy woodpeckers, nuthatches and hummingbirds. Even though the hummingbirds are at the bottom of the list enough remains for them that their occurrence in North America appears to be largely determined by the presence of trees with holes made by sapsuckers (Fig. 45). Naturally, there is a corresponding rank order among the invertebrates that feed on the sap. During the day wasps lead the list, followed by flies of the genera Calliphora and Lucilia and finally by small flies (Fig. 47). The large Mesembrina is chased away by all the others despite its size. And there are a number of small flies (primarily Sepsidae) that are clearly superior to the large calliphorids and the other fairly large muscids. Holding their wings spread out from the body and moving them through a circular path,

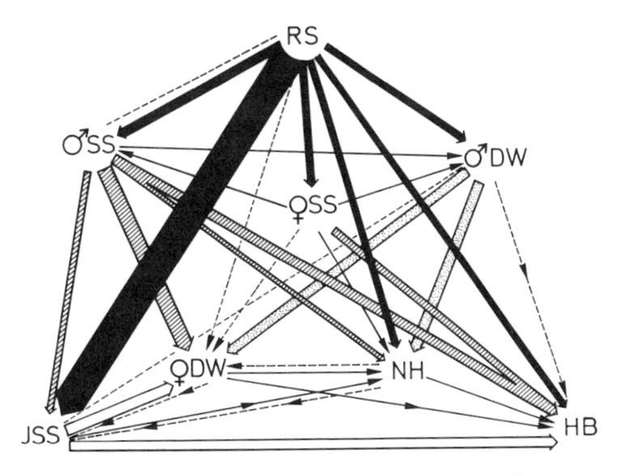

Fig. 45. Social hierarchy at sites where sap flow has been caused by sapsuckers. *RS* red squirrel; *SS* sapsucker; *JSS* juvenile sapsucker; *DW* downy woodpecker; *NH* nuthatch; *HB* hummingbird. (Foster and Tate, 1966)

Fig. 46. Moths of different species feeding on sap that has emerged from a tree

these small flies rush at the much larger ones; the dark spot at the tip of the wing makes the circular motion more distinct, and when faced with such an attack any large fly inevitably yields. Hornets are rare visitors to the sap source, but when they do come they are dominant over all others. At night there are three possible sets of feeders (Fig. 47). Either the site is entirely blocked by ants (Formica, Lasius), or earwigs (Forficula) achieve the same thing. In either case, the insects surround the sap in great numbers and allow no other animal to approach. But if only a few ants or earwigs are present, the predominant visitors are moths. As many as ten different species, represented by about equal numbers of individuals, can be found simultaneously at a single site. In general the large underwings (Catocala) dominate; the other moths are all about the same size, and give way to the wingbeats of the underwings. Only if the moths leave sufficient space do carabid beetles and bush crickets come to the food source. When tree sap is made available in this way, the number of species and individuals in an area can be distinctly increased.

We have mentioned several times the large differences in density of animal populations on poor (granitic) and rich (basaltic, calcareous) soils (cf. Fig. 116). Evidently when a substance is present only in limiting amounts in the soil, it can be supplemented if scattered exposed stones (or salt licks made by humans) provide the necessary trace elements. Gorillas in the African jungle migrate considerable distances to such sites, as do orangutans in southeastern Asia (Mackinnon, 1974). It has been postulated that elks can live where the soil

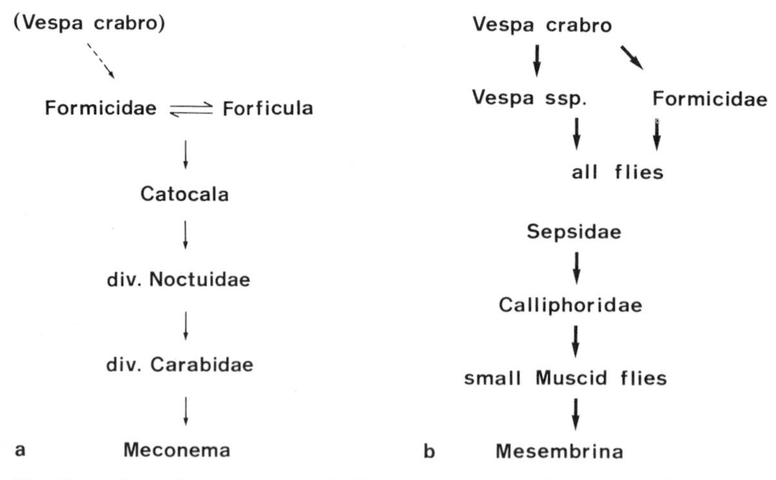

Fig. 47 a, b. Social hierarchy at sap-feeding sites. a At night, b during the day

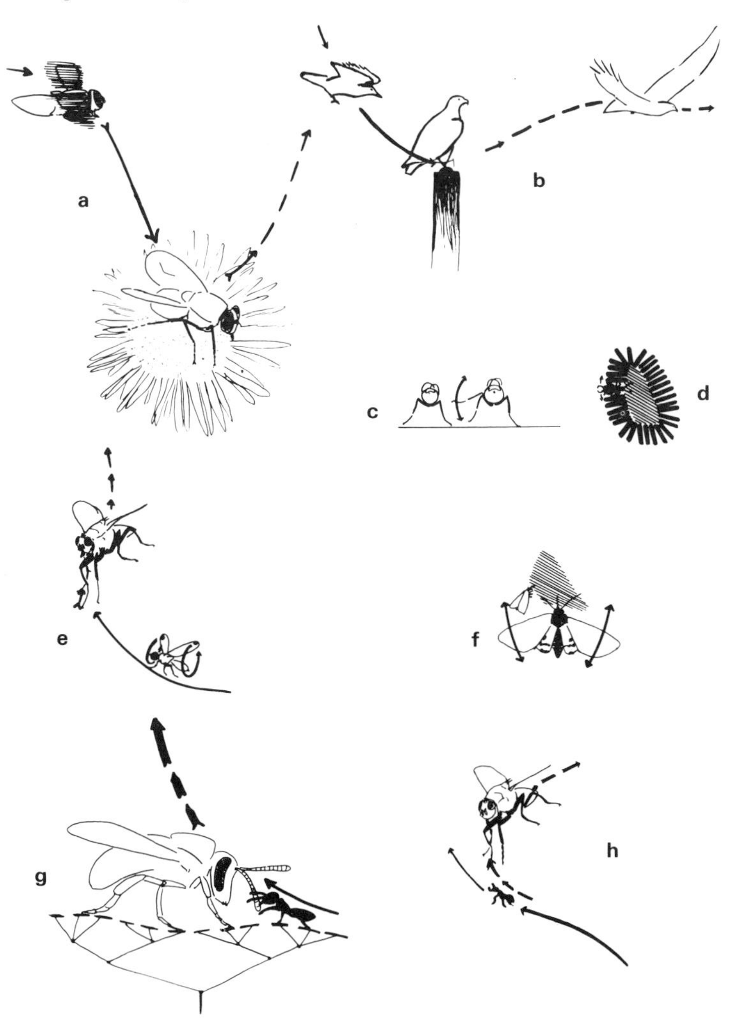

Fig. 48 a–h. Behavioral mechanisms in competition. **a** A fly hovers behind another that is on a flower and thus drives it off. **b** A similar situation, in which a crow attacks a buzzard, and then perches on the pole itself. This counts as competition only if lookout points are in limited supply. **c** Distance-keeping mechanism; one fly keeps its neighbor away by moving its middle leg. **d** Earwigs (Forficula) achieve the same thing by moving the end of the body back and forth. **e** A sepsid fly beats its wings while walking toward a calliphorid, and the latter withdraws. **f** Catocala drives other moths away from the vicinity by beating its wings. **g** An ant bites the antenna of a wasp, and the wasp flees. **h** A fly moves out of the way of an ant

is poor only by eating plants rich in sodium (Botkin et al., 1973). In central Europe it has been the practice since the Middle Ages to set out special licking stones for the wild animals living on poor soils; even at that time, this responsibility was assigned to forestry officials. In Africa, too, there seems to be a relationship between salt-containing material and animal density, as there is in Australia (Blair-West et al., 1968). There can be considerable competition for such salt sources; salty plants like Salicornia are a preferred food of rabbits and hares in inland habitats. These findings have not been universally accepted. The ability of animal kidneys to retain large amounts of sodium has repeatedly been raised as an objection. Probably salt content is closely related to population density only where the soil is very low in minerals – which in present-day agricultural countries, with winter salt-spreading and summer fertilization, practically never occurs (cf. Weeks et al., 1976).

We can think of the resources in these cases as "communal facilities" within the habitat, which are important to many species but used by each for a relatively brief time during its life cycle, and with respect to which there is a rank ordering. Water holes, salt-licks, lookout points, special feeding sites [flowers for insects, honeydew (Reichholf, 1973)], and overwintering shelters are examples of such communal facilities and are responsible for great diversity of species. Competition at an unexpected place can determine the distribution of a species.

The higher plants are in quite a different situation. The individual plant is not capable of selecting a suitable habitat; where the seed germinates, the plant must grow or perish. Only a few plants (e.g., Cuscuta) can change position slightly by growth and thus have a very limited ability to choose. For this reason, the distribution of vascular plants is largely the result of marked interspecific competition. Species better adapted to a particular environment crowd out others less well adapted, even though the latter are physiologically entirely capable of survival on the site (Fig. 49c). We exploit this situation when we cultivate plants in our gardens; a variety of species can thrive there, as long as we protect them from competition by others. The difficulties we encounter in raising animals are incomparably greater. The fact that the distribution of higher plants is determined by competition has an important consequence. A fairly large firmly rooted plant can survive in a habitat where it would not naturally occur, and can prevent considerably better suited plants from becoming established there. In parks everywhere, and frequently in forests, trees are growing that could not hold their own under ordinary circumstances. When they were young, humans protected them from the pressure of interspecific competition; when grown, they need no more protection and are fully competitive (though their seeds are not). Once an artifical plant community has become established it can be extremely stable, and can successfully resist displacement by the naturally occurring community. It may be possible for many of the dry grass regions created by people in central Europe to maintain themselves for many decades before they slowly, through intermediate stages, give way to the natural vegetation. With very little human assistance such grasslands can survive indefinitely and exclude the natural vegetation. The best adapted in this case is not a particular species, but the old plant well rooted in the soil.

This illustrates a further principle. Many plants are found in nature on sites that by no means correspond to their physiological optima. When competition is removed they can grow on physiologically optimal sites, where they provide far higher yields than they do in their natural habitats. Physiological and ecological optima are thus entirely different things.

Microorganisms resemble plants in this regard. Their distribution, too, is largely the result of interspecific competition. They too must germinate on the spot and then grow or perish; no real opportunity for choice is available.

All these discussions raise the question of the mechanisms underlying interspecific competition. What is its basis? There are various mechanisms, and these have been assigned to different categories – for example, interference by mutual disturbance, or exploitation when the food resources of one animal are destroyed by another. As important as such distinctions are in general, they are of little use in a small introductory text – especially since they have been developed for particular groups of animals and are thus not applicable to microorganisms and plants, and do not even apply to all animals.

A particularly impressive method is the elimination of competitors by means of chemicals, a familiar occurrence among

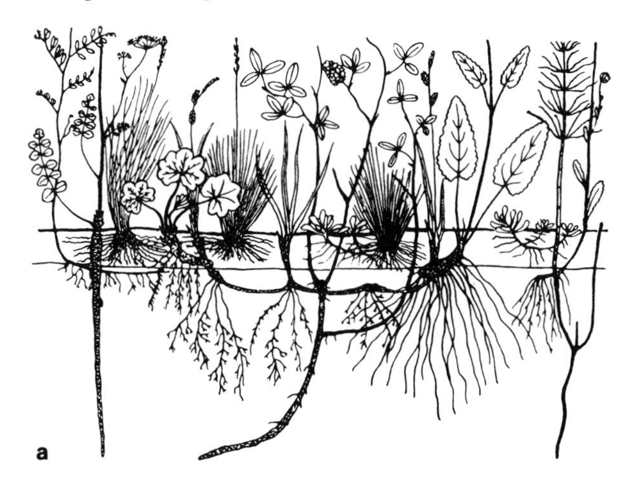

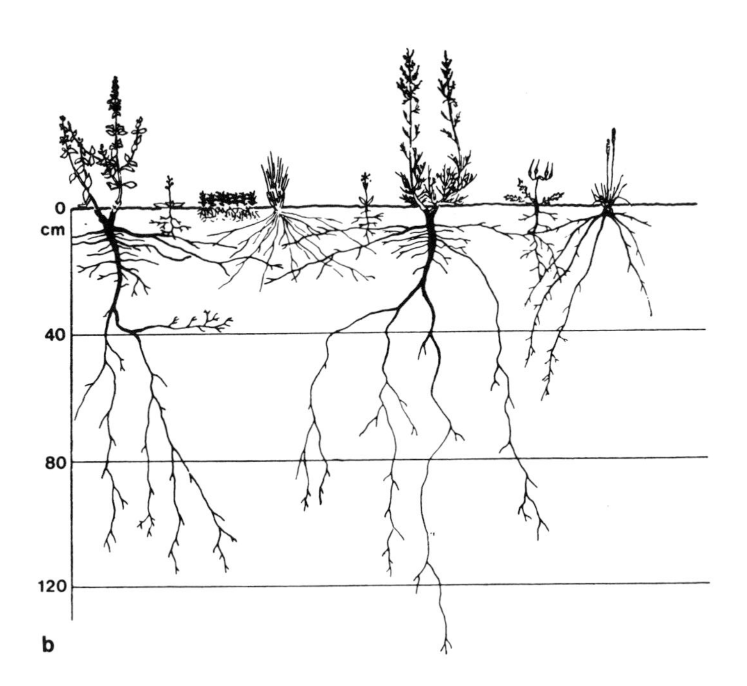

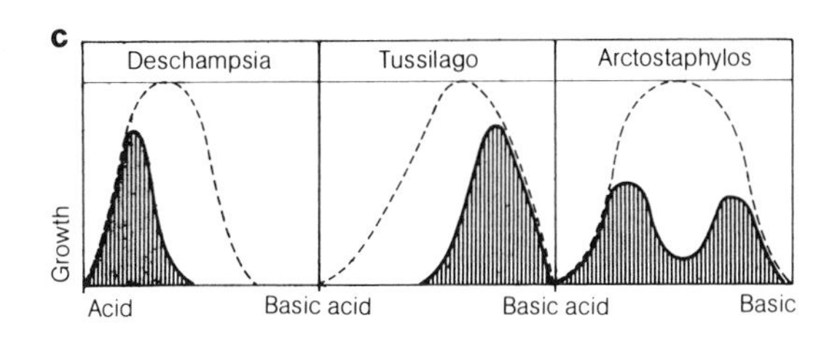

Fig. 49. a, b Reduction of root competition in stands of plants by stratification. (Walter, 1973). **c** Diagram of the pH-dependent behavior of different plants in monoculture ("physiological optimum curve", *dashed*) and under conditions of natural competition (ecological optimum curve, *solid* line with shading below). In the region between the solid and dashed curves each species can grow only if it is cultivated in rows and protected from the competition pressure of other, better-adapted species. (Ellenberg and Knapp in Larcher, 1973)

microorganisms (penicillin!). The famous algal poisons are probably in the last analysis a mechanism for competition. The highly poisonous toxin produced by the dinoflagellate Gonyaulax is one of the most lethal to humans. But marine animals can make use of it. Filter-feeding molluscs (e.g., Mytilus, Modiolus, and Mya) accumulate this toxin during a Gonyaulax bloom and then become extremely poisonous to many animals. In regions where Gonyaulax lives the molluscs do not lose their toxicity completely at any time of year (Krämer, 1978).

In all these forms allelopathic effects are the rule rather than the exception. In 1965 Vaartaja and Salisbury isolated fungi and actinomycetes from the forest floor, separated them by species and cultured them on agar, with two different species in each container. Of the 1205 different combinations that resulted, far more than half exhibited reciprocal inhibition. Only in a very few cases could stimulation be observed.

Chemicals also seem to play a central role in the competition among higher plants and between these and microorganisms (this interaction is usually called allelopathy in the botanical literature). Ling (the heather Calluna vulgaris) produces, in its roots and leaves, substances harmful to mycorrhiza fungi. Because all the trees that grow within the range of distribution of ling are dependent on such mycorrhiza, trees that come into competition with ling are injured. When the ling is removed, the trees grow significantly more rapidly and are healthier. The heaths in the northwestern Spanish province Galicia, which comprise several species of Erica, appear to prevent recolonization of the region by the endemic oak forest via an allelopathic action. Erica scoparia, for example, contains 10 water-soluble phenols; one of these, vanillic acid, has been shown to have a definite allelopathic effect in experiments on germination (Ballester et al., 1977). The allelopathic effects of the walnut (Juglans regia) and its relatives have become famous. In leaves, fruits, and other tissues of this tree there is a nontoxic quinone which is washed out of the dead tissues and fruits by rain. In the soil it is oxidized and prevents the growth of many plants that might be competitors. In the chapparal of southern California, two plants with an aromatic scent play a special role – a species of sage (Salvia leucophylla) and one of mugwort (Artemisia californica). On some loamy soils competing species cannot survive within a radius of one to two meters from these plants. A series of observations and experiments ruled out the possibility that these competitors might be suppressed by shade, dryness, nutrients or effects due to animals, nor are substances in the soil involved. Rather, the terpenes the plants give off into the air are toxic to other plants (Muller, 1967; Sondheimer and Simeone, 1970). A remarkable form of allelopathy is exhibited by the genus Tamarix. Even when these plants are growing on normal soil, with very little salt content, they excrete large amounts of NaCl from glands on the leaves. In contrast to most land plants, which take hardly any salt in through the roots, these take up salt from deep in the soil and transport it to the surface. As this process continues, the superficial soil under the crown of the tamarisk becomes increasingly saline, and the ordinary vegetation gives way to salt-adapted plants. This salt vegetation must in general be regarded as a less potent competitor than the plants that would normally exist here. The other bushes and trees endemic to the region are particularly affected by the salt. Tamarisks, then, while producing no toxins themselves, are able to accumulate and thus benefit from substances already available.

The extent to which such allelopathy is significant in the general field of plant ecology is a matter of much current debate. So far no final decision can be reached. It is to be expected that coevolution of allelopathic and resistant plants would occur, as has happened to poisonous plants and phytophagous animals; such coevolution would lead to very typical associations of plants. But as yet essentially nothing is known of such events. Whether the "root competition" that has repeatedly been described is based on allelopathic effects or some other processes is unclear; the term indicates the site of competition but says nothing about the mechanism. No one can deny that competition in the root horizon plays a very important role. Its importance is evident from an experiment done in the United States. In a dense forest a deep hole was dug and lined with rings of concrete. All the roots were removed from the soil that was brought up, and then this soil was put back into the hole and subsequent plant growth observed. The concrete lining ensured that no tree roots could penetrate the root-free soil. As time went on an unusually rich flora grew on this site. The absence of these plants under ordinary conditions had been ascribed to shading by the large trees, but the result showed that shade was not at all as effective as had been thought – the roots were the most important elements. Of course, the effect of shade should not be underestimated; it is certainly also a central factor in competition. The tallness of the trees may well have come about as an adaptation to other shade-casting competitors.

Leaves may also exclude competitors in another way; if they are hard to decompose, they can form a deterrent layer of decaying foliage on the ground. In each forest this layer must be kept down to a certain thickness, because most of the trees' own seeds must be able to germinate beneath the layer and grow through it. At the same time, alien seeds must be hindered from doing the same thing. If decomposition occurs so slowly that the layer becomes too thick the forest cannot naturally rejuvenate itself. If it is too thin because of rapid decay, other plants will form a dense undergrowth and in their turn will prevent germination of the tree seeds – which also blocks rejuvenation.

The competitive mechanisms of animals are still less satisfactorily analyzed than those of plants. Here, too, there are chemical mechanisms, effective particularly in small bodies of water. The spawn of frogs and toads can slow the growth of fish. But the studies of this phenomenon that have been done so far do not permit any confident interpretation, and there has probably never been a chemical analysis of the effective substances.

On the other hand, predation plays a central role in the competition among animals. The competition between Tribolium and Oryzaephilus described previously is an example – the young Oryzaephilus larvae are simply eaten by the Tribolium larvae. Indeed, this principle holds everywhere. There is almost no animal that would reject a protein-rich meal if it presented itself. The competition among the possible colonizers of the ocean floor is also based on this factor, as we have seen. Finally, there is competition by simple disturbance. The stronger, more aggressive opponent drives the others away (Fig. 48). The rank order established by this mechanism can often be reversed, for aggressiveness decreases or increases depending on an animal's nutritional state, thirst, and so on.

To summarize, too little distinction is made in the ecological literature between the historical and the current aspects of competition. In most cases the mechanisms underlying present-day competition are unclear. The ecological significance of competition must certainly be regarded as very high. Among animals it is most important when they are competing for common facilities within the ecosystem, resources that are broadly distributed and need be visited by the animals only rarely. Sometimes there is no sharp distinction between competition and predation by animals. Spatial competition, root competition and other such terms are not explanatory, but merely imply the site at which competition occurs in a way not yet understood.

8. The Conspecific as an Environmental Factor

We started with the notion that the organism becomes progressively better adapted to its environment in the course of evolution – that evolution is a process of optimization. Within this framework it seems difficult to explain some of the phenomena that we encounter again and again. For example, consider the reindeer on the island of Spitsbergen – the extreme limit of their range, a region with a growing season of less than a month. Why should these animals, when energy and mineral resources are so scant, each year grow a new set of antlers weighing as much as 8 kg, only to shed it again each year? Why should the inhospitable arctic, of all places, be inhabited by the only deer species of which male and female can afford the luxury of such antlers, the female even at the same time as she is carrying her young?

This question ignores the fact that conspecifics also form part of the environment of each individual organism – and evolution occurred under the selection pressure of these sharpest of all possible competitors. Let us continue with the reindeer. The dominant member of the group, and thus the first to feed, is the animal with the most massive antlers. In the spring those of the males begin to grow sooner than those of the females, and the male antlers are shed after the mating season in autumn, before the onset of real winter. Among the females not carrying young, growth of the antlers begins later and they are not shed until the end of the year. The latter are therefore dominant during the first half of the winter; they can assert themselves against the males and obtain enough to eat. Adequate food is particularly important to them at this time, because in general they have become pregnant during the autumn. Onset of antler development is most delayed among the mother deer, occurring long after the birth of the calves. The stags' antlers are fully developed while the females with young show hardly any sign of antlers. Eventually they do grow; they are relatively small, but are retained until just before the birth of the next calves. This timing allows the mother deer a distinct group dominance during the time of the greatest food shortage, toward the end of winter, when they are especially in need of food because they are in the last stage of pregnancy. So what at first seemed a paradox – that in a region so inimical to life even female animals can afford antlers – is explicable as a possible strategy for dealing with the environment. At the same time, it illustrates the significance of intraspecific competition.

Research on this and related problems has been emphasized by many workers in the last twenty years. One of the first reviews is given in the extremely stimulating – though many of the theoretical points are no longer tenable – book of Wynne-Edwards (1962); the fundamentals have been most recently summarized by Wilson, in his work on insect states, published in 1971, and in the 1975 synthesis of sociobiology. The central theoretical bases of this field, on the borderline between autecology and population ecology, can be found in Dawkins (1978). Because these studies build a bridge between population ecology and autecology, the analysis of how animals live is no longer an isolated area – it has become an inseparable element of ecology. A few examples will show how this occurs.

Leuthold investigated the strategy by which African antilopes adapt to their habitat. Those that live in the forests are mostly rather small, and tend to be nearer the ground in front than behind. They feed predominantly on buds and fruit (high-quality food), so that, like the roe deer, they have a small rumen and thus cannot live on food with a high cellulose content. When danger threatens they rarely flee; they try to escape notice by crouching, motionless. Most of them live singly or in pairs; either the buck or the pair maintains a territory, and the animals do not ordinarily move from this place. The territories are defended; they represent not only living space, but private food stocks. The animals need a steady supply of their special high-quality food throughout the year. Steppe antilopes, on the other hand, do not have such selective requirements. In the long term they can manage with plants that are hard to digest (i. e., of much lower quality). Accordingly, their rumens are large. When in danger they flee over long distances, and in some cases may actually attack. All of them live in nomadic herds; territories are taken up only occasionally, and only by the males. These territories have nothing to do with food availability, but are exclusively related to mating activities. The food supply of these animals can be quite nonuniform over the course of a year. Because they move about so much, it is necessary that the young antilopes soon after birth be able to wander with the herd and flee from predators. By comparison, the young forest antilope is almost like a nestling bird.

The antilopes of steppe and forest, then, experience selection in entirely different directions. This is not to say that all the animals are really diametrically opposite; evolution has progressed to a different extent in the different species, and just as there is a smooth transition between woods and open range, so there are gradations among the antilopes. But the strategies of selection are the same everywhere. We can apply this example to deer; the reindeer of the tundra form large herds, whereas the forest reindeer live alone. The tundra reindeer escape danger by long-distance flight; those of the forest make themselves inconspicuous and allow predators to approach closely. In fact, predators themselves fit into this scheme. The steppe-dwelling lion lives in groups, whereas the tiger and leopard hunt alone (cf. Hendrichs, 1978).

The way of life of Scandinavian owls has been examined from a quite different point of view. The long-eared owl is a strict food specialist, feeding exclusively on water voles. When the snow lies thick in the winter, these are hard to hunt down. For this reason long-eared owls are migratory birds; the pair bond is not a close one, and usually lasts for only one breeding season. In the spring, when the birds return to the breeding grounds, they find many possible nesting sites – in the nests of crows or others of appropriate size. Tengmalm's owl has a harder time. This owl also lives almost entirely on voles, so that one would expect it to behave like the long-eared owl. However, it breeds in holes in trees, and there is in general a scarcity of these; therefore only the females and young migrate south in the winter, while the male holds the territory with the nest hole. Apparently the food supply is sufficient for a single bird even in winter. Tengmalm's owl can also shift to feeding on birds more easily than can the long-eared owl. Development in this direction has been taken still further by the Ural owl, a bird which nests in very tall old hollow trees. The crucial limiting factor for the Ural owl is the availability of a large enough nest hole. The animals form permanent pairs, which spend the whole year in their territory, defending it from conspecifics and other animals interested in the breeding hole. Under these conditions food specialization is impossible. Ural owls are generalists, feeding regularly on mammals and birds that range in size from mice to pigeons and squirrels.

Considering the interactions among animals in this way, we can arrive at a possible explanation of the apparent paradox that animals form herds and colonies. Actually such a gregarious life style should hardly be possible, in view of the difficulties in acquiring food, the accessibility of such groups to parasites, and the ease with which parasites can be transmitted. Sociobiological considerations can clear up some of these problems (cf. p. 122).

9. Ecological Neurobiology

An animal must find and recognize its habitat, its food, and its enemies. It must be able to orient in its habitat. All this requires sensory abilities, which have been mentioned briefly in various contexts. Of foremost interest to the ecologist is not so much the way a sense cell operates and the primary processes leading to excitation – the focal points of modern sensory and neurophysiology. The ecologist is chiefly concerned with the adaptation of the sense organs to the way of life, the ecology of the animal. The publication of Karl von Frisch's famous work on bees (1965) initiated what has become a field of the greatest importance to ecology: neurobiology. Just as we consider it a matter of course that the biochemical aspects (Sondheimer and Simeone, 1970; Hochachka and Somero, 1973) and metabolic aspects (Larcher, 1973; Schmidt-Nielsen, 1975) of ecology are sufficiently fundamental to be treated here, so we must recognize ecological neurobiology as a fundamental area of research, essential to an understanding of zoological phenomena in the framework of ecology.

It is remarkable that this approach has often been disregarded by theoretical and general ecologists, whereas applied ecologists have paid increasing attention to it in recent years. Knowledge of the factors and mechanisms that govern or affect the choice of host by destructive insects and their parasites is now a self-evident necessity for plant cultivation and pest control. (An outstanding modern review of this field of research is given in the book of Chapman and Bernays, 1978.) For example, caterpillars of the silkmoth Bombyx mori will not consume food unless three factors are present – an attractant, a biting factor, and a swallowing factor. Each of these must be individually recognized, by specific sense organs. Evidently many plants contain substances that "stimulate the appetite" of insects, for they eat such plants distinctly more rapidly than plants lacking these substances (and hence grow more rapidly). Thus unexpected classes of substances, such as phenols, suddenly become essential to insects and acquire immense ecological significance.

If one examines the behavior of the three amphipods Pontoporeia affinis, Gammarus duebeni, and Gammarus oceanicus in humidity gradients, one finds that Pontoporeia dries up without attempting any active response. The two Gammarus species seek out positions where the relative humidity is about 100%; here they can survive for a long time. Pontoporeia is a deepwater animal, whereas the Gammarus species live in the littoral and at low tide are occasionally left dry (Lagerspetz, 1963). The behavior of the animals in this artificial situation is comprehensible only in the context of their ecological distribution; it is associated with the function of certain sense organs, and the existence of these sense organs in turn is explicable only in the ecological context.

Chemical sense organs enable an animal to find and recognize specific food, a sexual partner, and the boundaries of a territory. A juvenile salmon, migrating slowly downstream after its emergence from the egg, records the scent of its home brook and that of its home river before it moves on to the high seas. As an adult returning from the ocean to spawn in fresh water, it recognizes the mouth of the home river, and follows its tributaries up to the home brook, by chemical stimuli. An extremely complicated chemical language has been developed by the social Hymenoptera, ants in particular. There are specific odors for alarm, for attraction, for recruitment, and many others (Hölldobler, 1970, 1977; Wilson, 1971). The fact that all this could not prevent parasites in insect colonies from copying their hosts is another matter.

Sense organs that respond to mechanical stimuli exploit the properties of the medium (air or water) or the substrate (water surface, soil surface, tree trunk) (Markl, 1972; Wiese, 1972; Markl and Hauff, 1973; Markl et al., 1973). For example, a mosquito hanging at the surface of the water responds very precisely to the waves that an approaching backswimmer (Notonecta) produces on this surface; on the other hand, a Notonecta stationary at the water surface can detect and localize exactly a mosquito larva hanging at the water surface by the surface waves it sends out (Fig. 50). Lepidopteran caterpillars have special bristles responsive to airborne vibration, with which they detect the approach of a flying parasitic wasp; the caterpillar reacts by rolling up and dropping from the plant on which it was feeding (Markl, 1978). The lateral-line system of fish functions only in still water, and thus has not developed in species adapted to life in mountain brooks. The vibrissae on the heads of predators in many cases play a crucial role in measuring the width of a passage; if the vibrissae touch the sides, the animal is too large to get through.

In addition to these mechanisms for passive localization and orientation, there is the possibility of active position-finding, based in the last analysis on the same principles as the technology of radar and sonar. Bats send out ultrasonic vocalizations and orient by their echos. This behavior is treated elsewhere, in relation to the coevolution of moths (cf. p. 168). The South American oilbird, Steatornis, operates by the same principle; it breeds in extremely deep caves, emerging at night in search of palm fruits. In this case the pulses produced are not ultrasonic but merely very high tones, less advantageous technically. Ultrasound is used by whales and dolphins, for very accurate localization of fish as well as planktonic invertebrates. The large size of the sperm whale head appears to function at least in part in localization; layers of different substances act like a lens, to bundle the sound. Moreover, it seems likely that most whales and dolphins have developed a highly differentiated system of communication by sound in the audible range. If an animal is to make use of all these capacities, it must also be capable of directional hearing. But directional hearing has never been evolved by primary aquatic animals. The whales and dolphins, bringing the ability with them as an inheritance from their terrestrial ancestors, are the only organisms capable of directional hearing underwater. Can this be explained? Directional hearing is based on a difference in information at the two ears – with respect to either time of arrival or relative intensity of the signal. In water the

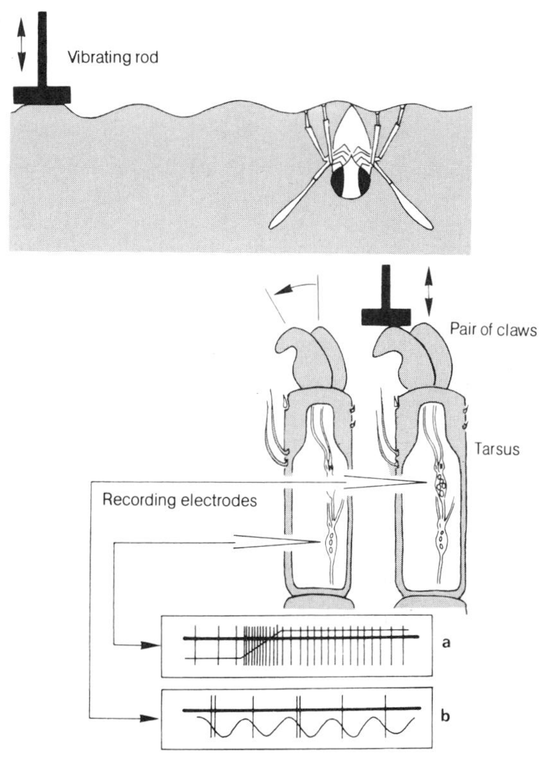

Fig. 50. Prey capture by the backswimmer. *Above:* diagram of the principle by which surface waves were produced experimentally. *Below: a* phasic-tonic responses of neurons of the proximal scoloparium when the pair of claws is displaced; *b* phasic responses of neurons in the distal scoloparium to vibration stimuli, when the vibrating rod is directly coupled to a claw. (Ewert, from data of Markl et al., 1974).

velocity of sound, about 1,500 m/s, is considerably higher than in air. Moreover, a sound originating underwater is transmitted with very little attenuation and can therefore be heard much farther away from the source than it could be on land. These features make it enormously difficult to detect direction underwater, for the signals reach both ears essentially at the same time and with the same intensity. And because organisms have about the same density as the water surrounding them, sound is poorly reflected, tending rather to pass straight through the body. Sound coming from the side, therefore, does not need to "make a detour" around the animal before arriving at the opposite ear;

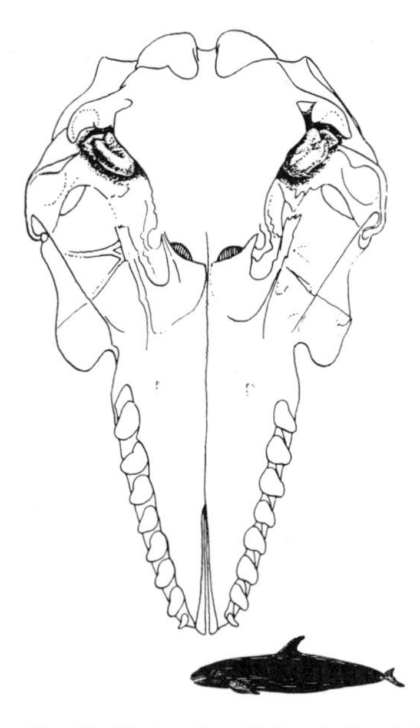

Fig. 51. Underside of the skull of the false killer whale (Pseudorca crassidens), showing the position of the bulla. (Slijper, 1966)

both the delay in arrival and the damping of the sound are exceedingly small. To compensate, the hearing apparatus of dolphins and whales has become specially modified. It is relatively broad at the base, because the animals are large, and the entire stimulus-receiving and -conducting apparatus is surrounded by a very conspicuous layer of bone. The skull bones of these animals in general are soft and fatty (and thus appear gray), but the armor enclosing the auditory apparatus consists of shimmering white, extremely hard bone (which, when an animal is cut up, ordinarily is lost and can no longer be found in the skull; cf. Fig. 51). Because of the difference in consistency of the skull and the material shielding the ear, practically simultaneous excitation of the two ears by sound passing through the body is prevented. (And now we know why fish are such an ideal target – their swim bladders are excellent sound reflectors. We also know, then, why it is a selective advantage for some fish of the high seas to give up the swim bladder.)

Another sort of active localization system has been developed by the weak electric fishes. They generate an electric field around themselves (Fig. 52) and by detecting distortions in that field can localize objects, predators or prey (good or poor electrical conductors) in the water. This mechanism can work only in fresh water, because the high conductivity of sea water would not permit such a weak electric field to be set up. (Electricity can also be used to kill prey animals and for defense, and in this case too the ocean presents special difficulties. The electric ray must fling itself upon its prey and cover it completely with its body before it can discharge a lethal current, cf. Fig. 53; Harder, 1965 a, b.) This mode of orientation has been brought to its highest development by fish of various families that live in quiet, muddy water. In such turbid conditions the lateral-line system becomes occluded and vision is impossible.

The degree to which acoustic signals can be informative is quite different in water and out of it, and it may vary among different

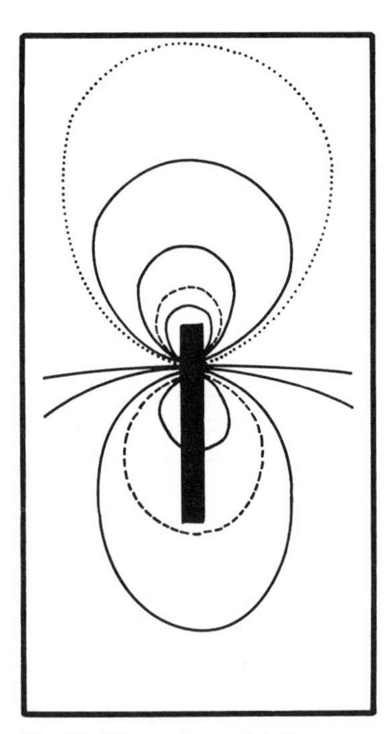

Fig. 52. The equipotential lines around an electric fish (Gnathonemus petersii) in an aquarium 30.4 × 58.5 cm in area, as seen from above. (Harder, 1965)

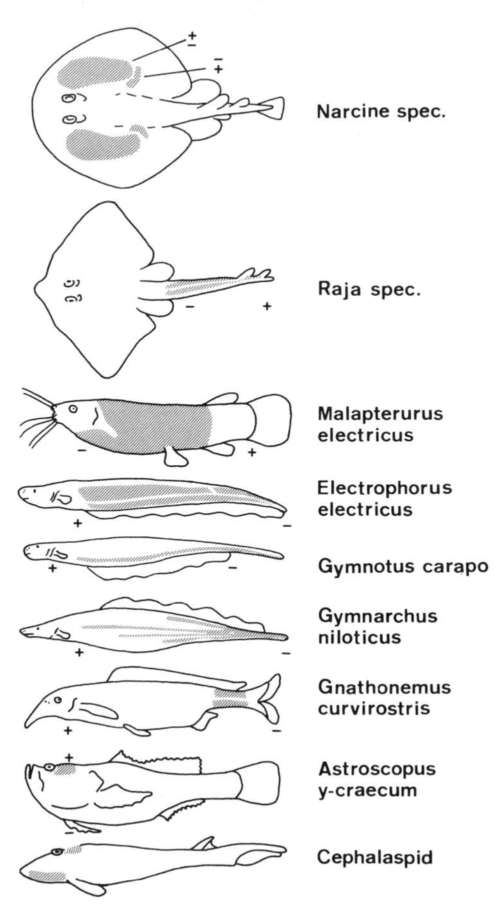

Fig. 53. Some typical electric fishes. The electric organs are indicated by cross-hatching (Harder, 1965). (The Cephalaspidae are known only as fossils, but the structure of the skull suggests that electric organs were present)

terrestrial habitats. Can a dense forest be equated with an open steppe? A good deal of effort has recently been expended on this question, because it may provide a key to understanding certain differences in bird songs. Some have assumed that a dense forest is penetrated particularly well by very specific sound frequencies, and that birds use these particular frequencies in their territorial songs. But the situation is obviously considerably more difficult and more complicated, and there seems to be no simple explanation (Marler, 1977).

The same distinction between air and water holds in the case of vision. Because the density of animal tissue is about the same as that of water, an eye designed to provide sharp images in both air and water must be capable of accommodating over a wide range. Diving birds, for example, have such eyes. Seals, on the other hand, have evolved an underwater eye, which forms blurred images in air. Whirligig beetles (Gyrinus) swim directly at the water surface; correspondingly, their facet eyes are subdivided into an above-water and an underwater part. It is generally true that eyes under water require higher light intensities. The compound eyes of aquatic crustaceans are all distinguished by an angular aperture larger than that found in their relatives living out of water (Thiele, 1968). Moreover, the eyes of aquatic animals (fish, for instance) are focussed in the near vicinity when at rest, while the eyes of land animals – for example, mammals and birds – are far-accommodated at rest. Finally, people frequently speak as though the colors we see in modern pictures of coral reefs are real; some have even speculated that the striking white-and-red patterns of certain animals represent warning coloration. Nothing could be further from the truth. Red light is strongly absorbed by water, so that it cannot penetrate very far; at a depth of only a few meters red appears gray or even bluish. The colors visible under water differ greatly from those to be seen in a color photo (made with a flash bulb) or in an animal recently brought to land. And just as one encounters such problems in comparing the two major media land and water, so one finds differences in visual perception during the course of the day and night and in different habitats.

When we drive along a tree-lined avenue when the sun is low, we have problems with the rapid alternation of sudden light and shade; our eye has not evolved to deal with such quick changes in light intensity. The eyes of flying birds and insects appear to be capable of much more rapid adaptation under such conditions. Particularly in dry tropical forests, where the contrast between light and shade is very great and is present not only in the morning and evening, adaptations would be expected. Naturally, there are also adaptations to dark habitats; the eye of the deep-sea ostracod Gigantocypris does not have a lens, but rather is a mirror eye, in which shortened structure is accompanied by increased light capture. Reflecting layers behind the light-sensitive layer of an eye are typical of many nocturnal insects, amphibians, reptiles, and mammals; this is the reason why such eyes "glow" in the dark. A ray of light that enters the eye is reflected and stimulates the retina a second time. This does impair visual acuity, but still provides the animal with information it might otherwise miss.

Further progress along the road of specialization leads to the sun orientation of beach animals (littoral amphipods, isopods, beetles, and spiders), which permits them to find their way within the habitat and to stay where there is the right amount of moisture (Pardi, 1960; Remmert, 1965a), just as the sun orientation of bees enables them to navigate within their home range (von Frisch, 1965). Further still, we come to a phenomenon as yet largely unexplained – long-distance orientation by animals (Galler et al., 1972; Schmidt-Koenig, 1978).

10. Other Ecological Factors

The distribution of animals and plants can be critically affected by small factors that have been little noticed. This will become evident in the examples that follow.

The operculate snail Theodoxus fluviatilis lives on the shores of lakes and bodies of brackish water and on stones in rivers and brooks that do not flow too rapidly; it feeds on diatoms. Its osmotic-pressure tolerance is relatively wide, ranging from fresh water to brackish water with a salinity of barely 10‰. Its temperature requirements are also fairly flexible. Nevertheless, it is absent from many sites where one would expect to find it. The meticulous experiments of Neumann (1961) have shown that this snail is not able to digest the diatoms it scrapes from the stones unless they are broken apart during the scraping. And this occurs only if the stone surface is very rough. In an experimental situation, when the snails are kept with diatom cultures on pieces of glass, they starve to death. But if the glass has previously been etched with hydrofluoric acid to roughen the surface, the snails thrive (Neumann, 1961).

Meinertzhagen (1954) took up the question why so many migratory birds leave the taiga in winter. All experiments indicate that many species should have no problem in surviving the prevailing winter temperatures. Food is as abundant as in the forests lying somewhat further south, so that a food storage can hardly be the reason. Nor is there likely to be any difference in the accessibility of the food. According to Meinertzhagen's analysis, it appears likely that when the snow cover is deep the birds are prevented from keeping enough stones in their crops. Without these stones, they cannot digest the food that they find. In such regions roads kept free of snow provide a winter refuge for a number of birds; even grouse may collect there.

11. Periodic Changes in the Habitat

Many of the ecologically significant factors are subjected to cyclic changes correlated with the earth's rotation. Such factors include light, temperature, relative humidity, and the activities of predators and competitors. It is evident, then, that in a normal habitat conditions are optimal for a species only at certain times of day or seasons of the year. This is as clearly true of plants, which carry on light-dependent photosynthesis, as of animal activities; feeding and reproductive behavior, as well as defense of territory and the search for a favorable locale, are affected by timing – some animals are active in broad daylight, others in twilight, and others in the dark, some in summer and others in winter. Almost any organism, then, must be capable of enduring the regular recurrence of unfavorable periods in its habitat. How long must the minimal daily activity period be if an animal is to survive an enforced rest during the unfavorable period? How long a season of activity must an animal have, to survive the winter or a time of drought? At northern latitudes, for example, the hours of daylight during the winter are too brief for many birds to find the food they need. Because their energy consumption is high at the low winter temperatures, such species cannot overwinter at these latitudes. Wallgren (1954) showed that the ortolan (Emberiza hortulana) is restricted by this factor.

In general, though, the actual situation is not so simple. Plants and animals have adapted to the cyclic fluctuations in environmental conditions, and in many cases they depend on them for normal life. Only very few plants in the tropics and the temperate zone are capable of existing under maintained light; they need the darkness in order to grow normally. Under constant light they develop necroses, just as many animals under constant light show behavioral disturbances or impaired reproduction. During the course of a year the weight of a bird or mammal undergoes characteristic regular oscillations, which serve to store up nutrients for the unfavorable season. The ability of insects, warm-blooded animals, and plants to tolerate cold also exhibits a regular annual rhythm. Birds and mammals put on winter dress; the winter plumage of the house sparrow is 30% heavier than the summer plumage. In conditions of diurnal rhythmicity egg production by ectothermic animals is increased, as is productivity in most plants (cf. p. 31). Some ants and snails seem to require the regular impetus to metabolism that the daily oscillation in the environment provides. Many species – both plants and animals – need to experience winter

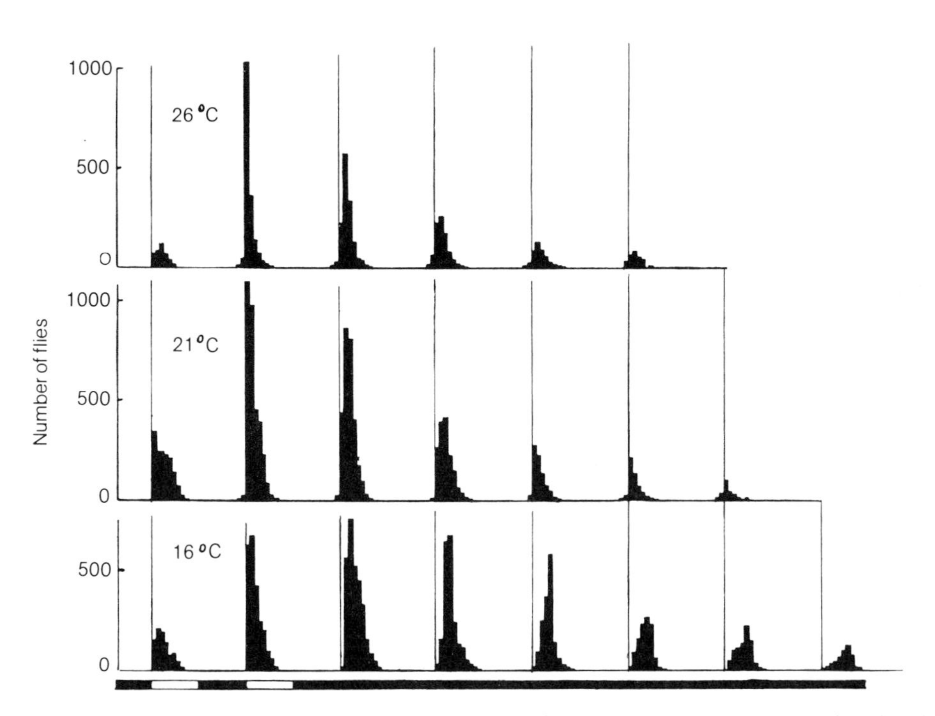

Fig. 54. The emergence of Drosophila pseudoobscura at different constant temperatures, under otherwise constant conditions. Despite the maintained darkness and the wide range of temperatures, the peak emergence times are separated by about 24 h in all cases; the animals have a physiological clock. (Pittendrigh in Remmert, 1965 a)

conditions in order that their seeds germinate, or their larvae hatch out, in the spring.

This close meshing of vital processes and the rhythmic environment is documented by the evolution of a physiological system for the various periodicities involved. Both plants and animals have a physiological clock with a period of about 24 h. If the organisms are kept under constant conditions the clock continues to run, but it is no longer "set" by external factors. Just as an old alarm clock gradually drifts away from the normal period of the earth's rotation, 24 h, and must be reset by its owner, the physiological clock of an animal or plant drifts away from the time given by the earth's rotation after a few days. Thus we speak of the physiological clock as having a period of about 24 h – it is a *circa*dian clock.

The internal clock that marks the time of year functions similarly – it is a circannual clock. There have been clear demonstrations of the existence of both kinds of clock in plants and animals. The seeds of most central European plants germinate especially well in certain months; if they are kept under constant conditions, they tend to germinate at about the same time in each successive year. Birds of the temperate zone come into a reproductive state fairly regularly every 12 months. This physiological clock evidently keeps better time in migratory birds that regularly cross the equator than in species that do not migrate or travel only short distances. Both the diurnal and the annual clock, then, seem to involve oscillations within the organism that persist for a long time (Fig. 54).

Instances of adaptation to the tidal rhythm are also known; there are semilunar clocks, lunar clocks, tidal and bitidal clocks. Some of these seem to be based not on oscillatory processes but on the hour-glass principle. Once triggered, they run down in a fixed time, and can then measure that time again only if they are triggered anew.

These physiological clocks are "set" by environmental factors; in all cases the alternation between light and dark appears to play the leading role. In the far north the light/dark alternation is evidently replaced by an alternation in the spectral composition of the light. During the very bright "night" the light contains more long-wavelength (reddish) components than during the day, which is very little brighter (Krüll, 1976 a, b; Fig. 55). Temperature can serve as a timing signal for some ectothermic animals, but never for warm-blooded animals, and there are some ectotherms that do not respond to temperature changes as a timing signal. There are other potential timing signals, but only a few (auditory signals such as bird song, humidity, turbulence) so far are likely candidates, and these are apparently effective only in isolated cases. The timing signal for all circannual clocks seems to be the change, over the course of a year, in the relative lengths of the light and dark periods. For a long time there has been good evidence of this in plants (there are long-day plants and short-day plants). The transition among seasonal forms of animals can be ascribed to this factor, as can the onset of sexual maturity and the beginning of the spring migration in birds (cf. Remmert, 1965 a).

The evolution of a physiological clock is an obviously plausible adaptation to regularly changing environmental conditions. It enables any organism at any moment to adjust itself to the situation it will encounter the next moment. But as plausible as this may be, there is practically no supporting evidence. The evolution of such a complicated physiological clock – operating by mechanisms still far from being understood – does not seem so likely when one considers that the same effect could be achieved by simple reactions to environmental factors, in certain circumstances involving an hour-glass-type mechanism to measure short times.

Nevertheless, it seems that all plants and animals have an internal clock, which op-

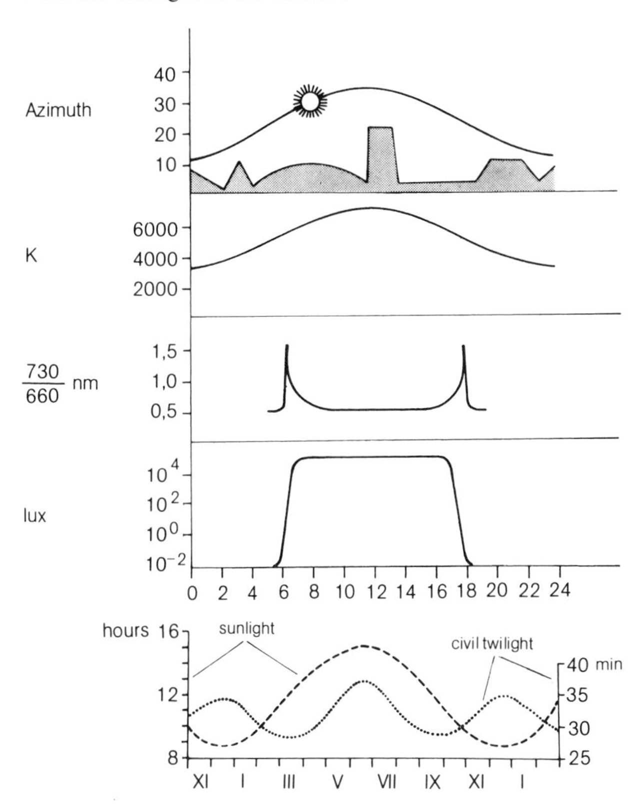

Fig. 55. Aspects of the timing signal provided by light/dark alternation. From top to bottom: azimuth position of the sun in relation to landmarks in the far Arctic (summer); color temperature (in Kelvin) of the light in the far Arctic (summer); far-red/red ratio of the light during the course of a day (June, Washington D.C.); brightness in lux during an extremely bright day at about 40 °N latitude in June, diagrammatic; theoretical length of the sunlight hours and of civil twilight at 45 °N latitude. (Remmert, 1976)

erates on the same principle throughout and is universally temperature-compensated. In the case of sun orientation, its function is quite clear. Animals orienting by the sun must take into account its movement and the differences in azimuth angle at different hours and at different latitudes, if the sun orientation is to function properly. That these are included in the calculation has been confirmed for a great variety of animal groups. Bees, for example, find their food plants and their hives by this principle; lizards find their territories; crustaceans, beetles, and spiders on ocean beaches use the sun to determine the direction in which to flee. Marine turtles and fishes are also capable of sun orientation.

Nearly all vital functions of plants and animals are coupled to certain times of day. The opening and closing of flowers, the production of flower scents, and the lifting and lowering of the leaves of many plants are examples, as are the emergence of an insect from the pupal case, which almost always occurs at a particular time of day (Remmert, 1963), and the timing of sessions of activity and rest by animals in general. Many species evidently engage in special forms of activity at certain times of day; copulation, egg laying, and feeding are among these. A number of insects have been shown to schedule such activities fairly closely. But on the whole we know very little indeed about this last phenomenon – which is precisely the aspect most likely to be of interest and significance in ecology. Planktonic animals carry out regular vertical migrations with a diurnal periodicity, coming to the surface only at night. As a result, they can consume planktonic algae only at the time when the algae are not engaged in photosynthesis and the production of matter. Hence secondary productivity is very high (Fig. 57). During the night algal respiration draws chiefly on the energy stored by day; by eating the planktonic algae, planktonic animals reduce the

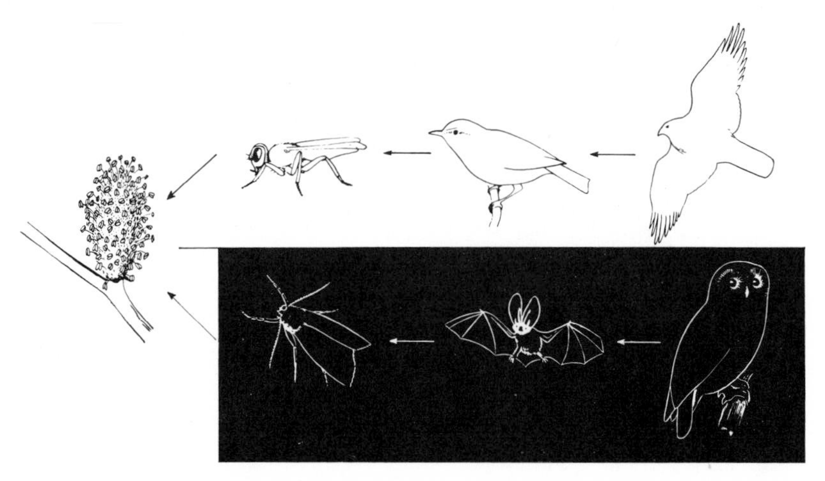

Fig. 56. The food chain involving willow (Salix) blossoms is different by day than by night. Daytime: fly (Egle), bird (Phylloscopus collybita), bird (Accipiter nisus). Nighttime: moth (Taeniocampa monima), bat, bird (Strix aluco). (Remmert, 1969 b)

algal population density and stimulate renewed growth and subdivision of the algae.

On more than one occasion it has been proposed that several species might be mutually synchronized by means of the physiological clock. At a gross level, when one simply compares day and night (Fig. 56), something like this does happen. But even here species active by both day and night – such as mice and shrews (cf. p. 44) – provide direct links between a nocturnal and a diurnal food chain. A really precise matching of the schedules of different species does not in general occur. At most, one might mention the staggered times of day and year at which birds produce their territorial songs; if all the bird species in a region were to sing at the same time, the territorial arrangement would be as indecipherable to the occupants of the territories as to the observer trying to learn bird calls. By singing only at fixed times during the year and day, birds reduce territorial fighting to a minimum and optimize the opportunities for demarcation.

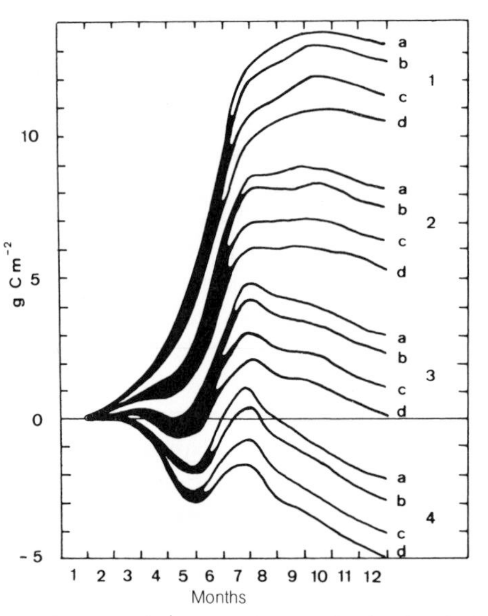

Fig. 57. Cumulative secondary production by planktonic crustaceans, assuming different rates of respiration (1–4) and different feeding times. *a* feeding during the early night hours; *b* feeding during the first half of the night; *c* feeding all night long; *d* feeding uniform throughout the day and night. Ordinate: production in grams of carbon per square meter. (MacAllister in Remmert, 1977)

But to return to the initial question – how short the activity time of an organism can be while permitting survival in the habitat – we must admit to an almost total ignorance. Because, as we have heard, the length of the daily light period has acquired a timing-signal function, experiments on this point are not easy to do; two effects of the light influence each other.

12. The Interaction of Environmental Factors

The autecologist or physiologist who wants data that justify predictive inferences is compelled to investigate each factor in isolation, under controlled conditions in the laboratory. Only if the physiological basis of the responses of organisms to environmental factors is known can such predictions be confidently made. The various stages in an organism's life cycle, however, respond to a given factor in different ways, as does an organism in different physiological states. Finally, analysis of even a single factor is hampered by the nonlinear dose-response relationship and by the possibility that there may be differences in principle between constant and fluctuating conditions (as we discussed in the context of temperature oscillations and the diurnal rhythm). In nature environmental factors never act in isolation; there are always many operating in concert. And this is the great problem of ecology. It is impossible to explore systematically all the conceivable combinations of factors in the laboratory. To what extent, then, can one apply the results of single-factor analysis to situations in which several factors are combined? Theoretically, given two factors, one can imagine three possibilities.

1. The two factors act entirely independently of one another. If this is so, there is no problem in using experimental data to interpret what happens in the field.

2. The two factors affect one another in a predictable way (e. g., temperature and water oxygen content, or temperature and humidity). In this case it is easy to do experiments that allow predictions.

3. The combination of the two factors involves special conditions that cannot be predicted.

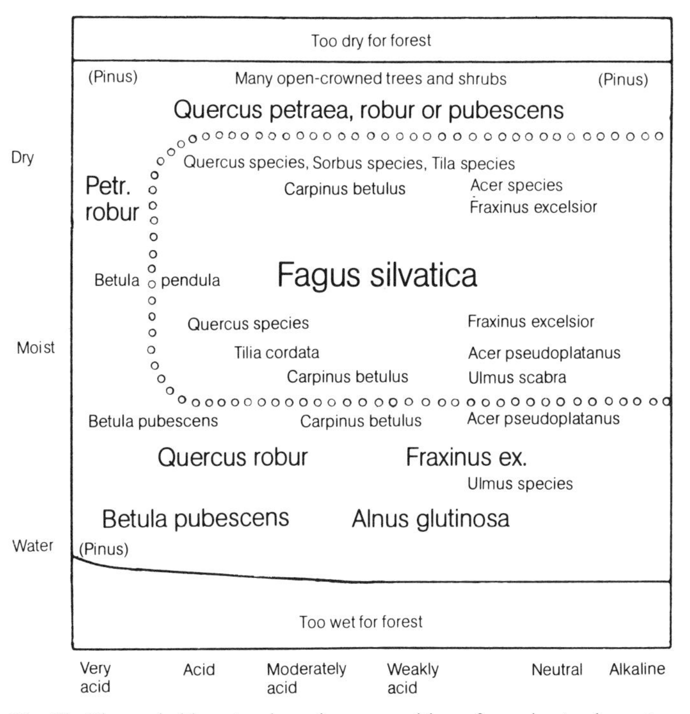

Fig. 58. The probable natural species composition of woods at submontane altitudes in western central Europe, as a function of water and nutrient balance. The size of the print indicates the relative number of individuals of the species in the tree stratum. The region enclosed by circles represents the habitats where beech is dominant. Two factor groups are included in this diagram. (Ellenberg, 1978)

There are examples of all three possibilities.

1. The distribution of many plant species can be directly related to climatic conditions; and of course, plants grow only on suitable soils. The effects of these two factors are independent (Fig. 58). Similarly, the occurrence of bottom-dwelling marine animals can be plotted as a function of salinity and of the grain size of the substrate. The cockle Cardium edule tends to be found in fine-grained sand, whereas the long clam, Mya arenaria, needs a certain admixture of silt. Both can live in a salinity range extending from just above 10‰ to just above 38‰. Here, again, one can easily predict where each clam will be found. The classical example of the independent simultaneous action of different factors is given by the production of organic matter in green plants. Photosynthesis requires light, water and carbon dioxide. As light intensity increases, the rate of photosynthesis rises to a saturation level, if the temperature is within the favorable range (Fig. 59). Variation over the course of the day is thus represented by a single-peaked curve (Fig. 62). Curves with two peaks indicate additional complications; either the water supply is insufficient, or the air temperature becomes so high that the stomata close, so that the plant conserves water and cannot take in any more CO_2 (Fig. 61). If the temperature remains high for too long a time, the plant operates at a loss. Photosynthesis cannot compensate for the energy used up in respiration (which naturally is greatly increased at higher temperatures). The economic significance is considerable in desert regions, where water scarcity and temperature restrict daily photosynthesis. True desert plants encounter fewer problems in this regard than agriculturally important species such as the apricot, which even when supplied with fairly large amounts of water can operate at a profit only temporarily (Fig. 61). Lichens in the Negev Desert dry out during

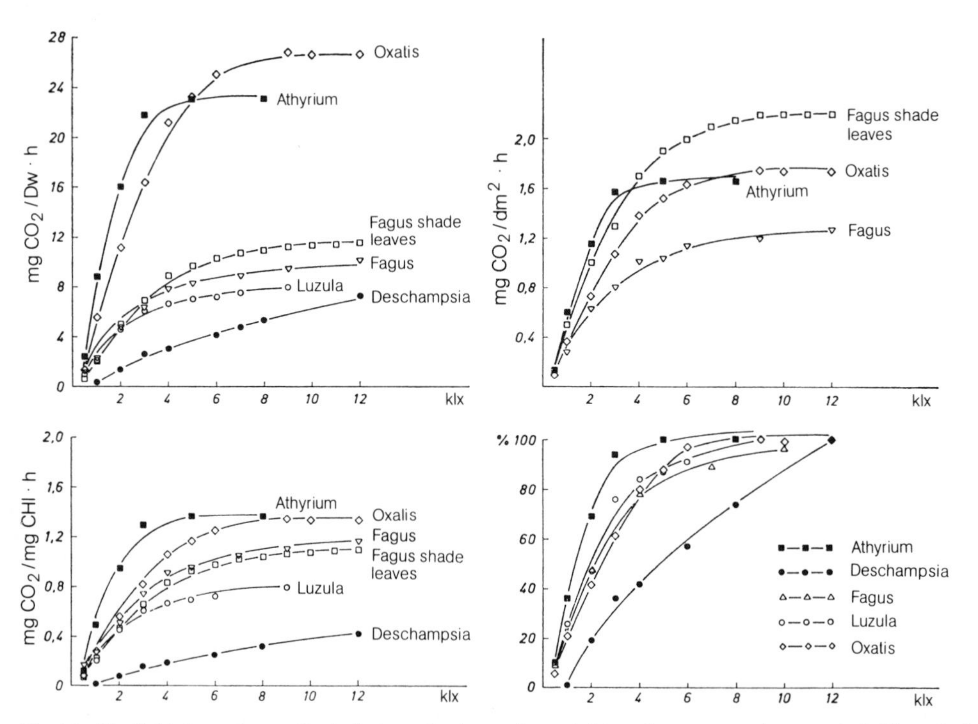

Fig. 59. The light-dependence of net photosynthesis per dry weight, surface area, and amount of chlorophyll of the assimilation organs, and the maximal rates of CO_2 uptake of various plants and of the shade leaves of beech trees. (Schulze, 1972)

the day and take in water at night. Photosynthesis is possible for these plants only during the earliest hours of the morning (Fig. 60).

Although more complicated, the following example is still readily interpretable. Respiration, and thus nutrient consumption, in the oyster rises linearly with increasing temperature. But the amount of water pumped through the gill apparatus is described by a single-peaked curve, which in principle has the same shape under all conditions although it is slightly dependent on adaptation temperature. That is, water flow is largely independent of temperature in the low-temperature range (pumping rate very low), suddenly rises to a high maximum, and then returns to the low level at temperatures above 25 °C. The amount of food the oyster obtains is therefore dependent on temperature. At some temperatures, whether low or high, the balance is negative. This deficit can be compensated, to some extent, by differences in the concentration of food in the water (Fig. 63). A fourth, even a fifth, dimension can be added to the picture. Despite the large number of long-term ex-

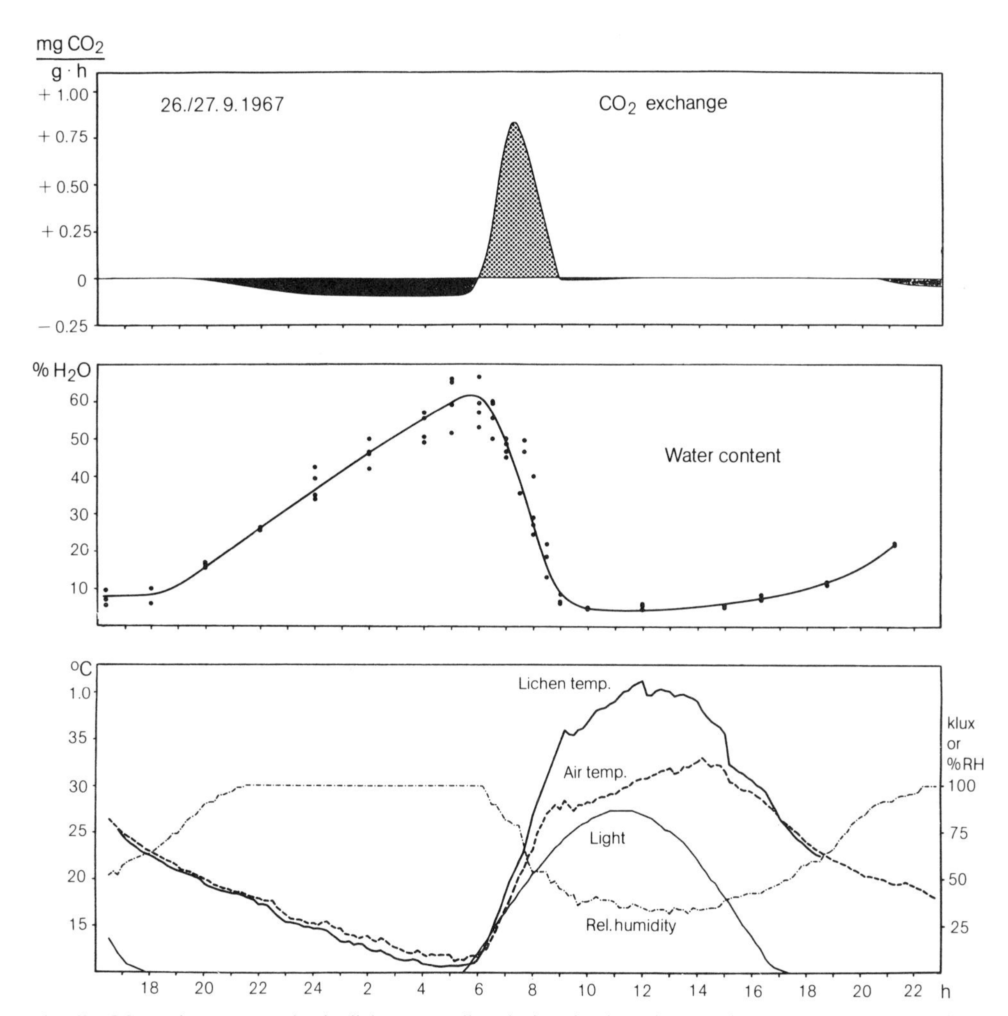

Fig. 60. CO_2 exchange *(top)* in the lichen Ramalina during the day when wet by dew; water content of the thalli *(middle)*; light intensity, relative humidity, and temperature of lichen and air *(bottom)*. In the Negev Desert, where these measurements were made, the lichens can carry on photosynthesis only during the early hours of the morning. (Lange et al., 1970)

periments that have been done, one soon finds oneself facing unpredicted or unpredictable special conditions. Even salinity and temperature often affect one another in an unpredictable way.

2. More oxygen dissolves in cold water than in warm water. Other things being equal, then, an animal in cold water has more oxygen available to it. This is a particularly weighty factor, because at higher temperatures the metabolic rate of an aquatic organism generally rises (all gill- and skin-breathing aquatic animals are ectothermic!). That is, it requires more oxygen as the amount available becomes less. The opinion frequently expressed by people interested in the technology of conservation – that increasing the temperature

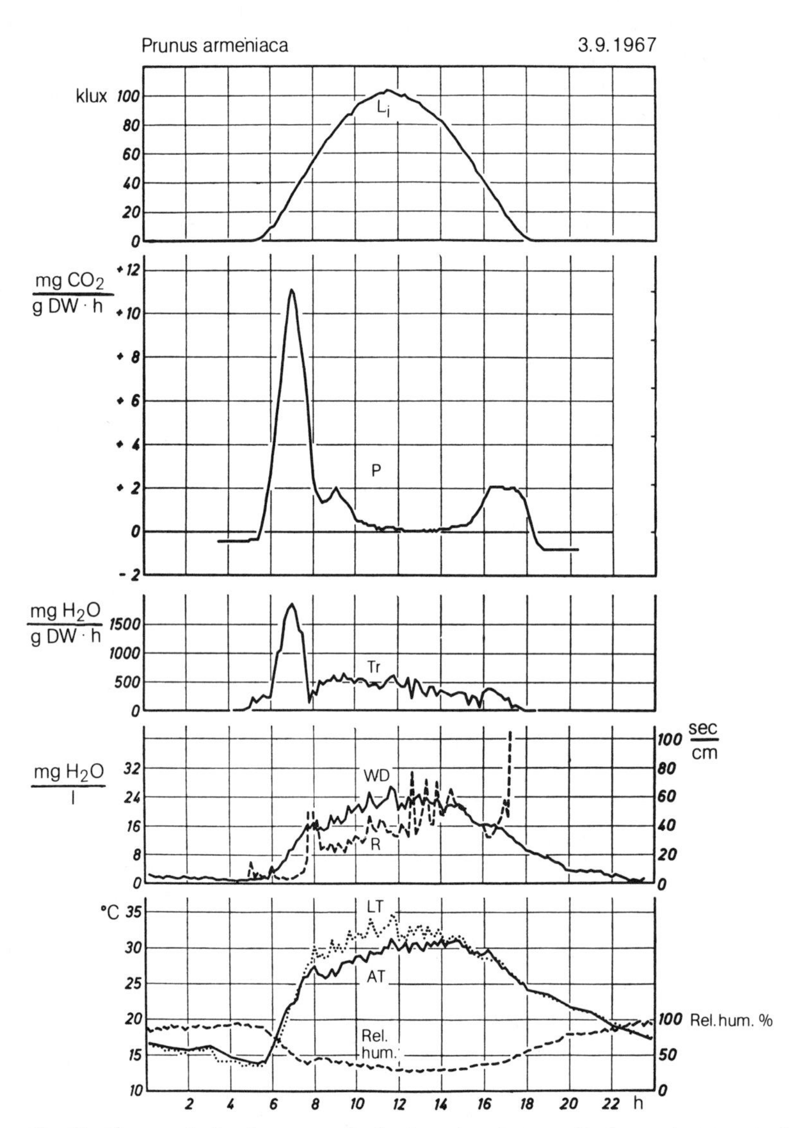

Fig. 61. Changes during the course of a day in various factors affecting apricot trees in the Negev Desert. *From top:* light intensity, carbon-dioxide uptake, transpiration, water-vapor difference between leaf and surrounding air *(WD)*, diffusion resistance to water vapor *(R)*, leaf temperature *(LT)*, air temperature *(AT)*, and relative humidity. When the leaf temperatures are too high transpiration and photosynthesis are stopped (cf. Fig. 62; Schulze et al., 1972)

of a river by one more degree should have negligible effects – is not valid as a general prediction. Any such claim must be supported by detailed investigation (Fig. 64). However, quite a bit can be inferred from the physical solubility of oxygen in water, and from humidity as well.

3. Finally, there are cases in which organisms respond unpredictably to factors in combination. These responses are of course explicable physiologically, so that in theory it should be possible to predict them. But in practice, as a rule, it is much too difficult. Moreover, in many cases prediction is hindered by the fact that the

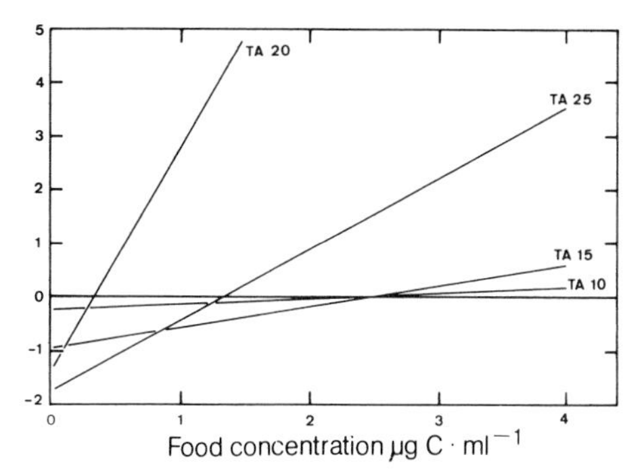

Fig. 63. Change in weight of oysters per day, as a function of differences in food concentration in the medium (*abscissa*, in µg carbon per ml) and at different acclimation temperatures *(TA)*. (Newell et al., 1977)

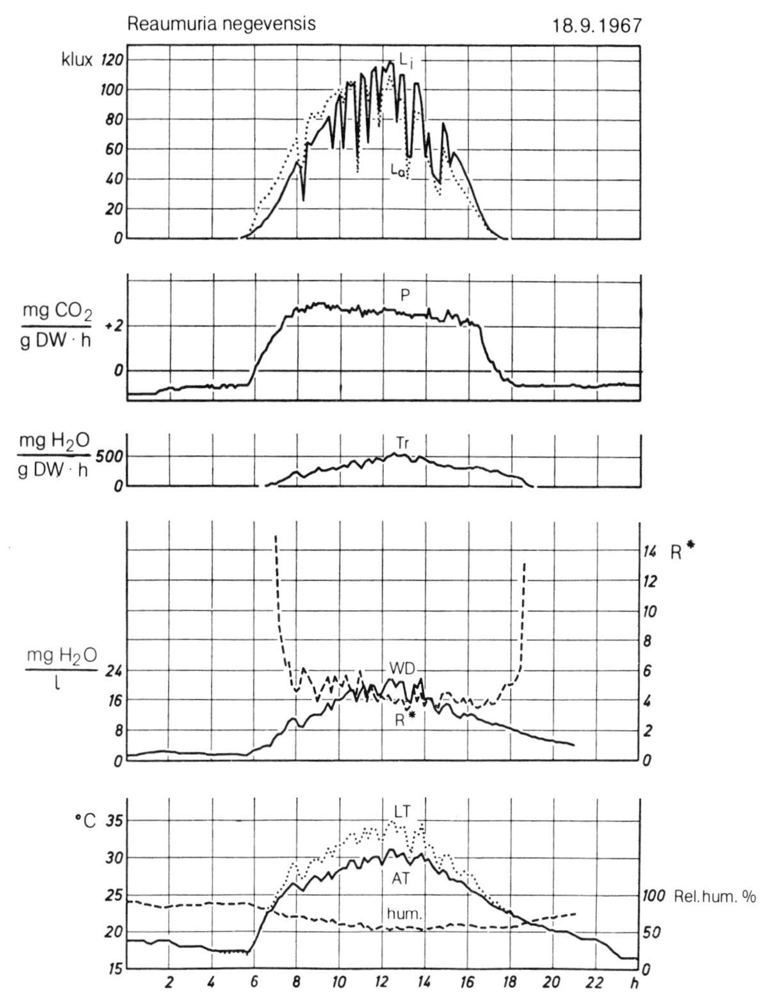

Fig. 62. Graphs as in Fig. 61, for a typical desert plant; it can carry on productive photosynthesis all day long, even though the leaf temperature reaches 35 °C (Schulze et al., 1972). *R**, relative diffusion resistance

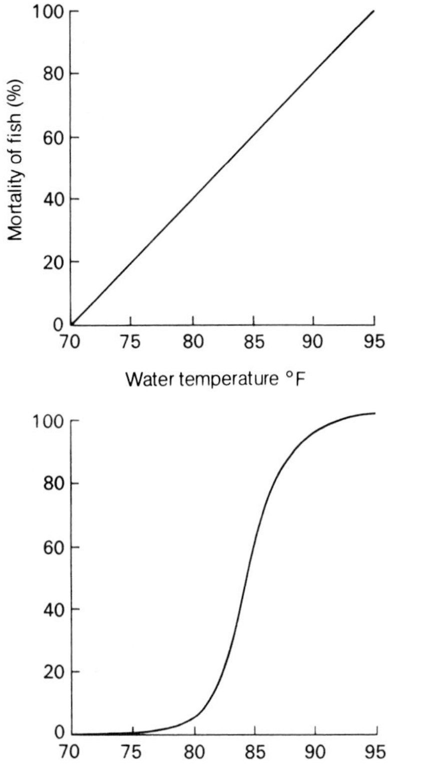

Fig. 64. Idealized linear and nonlinear dose-response relationships. In nature only the latter occur (Ehrlich et al., 1975). The upper curve is therefore false

physiological basis is not yet known. Fluctuating conditions offer an example; as has been described, temperature oscillation increases egg production by ectothermic animals, but we cannot yet say why.

The marine isopod Ligia oceanica can be kept for a long time – without reproducing – if at least two of the factors substrate, salinity, and food are in the optimal range. For example, it can live in fresh water if it is provided with food having high salt content (e.g., marine algae) and with coarse sand or gravel as a substrate. It can live on a mud substrate if this is kept moist with sea water and if it has marine algae or animals as food. Living on a substrate of coarse sand moistened with sea water, it can feed on oatmeal, apple, or nettle powder softened in fresh water (Figs. 65 and 66).

The action of carcinogenic substances on humans is multiplied, even exponentiated, when the air they breathe contains SO_2. Sulfur dioxide interrupts the movement of the cilia that clean the lungs, so that the carcinogens that are breathed in remain in the lungs for a long time and can exert their full effects.

The colonization of the Baltic Sea can be explained, in rough outline, on the basis of experimental findings. As the influx of fresh water into the eastern and northern basin proceeds, many animals shift their ranges from the superficial water levels to the depths, or colonize the depths as well as the surface regions. This phenomenon, known as brackish-water submergence (Remane, 1940, 1955) was not explained for a long time. Its origins are probably to be found in the following considerations:

a) Animals living in the upper water levels or in the tidal zone can tolerate broader fluctuations in salinity than deep-water

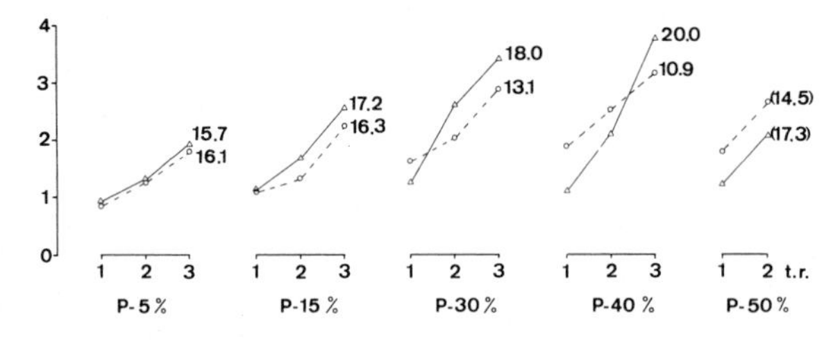

——— constant temperatures; – – – – temperatures fluctuating about the indicated mean during the day. The numbers next to the curves indicate the null point for development. L:D = 16:8, with corresponding variation in temperature. As protein content of the food rises, under fluctuating temperatures the null point for development of the species falls. (Merkel, 1977)

Fig. 65. Rate of development (100/duration of development) of Gryllus bimaculatus at different temperatures and with different kinds of food, identified by the percentage of protein (P 5% to P 50%). The temperatures are indicated by numbers, as follows:

1	2	3	Mode
29°/11 °C	35°/15 °C	37°/19 °C	Fluctuating
23 °C	27 °C	31 °C	Constant

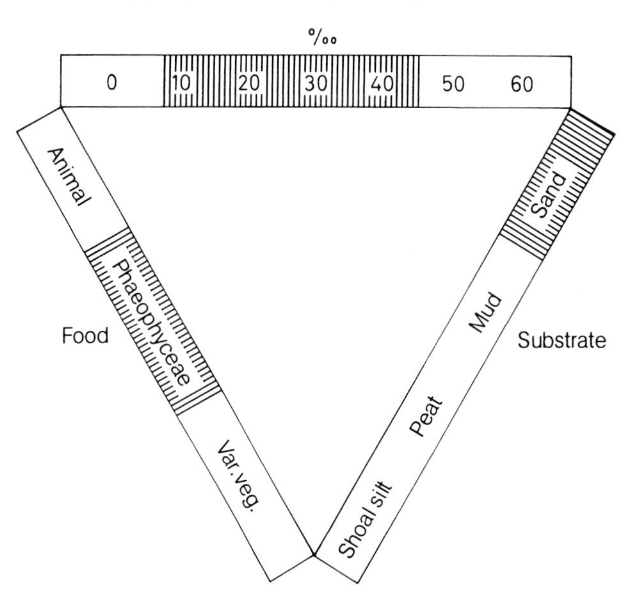

Fig. 66. Diagram of the ecological requirements of adult Ligia oceanica. Two of the three factor complexes must be in the optimal range *(shaded)* for adult animals to be able to survive. (Remmert, 1967)

animals. As the Baltic Sea gradually becomes more dilute, deep-water animals are affected more than those nearer the surface, almost all of which can invade brackish-water regions up to very low salinities – especially species capable of hypertonicity regulation.

b) As the deep-water inhabitants sensitive to brackish water are eliminated, their place, now devoid of competition, becomes available to all species still capable of tolerating this lowered salinity. Moreover, the salinity fluctuations in the deep basin are considerably less than at the surface. Dilution to the extent that occurs near the surface is practically ruled out. Species not dependent on surface conditions, able to penetrate the dark depths, can exploit this noncompetitive habitat; and those unable to endure the salinity fluctuations and the occasional extremely low salinity in superficial water do in fact move downward. Others, perhaps dependent on photosynthesis, on photosynthesizing symbionts, or on green plant food, cannot move away. Finally, a third group which by virtue of hypertonicity regulation or other mechanisms can survive long-term dilution, can become distributed over all depths. Thus even though conditions are becoming increasingly unfavorable, the habitat of a number of animals expands in an initially unpredictable way. In retrospect this is readily explained, but to predict it is difficult (Remmert, 1968 b).

It is even more difficult to understand this sort of factor combination on land. Movement of the air is known to have a cooling effect on moist surfaces, and thus on all plants and animals. The measured temperature in itself means relatively little, as illustrated by Table 6. The effects are enhanced at higher altitudes. Similarly, there is a relationship between temperature and moisture – whether in the form of clouds, fog, rain, or relative humidity. Some examples were presented on p. 22f.

It makes a difference, then, whether we regard the air as a stationary or a moving medium. Naturally, the same difference applies to the second major medium of plants and animals – water. Here, too, we must make clear distinctions between still water (found in nature chiefly in very small bodies of water), turbulent waters (all the larger lakes and the oceans; in nonturbulent water planktonic organisms sink relatively rapidly to the bottom), water flowing in one direction (brooks and rivers, with all the associated adaptations of the fauna and mechanisms to prevent their drifting away), and water that sweeps back

Table 6. Chilling effect of the wind. At a temperature of −29 °C and a wind speed of 40 km/h, humans are affected as they would be by −60 °C in still air (Weiss, 1975). The range of combinations potentially lethal to humans is enclosed by the line

Wind speed (km/h)	Temperature, − °C															
	9	12	15	18	21	24	26	29	31	34	37	40	42	45	47	51
8	12	15	18	21	24	26	29	31	34	37	40	42	45	47	51	54
16	18	24	26	29	31	27	40	42	45	51	54	56	60	62	68	71
24	24	29	31	34	40	42	45	51	54	56	62	65	68	73	76	79
32	24	31	34	37	42	45	51	54	60	62	65	71	73	79	82	84
40	29	34	37	42	45	51	54	60	62	68	71	76	79	84	87	93
48	31	34	40	45	47	54	56	62	65	71	73	79	82	87	90	96
56	34	37	40	45	51	54	60	62	68	73	76	82	84	90	93	98
64	34	37	42	47	51	56	60	65	71	73	73	82	87	90	96	101

and forth (for example, on shores where there is surf). Water movement of the last type provides the basis of the currents that flow through many aquatic animals (e.g., sponges) and carry their food. Similarly, the burrows of many land animals are constructed in such a way that air currents moving above ground set up a circulation of the air in the burrow (Vogel, 1978; Fig. 67).

The extraordinary complexity with which factors can interact is demonstrated by the distribution of North American deer. As humans invaded their habitat, building roads and highways, opening up the large unbroken stretches of woodland and destroying the wolf population, the relatively weak Odocoileus species (mule deer and white-tailed deer), previously restricted to the far south, could expand their range northward across the Canadian border. The regions they invaded had been occupied by caribou, wapiti, and elk. These large deer soon vanished from practically all the areas into which Odocoileus advanced – perhaps due to infestation by a parasite (a nematode) of the latter. Although the parasite had established a relatively innocuous functional relationship with its original host, it was lethal to the larger deer. Thus Odocoileus, otherwise an inferior competitor, by coevolution with a parasite gained the advantage in competition with the large deer. When humans provide mule and white-tailed deer with suitable conditions – they cannot thrive in very large uninterrupted forests with a large wolf population – they can drive the competing deer from their habitat with the help of the parasite. In some respects this example illustrates the limits of the field of physiological ecology. Only by combining the results of physiological experimentation and field observation can such a situation be explained.

The unpredictability of the effects of certain factors in combination impedes analysis in two senses – one cannot be sure in advance that an organism will be found in a particular place, and the knowledge that an organism does occur in a habitat does not necessarily justify conclusions about the conditions there. Although this uncertainty is always there in principle, in practice one can assume that within a circumscribed area the complex of interacting factors is uniform, and draw useful conclusions on this basis. Studies of plant distribution have demonstrated the practical value of such procedures. Within the limits of central Europe, for example, one can infer from the prevailing conditions which plants should be growing in a location, so that maps of the potential natural vegetation can be constructed. On the other hand, the presence of particular plants per-

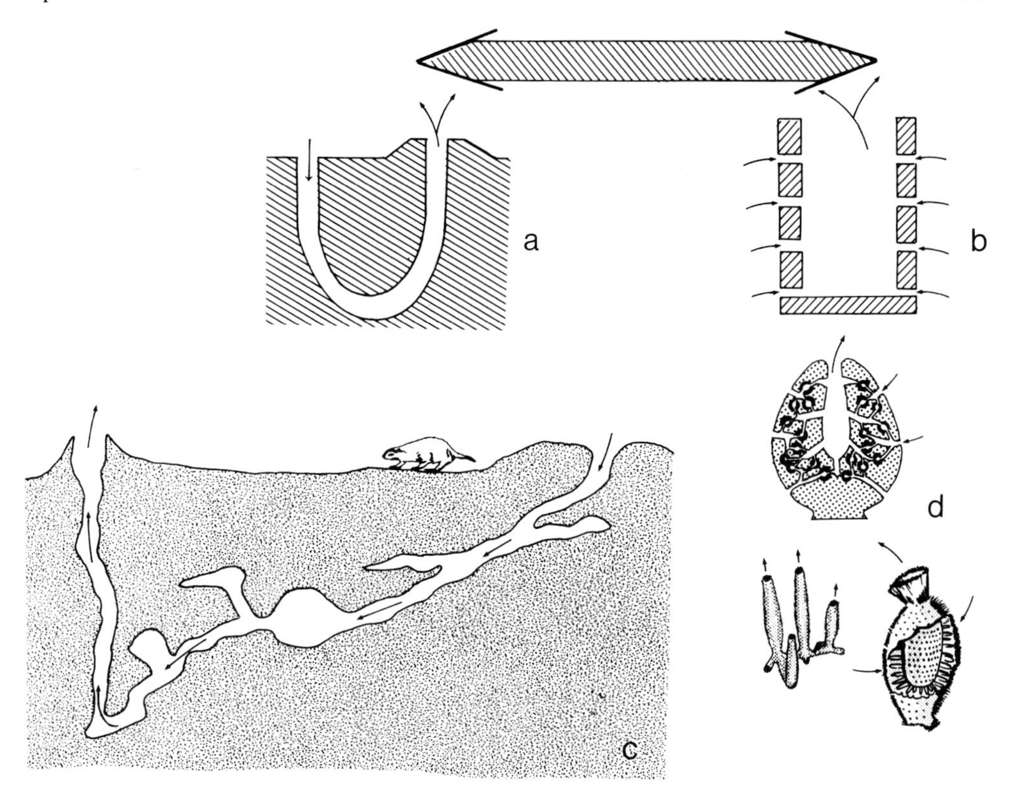

Fig. 67 a–d. Ways in which movements of air and water varying in direction (two-headed arrow, above) can be used by animals. **a** Diagram of the passage of wind or water through a U-tube with one end raised. **b** Diagram of the passage of air or water through a vessel with many small openings and one large opening (this principle is now being tested for wind-driven power plants). **c** A practical example of **a** – the ventilation of the burrow of a prairie dog (Cynomys). **d** An example of **b** – current flow through a sponge (Porifera). Combined and drawn from Vogel, 1978

mits accurate conclusions about water relations, the nature of the soil, and the local climate. The existing vegetation is a simpler guide than meteorological data to a decision about which crops will thrive in an area or about the kinds of parks that can be established there. Indeed, it has been possible to chart the vascular plants of central Europe in terms of their indicator value (Ellenberg, 1974a), and this information has been used on a large scale in both urban planning and the planning of conservation areas and national parks. Because they stay in one place and are relatively easy to identify, plants are more useful than animals for such purposes. Here is an area, then, in which autecological analysis has proved to be of considerable practical importance (Miyamaki and Tüxen, 1977).

13. Special Problems

It is probably true in general that association with a particular habitat is genetically fixed and unalterable by environmental influences. But there are a number of remarkable exceptions to this rule, which deserve further study. Outstanding among these is the wasp Trichogramma, which parasitizes the eggs of many insects. The shape and coloration of the wasp depend to a great extent on its host; indeed, wasps that have grown up in the eggs of certain hosts can be wingless. Understandably, this variability has produced chaos in attempts to classify the genus. Its systematics were clarified by Quednau (1957). In most cases a single species (evanescens) is involved, individuals of which can be "imprinted" on the host it parasitizes. Especially after several generations have used the same host species, the females show a clear tendency to choose this species for egg deposition. This is evidently a case of long-term modification of habitat choice. It is of economic interest, because the wasp has been and can further be used for biological pest control; it can be raised in large numbers under conditions that "set" it to a particular host target. The example shows how careful one must be in judging the extent to which an animal's biotope is obligatory. It may be that such cases are more common than has been thought. The roe deer (Capreolus capreolus) is known to follow clear-cut traditions in the choice of food plants; the fawns adopt the mother's preferences as to species and the part of the plant chosen. Hence one finds regional differences in the amount of grazing of particular plants (Ellenberg, 1974b). Further studies on this point are urgently needed.

When the ecological requirements of juvenile and adult animals are not the same, remarkable phenomena can result. Working independently, Ribaut in Lausanne and Erz in Dortmund and Kiel (both of whom published in 1964) showed that cities are a favorable environment for full-grown blackbirds. Their mortality is lower there than in the woods, and in particular the opportunities for overwintering are better. On the other hand, more nests, with or without eggs or young, are destroyed in the city than in the countryside. The population of urban blackbirds cannot maintain itself, but in many cities depends on a steady stream of immigrants from biotopes more suitable for growth of the young (it is tempting to see here a parallel to the human situation).

Areas of this sort, in which adults can thrive but cannot reproduce (or produce enough offspring), have for some time been called "sterile ranges" by zoogeographers. Many marine animals appear unable to reproduce in the Baltic Sea, with its lowered salinity, but great numbers of larvae are regularly swept in from the North Sea. These readily establish themselves, so the Baltic population remains large. Similarly, many lepidopterans arrive in central and northern Europe in the summer; some common species never succeed in reproducing there. Hardly any ecological approaches to this complex of problems have been attempted.

Even within a stage of an animal's life cycle that appears uniform, there can be distinct alterations in ecological requirements. Caterpillars of the lepidopteran species Euphydryas maturna feed gregariously on ash, poplar or beech leaves, but after overwintering they change both plant and habit; they are then found as individuals on plantain, scabious or veronica. Such sudden change in social behavior and food requirements is no rarity in the genus Euphydryas, and it can be even more pronounced in another butterfly group, the blues. Most blue caterpillars secrete a fluid attractive to ants; the ants protect them from predators. The caterpillar of the European large blue (Lycaena arion) feeds on thyme up to the third stage; then, when licked by an ant, it transforms itself into a packet resembling an ant pupa. The ant carries this packet into its nest, where the caterpillar continues to live – now as a predator, feeding on the ant larvae. When its growth is complete the caterpillar pupates and spends the winter in the anthill. Then the imago emerges and leaves the ants' nest (and no one knows why the imago is not eaten by the ants).

V. Case Studies in Autecology

1. Surface Chemistry and Choice of Biotope

How do barnacle (Balanus) larvae find a substrate suitable for metamorphosis? These crustacean larvae are free-swimming marine plankton which eventually pass through a cypris stage and finally turn into the sessile adult form. Barnacles are hermaphrodites, but must be fertilized by another individual, so that for reproduction to succeed, many individuals must live in close proximity. The cypris is the first sessile stage, and as such plays a key role in the barnacle life cycle. Crisp (1964), who has been particularly concerned with this problem, writes that it seems improbable enough for the barnacles Conchoderma and Coronula, which attach only to whales, to find a whale in the waters of the open ocean; that two larvae should find the same whale and settle on it side by side seems quite incredible. But this is just what happens. How do the barnacle larvae manage it?

If one keeps barnacle nauplia in a container with some hard objects to which balanids are already attached and others, as a control, without balanids, after metamorphosis to the cypris stage the larvae settle exclusively (as long as there is enough room) in the vicinity of the adult barnacles – the vacant objects at first remain so. The result is the same if dead barnacles are used rather than living ones, and even with old barnacle shells from museums, provided that they are not appreciably contaminated by poisons such as formalin. Some factor in the barnacle, then, which is not destroyed even by desiccation and prolonged storage, is recognized by the larvae. It would seem to be a chemical stimulus produced by the barnacle, but this assumption appears to be contradicted by the fact that even balanids that have been soaking in water for years are equally attractive; a chemical stimulus ought eventually to be leached out. And one becomes still more sceptical on learning that the cypris larva must make direct contact with the balanid shell before it is induced to settle. Crisp extracted from balanid tissue a soluble material that he identified as a cuticular protein, "arthropodin." This protein is particularly abundant in newly formed cuticle. When the cuticle hardens it is changed into "sclerotin." Plates covered with this protein are just as attractive to the larvae as Balanus shells – even if they have been dried out in the meanwhile. Now, if arthropodin is dissolved in the water that fills the container, and untreated shells are put in along with shells treated with the same protein concentration, all the larvae settle on the treated shells. This result is further evidence against a chemical stimulus in the sense of odor or taste. The molecules dissolved in the water evidently have no effect on larval metamorphosis. Rather, the larvae test the surfaces of objects until they find a structure that triggers settling. This observation implies that the protein must be oriented in a particular way on the object, a way that is detectable by the larval sense organs. In that case it ought to be irrelevant whether the arthropodin is applied to the object in a monomolecular layer or in a thick coating. This proposal was tested, by measuring the attractiveness to the larvae of objects covered with different numbers of arthropodin layers (Fig. 68). It turned out that if there is less than a single uninterrupted layer – that is, if there are relatively large "holes" in the coating – very few larvae settle nearby. As the number of layers is increased from 1 to 8, the number of larvae that settle rises steeply and saturates; no more larvae are attracted by objects bearing more than 8 layers, even as many as 1,000. In view of the difficulty in producing true monomo-

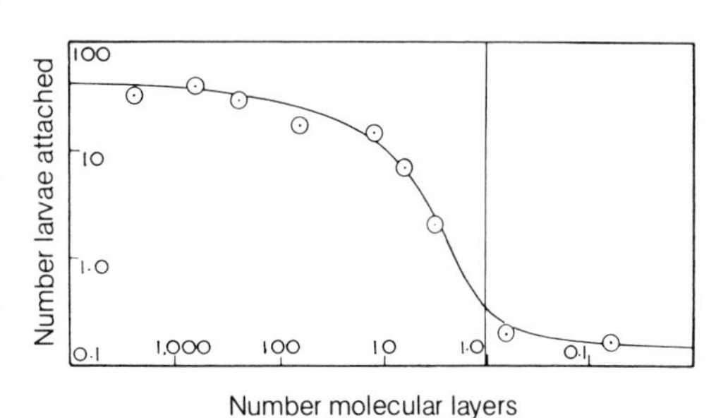

Fig. 68. The attractiveness of plates coated with different numbers of layers of the protein arthropodin, as places for cypris larvae of the barnacle Balanus balanoides to settle. They attach when at least one uninterrupted layer is present. (Crisp, 1964)

lecular layers, it must be assumed that the arthropodin coating does not become really uniform until five or six layers have been applied. This would mean that the crucial element is not the amount of arthropodin, but actually the surface structure itself, which is detected by the animals and which triggers metamorphosis.

It is clear that this sort of surface-structure recognition plays a central role in the metamorphosis of planktonic larvae. There is hardly any planktonic larva – whether polychaete, echinoderm, or cnidarian – that can be brought to metamorphosis in the laboratory without special handling. But it is often easy to induce metamorphosis if the culture tank contains specially conditioned sand. This is simply sand taken from the natural habitat of the animal and coated with a film of bacteria. Müller (1969), in his study of the hydroid polyp Hydractinia echinata, showed that quite specific bacteria are necessary to bring about metamorphosis. Furthermore, they must be cultured in specific nutrient solutions, and bacterial cultures in the stationary phase of growth give better results than those in the exponential phase. In this case, too, the surface of the bacterial coating is tested by the planula larvae before metamorphosis occurs. Again, a very thin film seems to suffice; the effect is no different when large quantities of bacteria are used. It is true that the metamorphosis of Hydractinia can be induced in other ways – for example, by lithium ions – but in nature it is most probably the chemistry of the bacterial surface that is decisive.

Contact receptors of this sort, which obtain information from surface structure, appear to be intimately involved in the choice of food plants and egg-laying sites by insects (Chapman and Bernays, 1978).

2. Synchronization with Habitat Conditions

Along the ocean coasts a large number of very precise environmental rhythms coincide – the diurnal rhythm, the ebb and flow of the tides, the regular alternation between spring and neap tides, and finally the annual rhythm. All these rhythmic changes in their habitat profoundly affect the lives of tidal-zone organisms. Neumann (1962 ff.) studied the way in which they affect the nematoceran insect Clunio, which lives on rocky substrates. The imago emerges from the pupa only at extreme low tide, and lives for a very short time; the male emerges about 20 min before the female, finds a female pupa, and sets the female free (Fig. 69). Copulation follows soon after, and the eggs are attached to stones.

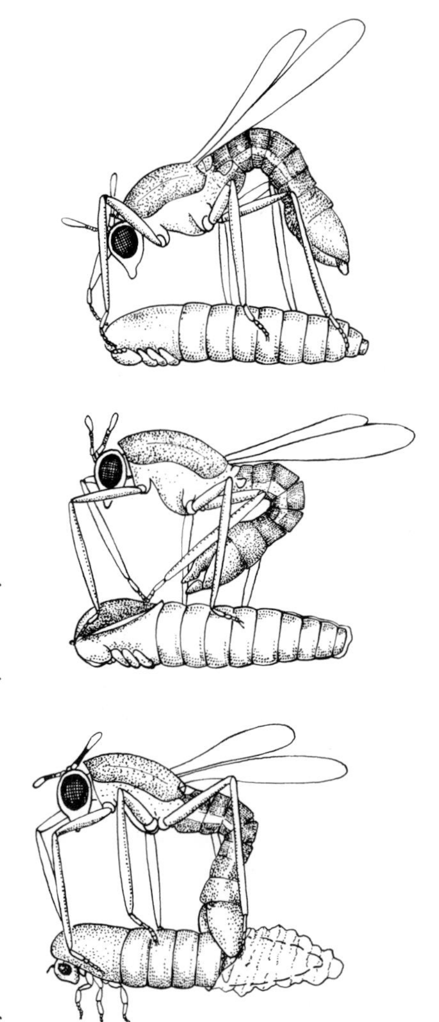

Fig. 69. A male Clunio helps the female out of the pupal case. (Hashimoto in Remmert, 1962)

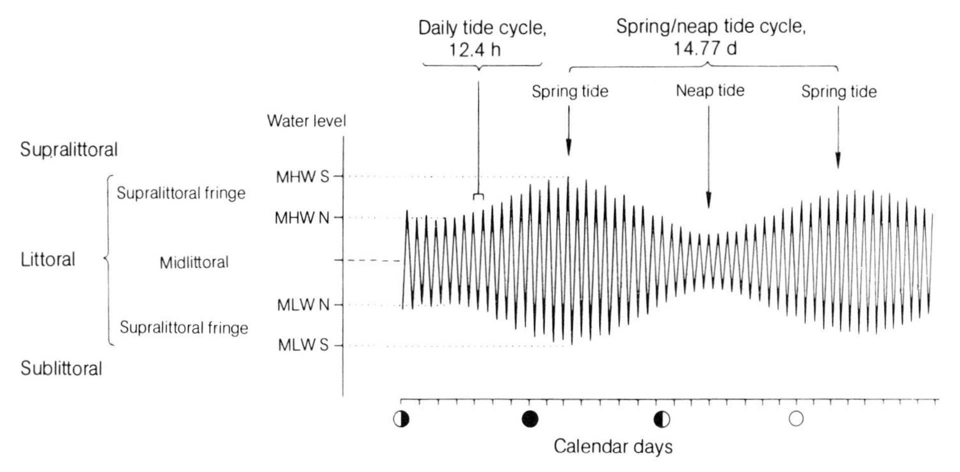

Fig. 70. The ecological subdivision of the tidal zone, in terms of the tides and phases of the moon. *MHWS, MLWS:* mean high and low water at spring tide; *MHWN, MLWN:* mean high and low water at neap tide. (Neumann, 1976)

About an hour later most of the adult animals are dead; the water rises and again covers the habitat. The development of Clunio follows an extremely precise schedule. The imagines must emerge at the time of spring tides, when the ebb tide moves the water line far out and briefly leaves the habitat dry. Then all the males in a condition to emerge from the pupa must do so simultaneously, within a few moments, in order that there be time to release the females, copulate, and secure the eggs.

Theoretically such synchronization could be achieved by a semilunar rhythm that preprograms the animals to the days of lowest low tide. A diurnal rhythm would have to be superimposed; the daily and monthly tidal rhythms interact in such a way that a particular water level – for example, extreme low tide – occurs at the same time of day every 14 days. Semilunar rhythms and diurnal rhythms together, then, could in theory account for the performance described. The animals need not have developed a tidal rhythm (Figs. 70 and 71).

This is in fact the case. If Clunio is kept in an aquarium, the imagines always emerge at a particular time of day, depending on the light/dark cycle. The existence of a semilunar rhythm can also be demonstrated. Larvae in aquaria under conditions of alternating light and dark are ex-

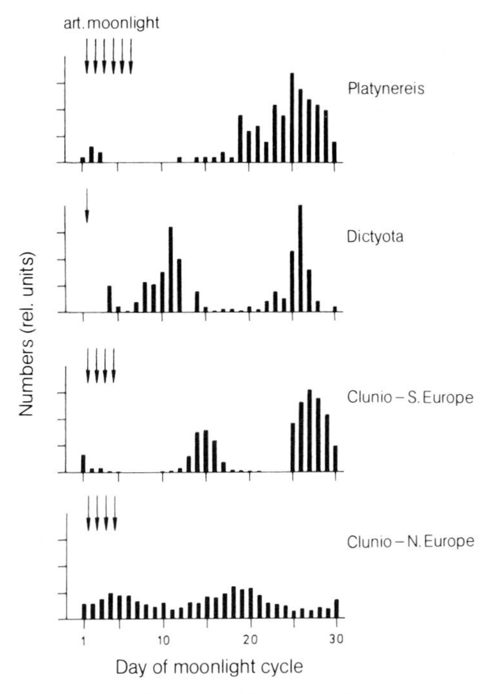

Fig. 71. The effect of artificial moonlight on the reproductive times of marine organisms. The arrows mark the nights with artificial moonlight (ca. 0.3 lux), presented several times at 30-day intervals, during either one or several successive nights. (Neumann, 1977)

posed to an additional "artificial moonlight" for a few nights. This weak extra illumination effectively reduces the amplitude of the diurnal light/dark oscillation. Under these conditions the semilunar clocks evidently present in the larvae and pupae are synchronized with one another; now emergence occurs every 14 days at the same time of day, even if the lighting conditions are returned to a normal, constant-amplitude light/dark alternation or to maintained light (Fig. 71).

It would seem, then, that the mechanism that causes concentration of the imago population by linking the life cycle to the temporal structure of the habitat is clear. But it conceals many problems. High water and low water do not occur simultaneously all along the coast (Fig. 72). The sole inference to be drawn from this fact is that very closely adjacent Clunio populations can emerge at different times of day – and this is precisely the case (Fig. 73). For each population the time of emergence is genetically fixed. Crossing two populations with different emergence times results in a second generation with an intermediate time. Evidently only a few genes are responsible.

The same programing could be achieved by means of an endogenous lunar rhythm, one with a period of just 30 days. One of the Clunio populations studied – that off the Basque coast – does in fact have such a lunar rhythm.

A prerequisite for effective synchronization by these rhythms is that the moon be a reliable timing signal. Is it really? Or are the bright nights that recur every month when the moon is full to some extent

Fig. 72. The sites from which the tested Clunio groups were taken, showing the distribution of flood-hour lines (the lines connecting locations with the same mean high-water time differences from the meridian passage of the moon in Greenwich). (Neumann, 1965)

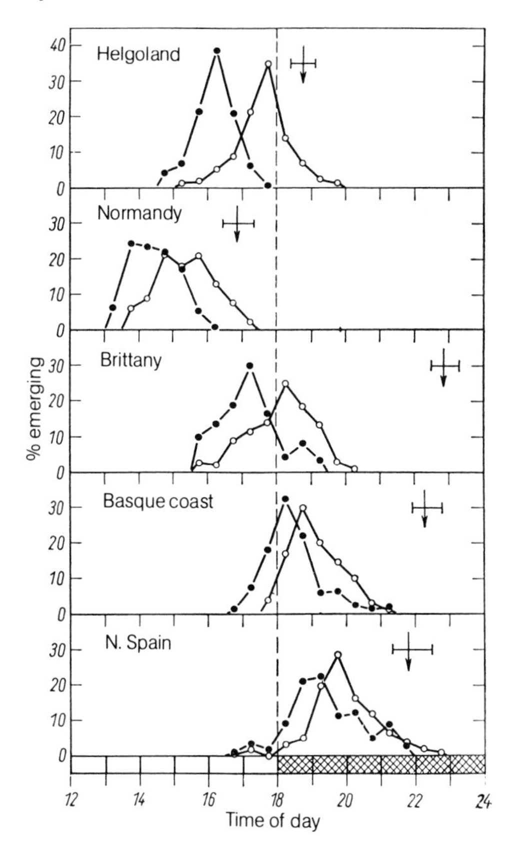

Fig. 73. Clunio marinus. The distribution of time of imago emergence over a day, for members of 5 geographical races kept in artificial light (L:D 12:12, light period 6 a.m. to 6 p.m., 20 °C). ● , males; ○ , females. In the part of the day not shown (midnight to noon) no imagines emerged. *Arrow:* the mean local low-tide time on the first day after full and new moon (second low tide of the day, by local time, average for 1963 with range of variation indicated). (Neumann, 1965)

masked by wind and weather or by the generally increased nighttime brightness during the summer? At southern latitudes, the moon actually does seem to provide a reliable signal . But further north, where the summer nights are bright, it does not; even as far south as Helgoland, the native animals give hardly any response to the signal "artificial moonlight." It was shown that in Helgoland the turbulence of the rising and receding water gives the timing signal (Fig. 74), whereas on the coasts of France and Spain it is not effective. Under experimental conditions, the Helgoland Clunio could be synchronized during larval development by turbulence just as the Spanish ones are by artificial moonlight. Yet another kind of behavior is found in arctic populations (from the region of Tromsö, Norway, about 300 km north of the polar circle). Here the animals occupy a rather different habitat, sandy beach flats over a granite stratum. The imagines emerge in summer at each ebb tide. Each time the water recedes from their habitat the temperature of the sand rises by about 2–3 degrees. If the animals are kept in constant temperature in the laboratory, they emerge at random times. If the temperature is raised once by the appropriate amount a few of the insects emerge immediately, whereas very many emerge 12 h later (Fig. 74). Then the times of emergence become scattered again. In this arctic

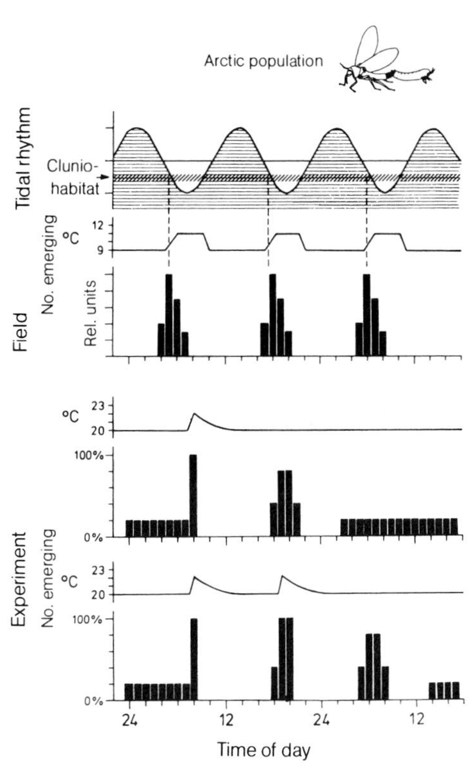

Fig. 74. Tidal periodicity of the times of emergence and reproduction in the arctic population of Clunio marinus (Tromsö, Norway), in the field and in an experiment in which the temperature was raised briefly as a timing signal. (Neumann, 1977)

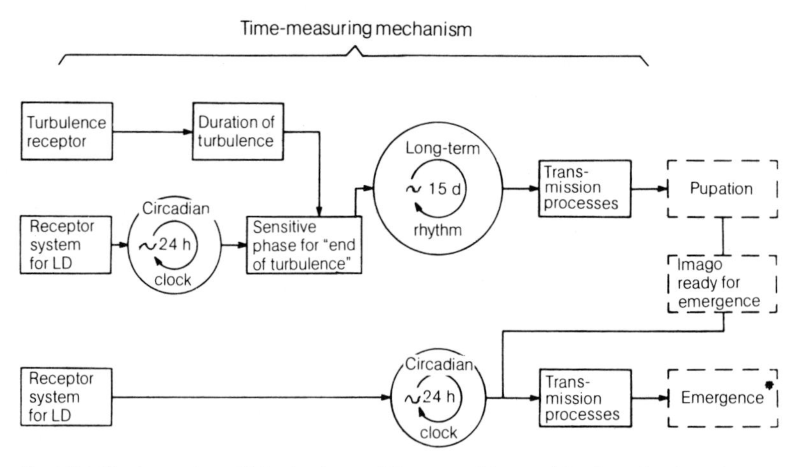

Fig. 75. Clunio marinus, Helgoland race. Diagram of the combined mechanisms for measurement of time, and their components. The mechanism for semilunar periodicity (upper half) controls the onset of pupation; the pupal stage lasts 3–5 days. The circadian measurement mechanism (lower half) controls the time of day at which those imagines ready to emerge do so. The time of emergence is directly adapted to the environmental situation (spring low tide). (Neumann, 1977)

population the clock is quite different from those previously considered, in that the latter continued to run under constant conditions – they are driven by some sort of oscillation. In the northern population a single stimulus, one brief increase in temperature, programs the animals to emerge 12 h later. This principle of synchronization is comparable to that of an hour-glass. The Baltic Sea is inhabited by a species that evidently differs from those previously mentioned, for there are no tides there; the animals live among red algae at depths of 5–10 m. Their eggs do not need to be attached to stones, because when they are laid they sink to the bottom rather than floating. In the vicinity of Bergen (western Norway) the Baltic Clunio species are found together with the Norwegian population of Clunio marinus. The two emerge here at different times. The mechanisms are summarized in Fig. 67. An extraordinary variety of mechanisms has been developed within a single species to match the life cycle to the temporal structure of the habitat. Different populations behave differently, in a way that is under genetic control. This sort of adaptation may well operate, in similar ways, in a great many tidal-zone organisms.

3. The Biology of Game Animals: Capercaillie and Roe Deer

One of the best known animals in the European forests is the large grouse called the capercaillie. Populations of this bird are currently falling off drastically – a tendency which conservation organizations and hunters alike are making efforts to halt. If they are to be successful, they must have detailed knowledge of the bird's habitat requirements. The information so far available reveals a complicated picture (Müller, 1974; Scherzinger, 1976; Brüll et al., 1977).

The cock needs a more or less isolated space in which to perform his courtship dance. This grouse is a relatively poor flyer. Where the tree crowns form a dense, level layer it cannot penetrate them; the trees must be of different heights. The tree selected for the courtship display should provide an extensive view (Fig. 76), and thus ideally is located on a slope (Figs. 77 and 78). When his display in the tree is completed the cock flies down to the ground and continues the courtship performance; this requires a clearing where the vegetation is low and the ground is level. Although the tree is on a slope, then,

Fig. 76. Diagram of a capercaillie biotope, in a montane mixed forest in nearly natural conditions. *1* young spruce, the winter food of the cock; *2* courtship tree; *3* bilberry bushes; *4* courtship area on ground; *5* picking out stones from roots; *6* sleeping place under cover; *7* sheltered breeding place; *8* spruce twigs, the winter food of the hen; *9* sand bath; *10* exposed sleeping place; *11* anthill; *12* beech foliage, the food in summer and autumn. (Scherzinger, 1977)

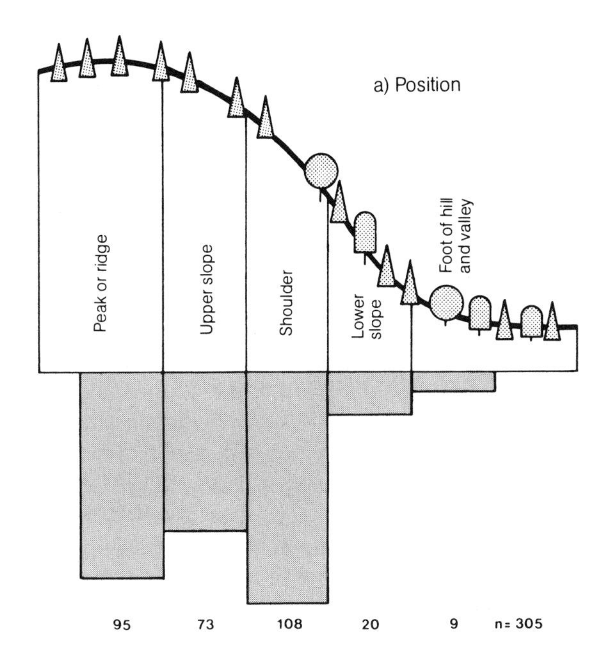

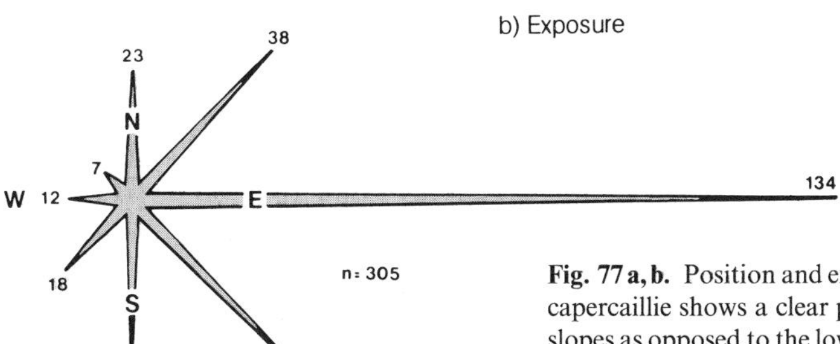

Fig. 77a, b. Position and exposure of the courtship trees. The capercaillie shows a clear preference for the peak and upper slopes as opposed to the lower levels. The tree selected is most likely to be on a slope with eastern exposure. (Scherzinger, 1977)

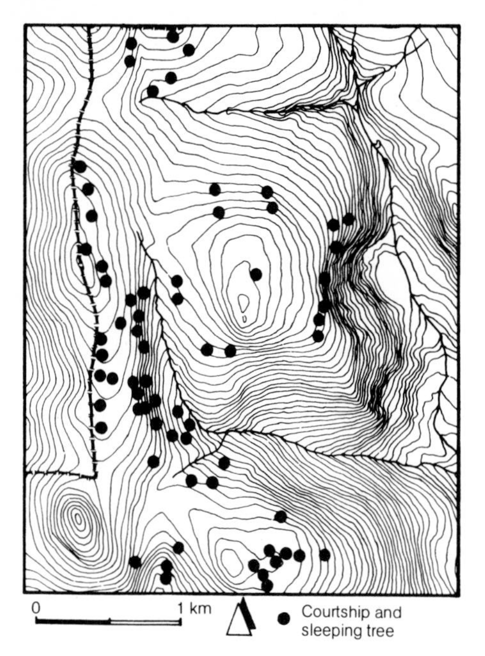

Fig. 78. Geographical distribution of the courtship trees. Because slopes with eastern exposure are preferred (in the Bayerischer Wald) there is a striking concentration of trees used for sleeping and display on all the ridges, hilltops, and open ground facing in this direction. (Scherzinger, 1977)

it must have a flat area of at least 20 × 30 m at its foot, where nothing very tall is growing. But at the edges of the clearing there must be dense undergrowth in which the hens can hide, and where they can build their nests not too far away from the courtship site. These nests are always located in the protection of thick shrubbery. Soon after hatching the young are led out of the shrubbery, but they always return to the bushes to rest. The young birds are sensitive to rain; rain in their first days of life frequently causes the death of the whole brood (Fig. 79). Moreover, during their first days the young must be fed on isects

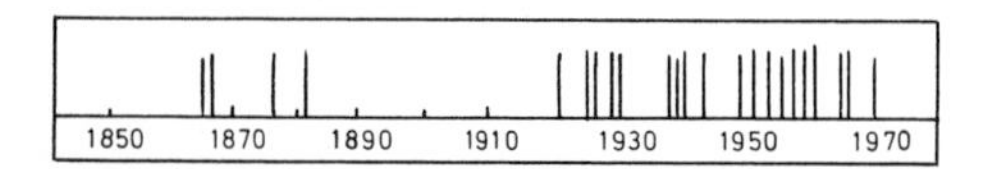

Fig. 79. Distribution of years with surplus precipitation, at least 100 mm (Brüll et al., 1977). The increasing frequency of such years was accompanied by a decline in the stocks of capercaillie, for the chicks cannot tolerate rain

that are not too small, whereas later they change to plant food. In winter they eat chiefly the needles of conifers; in spring leaf and flower buds and willow catkins make up a large part of the diet. In summer large amounts of berries from ericaceous plants (cranberry, bilberry, and their relatives) are consumed. This variety of demands cannot be met everywhere. Capercaillie can survive only in primeval mixed forest or in carefully planned woodland. A widespread system of forest management, in which each subdivision contains trees of the same age, is lethal to the capercaillie for several reasons. First, the crowns form a solid layer through which they cannot fly. This affects not only the courtship display, but survival in winter, when they must stay in the crowns of conifers to feed. Second, such woodland usually lacks the herbaceous and bush stratum necessary for nesting. And finally, the red deer population in these forests is generally excessive, so that even though the deer are provided with hay in winter they make severe inroads on the buds of the bushes the capercaillie needs to protect its nest. In the spring the roe and red deer compete with the capercaillie to a certain extent for food; the deer find willows just as palatable as the grouse do, though the former do not depend on them. A given area cannot support large populations of deer and capercaillie at the same time. Mass culling of the deer in certain areas was followed by a sharp increase in the capercaillie population.

This example serves as an illustration of the difficulties encountered in autecological research, even on well known animals. To the birds' highly specialized requirements with respect to climate, food, topography, and features of the vegetation is added competition by the deer, which affects the availability of both food and shelter. And this bird occupies more or less the same place all the time. If we had chosen to study an animal that either migrates south in winter or – as many insects do – requires very specific conditions for overwintering, the situation would be considerably more complicated. But all these separate factors and separate facets of the interactions between organism and environment must be known in detail if advances are to be made in either theoretical ecology or conservation methods. We shall return to these problems when discussing the division of resources among the members of a species (for territory establishment see p. 122ff.; cf. p. 74) and predator-prey systems.

Unlike the capercaillie, roe deer have found the agricultural regions of central Europe increasingly better as a habitat over the past 150 years. The population that now occupies central Europe is certainly a great deal larger than it would have been without modification of the countryside by humans. Nevertheless, important aspects of the biology and ecology of the roe deer have been studied only in the last few years (Ellenberg, 1974b, 1978). This association of the roe deer with cultivated areas results from its specialized feeding habits; these deer select food according to digestibility, choosing from a stand of plants only what is most easily digestible – chiefly fresh shoots. Older leaves are of very little value to roe deer, and when fed with hay of even the best quality they starve. Delicate, fresh greenery is found more regularly, and in larger quantities, in a park-like landscape heavily influenced by humans than in a natural landscape. Multiple sowings in a year, forest management, and other measures ensure that fresh sprouts are available throughout the period April to August (and perhaps even a bit longer).

But over an entire year, for the same reason, roe deer face enormous physiological problems in maintaining adequate nutrition. Their physiology is precisely matched to the course of the seasons. From September to the end of March their metabolism consumes relatively little energy. They eat comparatively little, and when it is very cold and the snow is deep they stop feeding almost entirely. As soon as the new leaves appear in April they begin to feed voraciously, and they continue until August whenever fresh green food is available. The food consumed in the winter months is limited to the buds and bark of softwoods and similar plants, and seeds (acorns!). This metabolic rhythm is maintained even under the conditions of captivity, when plenty of food is provided; in the winter the deer eat distinctly less than in summer. The rhythm in feeding is associated with a rhythm in weight, with high values in autumn and low values in spring. The same rhythm is reflected in the reproductive cycle; the egg is fertilized in June/July, but then its development is interrupted, and it does not continue to grow until spring. The fawn is born in May, when food is most abundant and both the mother and the young deer, which soon becomes independent, can eat their fill. The survival rate of the fawns is particularly high when they are born and raised in the territory of a strong buck.

The bucks begin to mark and defend their territories in February (Figs. 80 and 81). The old bucks begin earlier than the younger ones, and thus remain territorial over the years. As the bucks grow older they shift their territories progressively into areas where the best food is to be found, so that the territories of old bucks are smaller than those of bucks that are weaker and younger. For this reason the latter animals must consume more energy – to defend the long boundaries of their extensive territories and to find food in a territory where it is not so plentiful – than an old buck in an optimal, very small territory. But while the fawns are growing up the bucks are largely occupied with preparations for the mating season and with a strict demarcation of their territories; since the winter they have eaten enough to withstand the exhausting season of rut, so that at this time they do not offer the mother deer and fawns much competition for food. This is why the young in the territories of strong bucks (Fig. 82) grow so extremely well.

Another feature of the life of the roe deer that is synchronized with this annual cycle is the dispersal of the young, which is con-

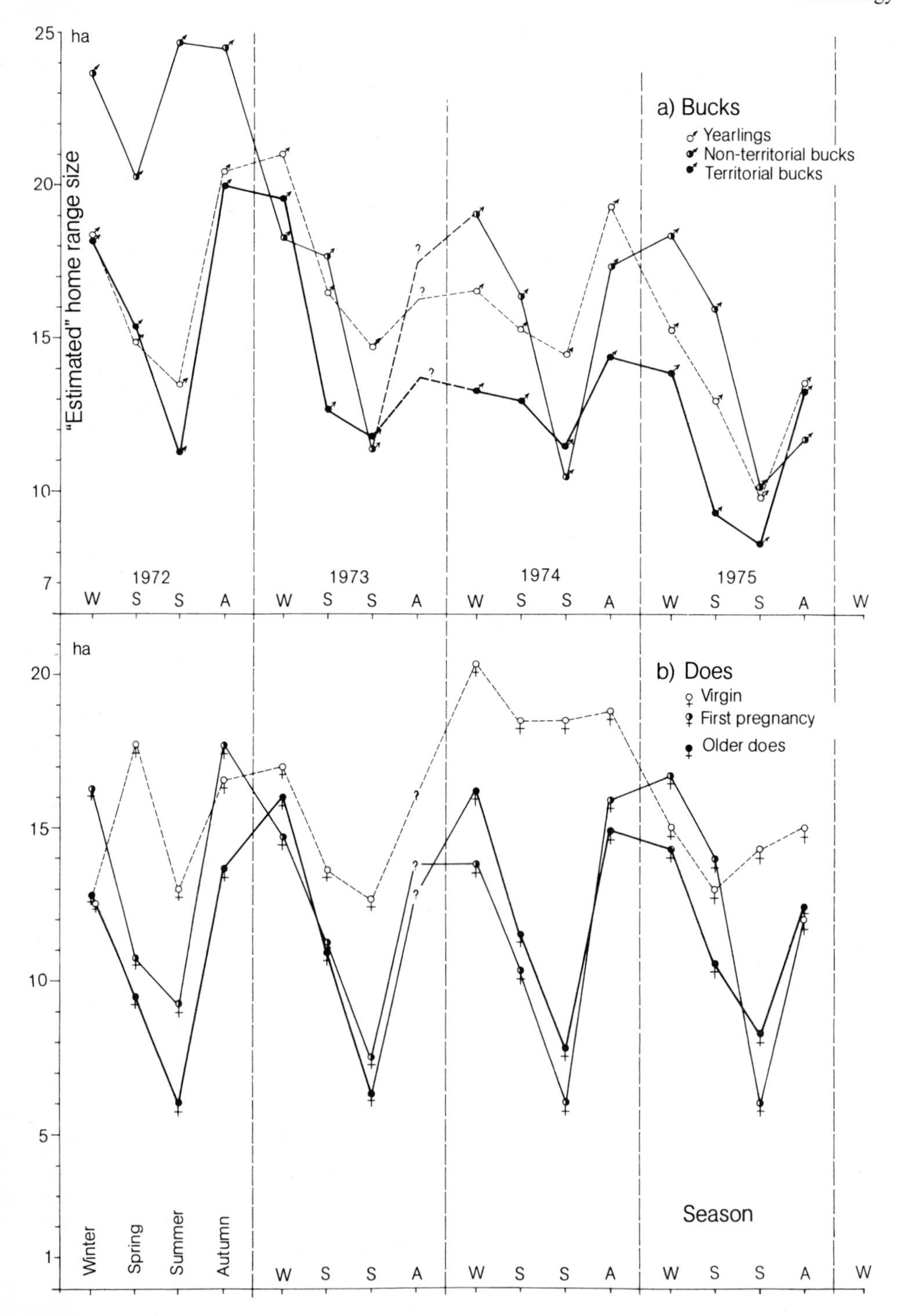

Fig. 80. Size of the home ranges, within the Stammham enclosure, of roe deer of different sexes and ages. (Ellenberg, 1974)

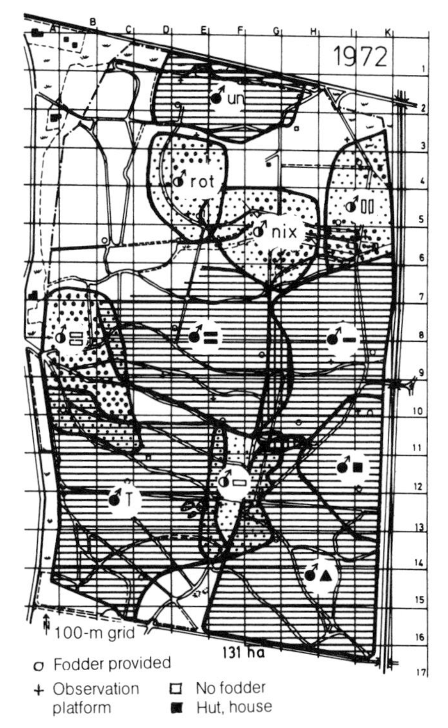

Fig. 81. The buck territories in a large deer enclosure (Stammham). (Ellenberg, 1974)

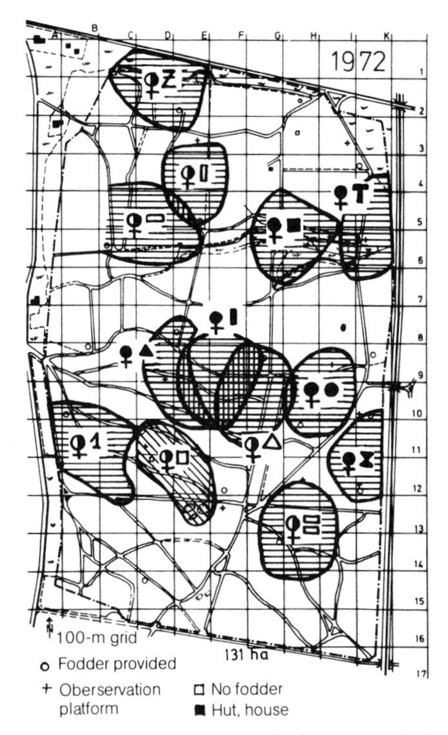

Fig. 82. The regions where the fawns are raised are in the same area as the buck territories. (Ellenberg, 1974)

centrated around the onset of territory marking and after the period of intense summer feeding, in August.

Depending on the availability of food, there is a pronounced autoregulation of roe-deer population size. Three mechanisms are primarily involved; though they operate in synergy, it is analytically useful to distinguish them.

These are:

1. The perinatal and early postnatal *infant mortality*. This is affected chiefly by the nutritional state of the mother during late pregnancy and in the first days after giving birth. In unfavorable biotopes as much as 75% mortality of newborn and small fawns has been observed. In addition, there is considerable mortality of the young before birth.

2. The variability of *ovulation rate*. When the mother-to-be has sufficient high-quality food in the 10–14 days preceding the rutting season the rate of ovulation is high, but when food is scarce it is low. This rate determines, to a great extent, the birth rates in the following spring. Young does in poor condition can fail to ovulate altogether. When conditions are favorable, each young doe develops an average of two eggs. Among older animals the rate may vary between 1.2 and 2.5 eggs per doe.

3. The modulation of the *sex ratio*. The ratio of males to females in the coming generation of fawns is fixed at the time of fertilization. A mother deer well nourished during the mating season bears predominantly female fawns, whereas those that are undernourished bear mostly males. Under controllable conditions and in the wild, sex ratios varying from 1:3 to 3:1 have been observed. In poor biotopes where the does are weak, so that in any case they would have few progeny, the discrepancy in number between males and females is further enhanced. Finally, the disproportion becomes even greater because of the higher mortality of small female fawns, as compared with the males.

C. Population Ecology

C. Population Ecology

I. Theory of Population Ecology

The "population" – formed of individuals of a single species able to exchange genetic material – is the basic unit of ecological phenomena. As far as the degree of genetic exchange is concerned, the definition is not strict. One can speak of the owl population of the Frankish Alb, in which case regular genetic exchange is emphasized; one can also speak of the central European owl population, taking into account that exchange with the Thuringian and Alpine populations, though rare, does occur to a certain extent.

All organisms normally produce more offspring than would be necessary to keep the population constant. Under such conditions the growth curve is exponential, so that after a few generations this particular organism would cover the face of the earth – unless some regulatory process intervenes. Regulation can take the form of a density-dependent effect on the rate of birth or death. When birth and death rates are the same, the population stays constant. Many different factors and mechanisms can bring about this density-dependent regulation; the term "density-dependent" is a logical postulate, not an explanation of the process.

There is a considerable advantage associated with population control by way of effects on mortality; ongoing selection of the best-suited genotypes can continually control and improve the adaptations of the individuals to their environment. The trick of diploidy enables plants and animals to "test" mutations, which as a rule initially appear as recessive traits; they can be carried through many generations as a "genetic load" before they are eventually eliminated or taken on by the whole population. For example, consider human sickle-cell anemia, an alteration of the hemoglobin in the red blood corpuscles. When it is represented in both chromosomes, the carrier either is born dead or dies as an infant. In the heterozygotic state, however, it conveys resistance to malaria and thus makes it possible for humans to settle in regions where malaria is prevalent. In regions where it proves valuable, a gene carried along as a genetic load is taken over by almost all members of the population.

But mortality as a population regulator has a disadvantage. If there is imprecision in the control of mortality level, individuals important to the population at a particular time (those able to reproduce, during the reproductive season) can be eliminated. The risk is particularly great in cases of density regulation by predators, in the broadest sense. Here specific adaptations have been evolved to avoid such accidents.

At this point we may well ask at what level a population is to be held. That there is some optimum is undeniable – neither a too low nor a too high density is desirable. But this optimum is difficult to define. Is it specified by the range within which the organisms are maximally productive, or that in which mortality is lowest? The two need not coincide. Is a constant population size optimal, or is it better for the size to fluctuate?

In recent decades the field of population ecology has made rapid progress owing to the development of mathematical models. To a much greater extent than autecology, which is based largely on physiology and biochemistry, population ecology today is a mathematical science, with a correspondingly universal approach. (As someone has said, "All biological problems concerned with more than a single organism are the legitimate subject matter of population biology.") But there is an associated danger: phenomena that appear similar but have quite different causes are subsumed under the same mathematical description – and one may overlook the fact that it is only a description, and not a causal explanation.

II. Population Genetics

The organisms in a diploid population have a common gene pool. Not every individual has available the total genetic information stored in the population; the individuals within a population are not genetically identical. When the total genetic information is considered, it turns out that different genes vary in frequency. On the assumption that the individuals mate randomly, with no preferred pairings, on the assumption that selection does not occur in any direction, and on the third assumption that we are dealing with a large population, these different gene frequencies are maintained indefinitely (Hardy-Weinberg Law). This rule holds for diploid as well as for haploid organisms. We can easily see how it comes about. Supposing we take six mice, three male and three female; one has the genetic information AA, three the genetic information Aa, and two the genetic information aa. If each of these individuals produces two gametes, we have six eggs and six sperms. Three eggs have the information A, and three have information a; two sperms have A and four have a. With random mating, the next generation again consists of six animals, one with AA, two with aa and three with Aa. In reality, though, with such a small population chance will not necessarily work out exactly like this. There could also be two individuals with AA, three with aa and one with Aa. But such an outcome is found only in small populations; it is called genetic drift. In the large populations with which we are ordinarily concerned, shifts in gene frequencies are the result of selection.

In changing environmental conditions, different individuals have the advantage – and thus better chances in the struggle for existence – at different times. The result is that as generations follow one another in a particular locality, the gene frequencies shift. With organisms that have many generations per year, during the course of the summer the number of individuals adapted to summer conditions steadily increases. When the temperature falls in the autumn their number declines again, to make room for individuals adapted to cooler conditions. During a year, then, the genetic information contained in the average individual of a population with many generations per year changes. This phenomenon has been demonstrated in Drosophila species in many parts of the world (Fig. 83). The situation is the same for the ladybird Adalia bipunctata. In this way optimal genetic adaptation to the changing conditions is ensured. Under changing climatic conditions the population can

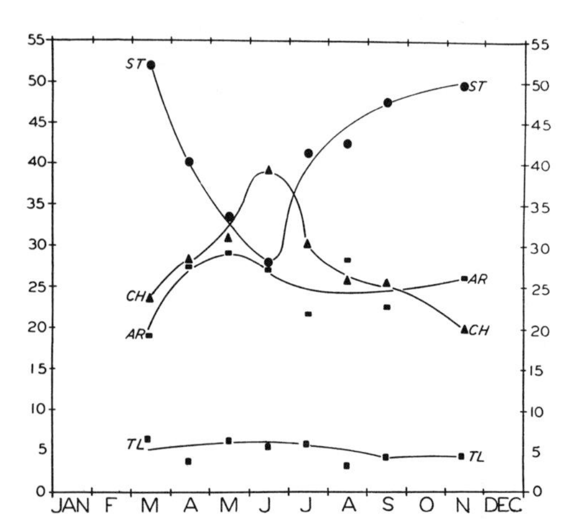

Fig. 83. Changes in the genetic information of a population of Drosophila pseudoobscura, in space and time. The gene frequencies observed at a particular location (California) change during the course of a year, in a way that resembles the geographical changes along a line crossing the United States from Texas to California. (Dobzhansky, 1947)

achieve a high rate of reproduction. The same observation has been made in studies of organisms with only one generation per year or less. It is a regular feature of the leafing-out of woods in spring that certain individual trees are ahead of the main group and others far behind. In favorable years, without late night frosts, the leaders are genetically at an advantage, because they have the opportunity for greater production. In years with late frosts they are in danger of producing nothing at all, for all their buds and leaves are irreversibly damaged. The extreme individuals also represent the genetic reservoir upon which the species draws in adjusting to long-term changes in climate. As the growing season shortens or lengthens over centuries, the population can respond without requiring a single mutation. The probability of a favorable mutation is so low that for long-lived organisms such as trees this is a crucial basis for adaptation.

All our animal populations also include individuals that depart from the norm. The European field cricket normally enters diapause in the penultimate larval stage, in autumn. The two final molts are postponed until spring, when the crickets come out of diapause. But in very warm summers a very small fraction of the central European population – less than one animal in a thousand – skips diapause and metamorphoses to the imago stage directly. These animals do not survive the winter under present-day conditions, but they represent the reservoir with which the population copes with long-term climatic fluctuations. This principle is illustrated even more clearly in the case of species distributed over many degrees of latitude. With respect to migratory behavior, hibernation or diapause their lives are closely meshed with the changing photoperiod as the seasons pass. A famous example is the dagger moth Acronycta rumicis, with a range that extends through Russia from the Black Sea to the Finnish border. In Leningrad this species passes through only one generation per year. It goes into diapause in the autumn; at this northern latitude the days are still very long, although it is cold and the caterpillars can no longer find food. On the Black Sea there are three or four generations per year, and the animals enter diapause late in the year, when the days are very short. They also emerge from diapause while the days are very short in the spring, whereas the Leningrad population requires long day lengths to emerge. The populations located between these two show all the progressive stages of intermediate behavior. The hybrid offspring of members of any two different populations have behavior that is intermediate between the two. It follows that this synchronization with the seasons is under polygenic control. Of course, this description refers only to the majority of the population; in each population there are certain individuals with photoperiodic responses characteristic of a quite different population.

The powerful effects that differences in genotype can exert are well illustrated by a human example. To central Europeans, milk and milk products are an important and regular part of the normal diet. But this does not hold true for almost all humans of other than central European ancestry; most humans can digest lactose only in early youth. In central Europe 2%–10% of adults are also unable to digest lactose, but the figure in Africa and Asia is 90%–98%. When these people consume lactose, in milk and its products, it diffuses into the circulatory system, where it is decomposed in an uncontrolled manner. The products of decomposition upset the acid/base balance in the blood and cause illness. In the gut, too, such lactose as is present is broken down in such a way as to produce substances that cause uncontrolled fermentation and ultimately intestinal disorders. A similar story is associated with dependence on Vitamin D (calciferol). It is common knowledge that it is the precursor of this vitamin which we obtain in food; this substance is converted to the actual vitamin in the skin, under the influence of ultraviolet light. Large amounts of skin pigment block the penetration of these rays where the sunlight is weak – in the lowlands and far from the tropics – and thus prevent this conversion. Therefore people with heavy pigmentation who live at temperate latitudes suffer from rickets on a normal diet; populations of such people cannot normally extend very far north or south of the equator. On the other hand, in very light-skinned people exposed to strong sunshine too much calciferol is produced, which can lead to calcification.

The differing genetic structure of individuals within a population thus permits adaptive modification of the reaction norm of the population when environmental conditions change – that is, it is a basis of evolution. This kind of evolution occurs entirely without mutation. Responses to altered conditions can be brought about simply by sampling the genetic information in successive months or years or at different places (Figs. 83 and 84). For this sort of local adaptation of the genome to penetrate the population, a certain amount of inbreeding is necessary. But as is well known, inbreeding has less desirable aspects. Theoretical considerations demand that there be a balance between inbreeding and its avoidance within a population. So far, however, no experimental evidence is available.

This principle is theoretically valid only for infinitely large populations – in practical terms, for very large populations. Where very small populations are concerned, a change in a population in the same direction over several generations does not necessarily imply adaptive modification of the sort just discussed. In small populations genetic drift can occur; that is, a chance combination of genes can accidentally persist over several generations and thus bring about what looks like a progressive change in the population for this peri-

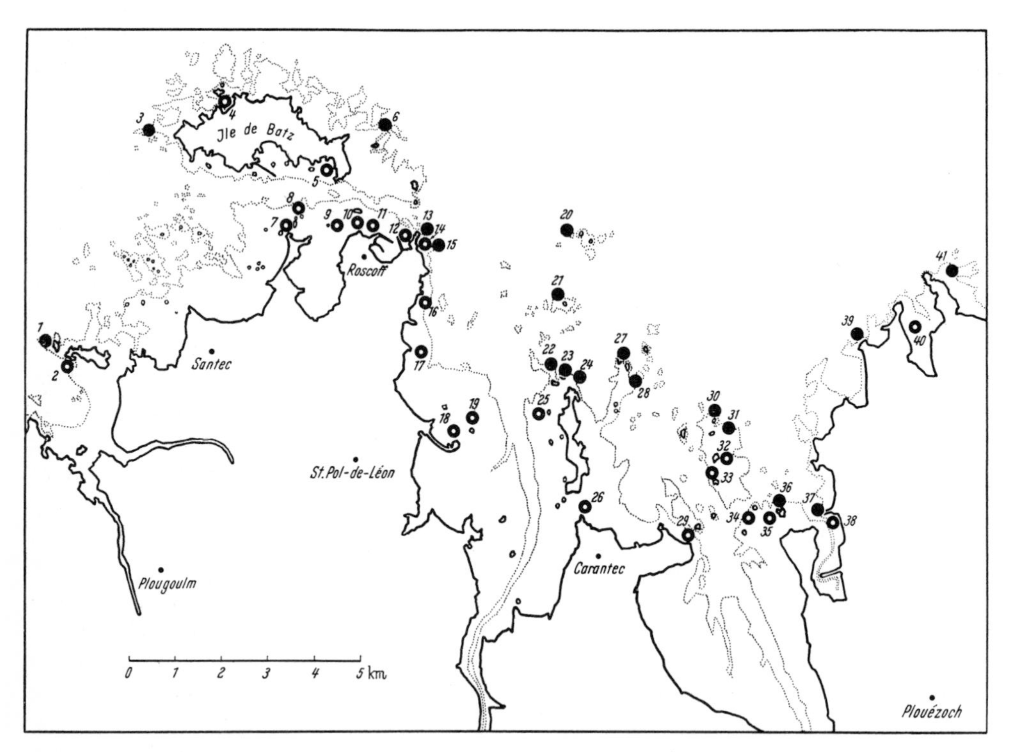

Fig. 84. The distribution of different genetic forms of the snail Purpura lapillus along the French coast. ● colony with 13 chromosomes, on sheltered surfaces; ○ colony with 18 chromosomes, on surfaces exposed to surf. (Staiger, 1954)

od. It is most unlikely – essentially impossible – for such a chance event to become evident in very large populations. A classical example of genetic drift in a small population is found among the ravens of the Faroe Islands. Around the turn of the century this population became famous for the relatively common occurrence of white-spotted individuals, a recessive characteristic. Today the Faroe ravens are black, like their relatives in Scotland and Iceland. We can be quite certain that the white ravens found in the Faroes around the turn of the century did not have a selective advantage – that their appearance was due to genetic drift in a small population (cf. Salomonsen, 1935).

The existence of genetic drift is of considerable significance to modern conservation. Areas in which the natural biological situation is preserved are steadily shrinking, and the rare plant and animal forms that survive there are but small residuals of a once extensive population. When genetic drift is observed in small populations one can never rule out the possibility that the populations may collapse and die out; in this case the alteration that has occurred by chance involves loss of genetic information essential to survival. In a large population, on the other hand, loss of genetic information is practically impossible. Natural large populations have a large reservoir of genetic information that normally is not used, and may even seem deleterious – hence the term "genetic load." But under certain environmental conditions this load may have a decisive positive value. In any population the genetic load contains a great deal of hereditary information that is evidently extremely injurious; its influence is normally not exerted because the traits are recessive and almost always suppressed by a dominant gene. Examples of such deleterious – in fact, hazardous – hereditary traits in humans are color blindness, hemophilia, and galactosemia (cf. Sperlich, 1973).

But if all individuals in a population are different, we may be sure that this diversity becomes significant when a new habitat – an island, perhaps – is colonized. Colonization of such a region is carried out by only a few individuals. As a result, the new island population that develops contains only part of the genetic information of the original population. A priori, then, the island population is likely to differ in appearance from the original form, or if not in appearance, in its physiological responses. This founder effect explains the marked tendency for distinct races to develop on isolated islands not widely separated from one another.

In the last analysis, every population confronts us with an ordinarily invisible polymorphism, which remains relatively constant as long as the environmental conditions do not change. This polymorphism can also be expressed in the external appearance of the animals and plants. The great variability among buzzards and ruffs, as well as certain snails (Cepaea) and cicadas (Mocydiopsis), offers an example. The important things are that mating is purely random within the population, not affected by the appearance of the animals, and that appearance does not convey any other sort of clear superiority or inferiority in the struggle for existence. Among the snails mentioned, it may be that on cultivated fields certain types are particularly easy for birds to spot, and are therefore at a selective disadvantage. But with the rotation of crops commonly practiced in central Europe, the disadvantage is divided up among the population – some types in one year, and others in the next.

The inhabitants of biotopes that remain constant for long periods are selected for the maximal uniform utilization of the habitat that can be achieved without affecting it. On the other hand, in habitats that rapidly come into being and vanish again organisms are selected for the speed with which they colonize the habitat, exploit it completely, and conduct a widespread search for a new favorable site. We can call organisms subject to these two types of selection K strategists (because

Table 7. Some consequences of *r* and *K* selection. (Stern and Tigerstedt, 1974; modified)

	r selection	*K* selection
Climate	Variable and/or unpredictable, uncertain	Fairly constant and/or predictable, more certain
Mortality	Often catastrophic, random, density-independent	More selective, density-dependent
Population size	Variable in time, no equilibrium, normally far below *K* of the environment, unsaturated ecosystems or partial systems, ecological vacuums, annual colonization	Fairly constant in time, equilibrium at or near *K* of the environment, saturated ecosystems, recolonization unnecessary
Intra- and interspecific competition	Variable, often lax	Normally vigorous
Selection favors:	1. Rapid development 2. High r_{max} 3. Early reproduction 4. Small body weight 5. One-time reproduction	1. Slow development 2. Greater competitive ability 3. Lower thresholds of resources 4. Delayed reproduction 5. Greater weight 6. Repeated reproduction
Life span	Short, usually less than a year	Long, usually more than a year

they are adapted to the capacity of the habitat, for which the symbol K is used) and r strategists (because the emphasis is on rapidity of development and reproduction and on a strong tendency to relinquish the habitat). Of course, there is practically a continuum of intermediate forms between the two, but in general large, long-lived animals and trees count as K strategists, whereas small animals and pioneer plant species are r strategists. The r strategists can be found primarily in evanescent habitats such as puddles of rain water, piles of earth, and the holes made by burrowing mammals; in many cases, they reproduce parthenogenetically (Daphnia, rotifers, aphids). All these forms can build up a large population in a short time, but in contrast to the K strategists they cannot maintain themselves against vigorous competition. They are really capable of living only in a habitat where competitors are scarce or nonexistent. The rapid colonization of habitats favorable to ectothermic animals in the spring is in many cases also done by r strategists. K strategists are distinguished by high competitive ability, a long life span, and few progeny. The number of such species is especially high in habitats that have been in existence for centuries. Table 7 summarizes these differences; from the characteristics given there we can infer that long-lived plants dominate chiefly in forest and steppe regions, whereas organisms that live only a year and thus colonize rapidly are heavily represented in deserts and semideserts, where sudden rainfall must be followed by rapid growth.

The notions of short-term and long-term habitats are of course relative. In regions very inimical to life – the spray zone along the seacoast, high-mountain and polar regions, and extreme desert environments – severe mechanical weathering occurs, exposing rocks upon which r strategists can rapidly settle. The lichens that do this exhibit the characteristics typical of r strategists; they have a very high rate of reproduction and are practically ubiquitous. As a result, they can appear everywhere as the first colonizers, the pioneer plants. Many different species can coexist on the same rock, with no discernible ecological differences among them. But in these inhospitable regions the lichens grow extremely slowly, and the pioneer plants continue to exist for very long periods – individual lichens as much as 4,000 years old have been described.

But the fact must not be overlooked that even in very constant habitats subsystems housing r strategists can exist. The feces and corpses of animals, for example, are such subsystems in the primeval forest. Moreover, a species can change from the r to the K strategy; the water fleas in our large lakes undergo massive reproduction in the spring, by parthenogenesis, and when the capacity of the habitat is reached they switch to bisexual reproduction (K strategy). Rotifers behave similarly. There is a graded difference, then, between r and K strategists. The extreme cases are useful as illustrations of the problems involved, but they do not represent any absolute quantities (cf. also p. 131, 245; cf. Sperlich, 1973; Stern and Tigerstedt, 1974). For this reason we speak of an r-K continuum.

The speed with which evolution can proceed under certain circumstances is apparent in the modification of plants that become weeds in cultivated fields. The cruciferous plant called gold-of-pleasure (Camelina sativa linicola) closely resembles flax (Linum usitatissimum, Linaceae), with its slender unbranched stalks and the small pale leaves. The two plants occur together, the gold-of-pleasure as a weed in fields of flax. The former is descended from a wild plant of shorter habit (C. gabrata) that when growing as a weed in the flax fields was not very conspicuous and thus, coincidentally camouflaged as a protected crop, benefitted from human care. It was treated just like the flax. The development of longer stems with fewer branches is important in a field of tall plants; gold-of-pleasure exhibits this characteristic as an inherited feature even on sites where it grows alone. But this similarity is not a critical factor in situations involving human care. Even conspicuous weeds (for instance, poppies or cornflowers) cannot be removed from fields of grain, for the fields are too large and eliminating the weeds would cause too much damage to the crop. Something else is more important. Originally the wall of the seed pod sprang smoothly and easily away from the supporting partition, a necessary feature under natural conditions to ensure scattering and dispersal of the seeds. But in cultivated fields the seeds harvested along with the ripe flax were those still in the pods. These seeds were mixed in with the linseeds during threshing and thus came under human protection. Their chances for broad distribution were much greater than those of seeds scattered by natural pod opening. Thus was evolved the characteristic of keeping the seed pods closed. Moreover, the sowing process exerted a further selective effect. Winnowing machines are used to separate the grain from the chaff, on the basis of a combination of size and weight. And this combination is the same in gold-of-pleasure seeds as in those of the flax plants. The winnowing machines cast the two kinds of seed equally far, so that the seed to be sown on the flax fields automatically contains those weed seeds that most closely resemble the crop seeds. Because of this interaction of factors similarity of characteristics selected for is produced without any intention on the part of plant, machine, or man. Backcrossing between gold-of-pleasure and its relatives has shown that in this case, too, the mimetic characteristics are genetically based, with several genes responsible.

Corresponding processes have led to the evolution of varieties in other grain species, and similar examples could be cited among other species of weed (Wickler, 1973).

III. Demography

If we record all the members of a species of organism within a given space, we find quite generally (as long as the species does not move among different habitats during

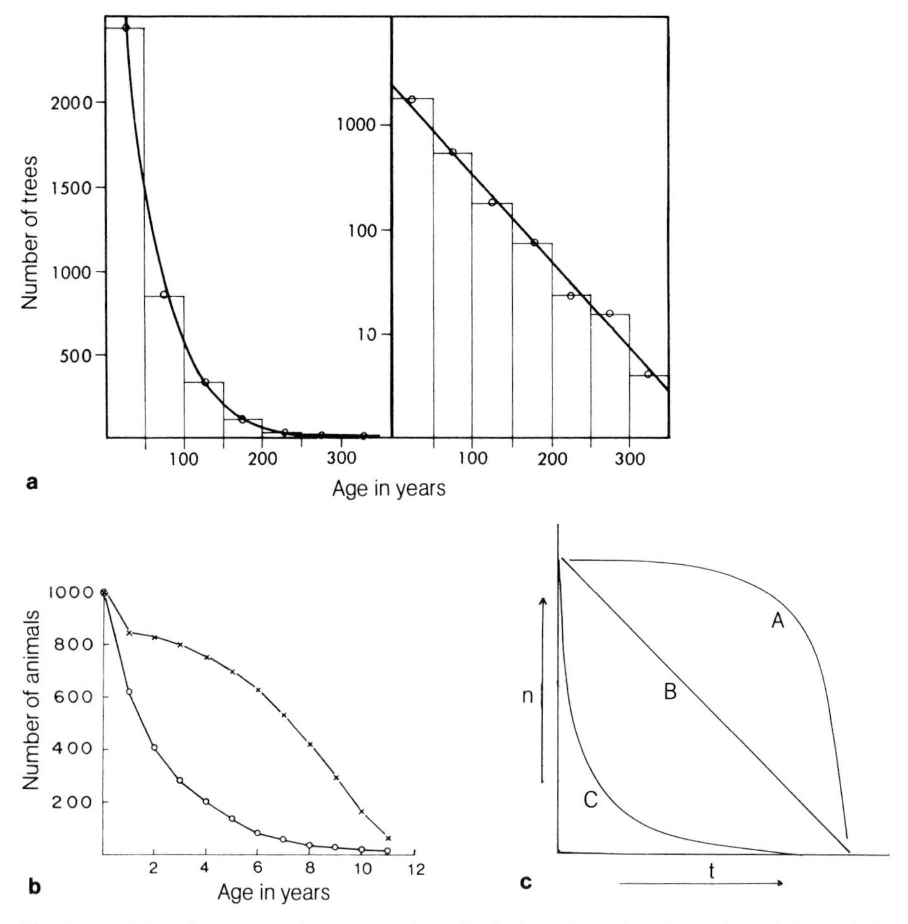

Fig. 85 a–c. Mortality curves. **a** Age composition of oaks in a North American primeval forest; *left* arithmetic scale; *right* logarithmic scale. (Whittaker, 1970). **b** Survival curves of lapwings (x) and domestic sheep (o). (Pielou, 1974). **c** The three mortality curves usually shown, schematic

its life cycle) that there is a characteristic distribution of age classes (Figs. 85 and 86). Young organisms are present in far greater numbers than old. The predominance of the young stages is more pronounced, the higher the rate of reproduction of the organism. From this age-class distribution it can be concluded that mortality is not uniform throughout the life cycle; juveniles have a particularly high death rate. Figure 85c shows the relationship between age and mortality as it is usually presented. Only curve C is likely to be realistic. Curve B is obtained only if – in the case of birds, for example – eggs and nestlings or still unfledged young are left out of the discussion. So far no example of curve B has been reliably confirmed in which all age classes were properly taken into account. In nature, curve A is presumably very rarely represented, although it can be taken (to a certain extent) as a description of humans living in a technological environment. Under these conditions the high infant and child mortality found in less developed societies is greatly reduced, and the same may well apply to social insects. In general, though, we are left with curve C (Fig. 85c).

Modifications of this type of distribution are not exactly rare. Frequently particular age classes predominate, as happens in

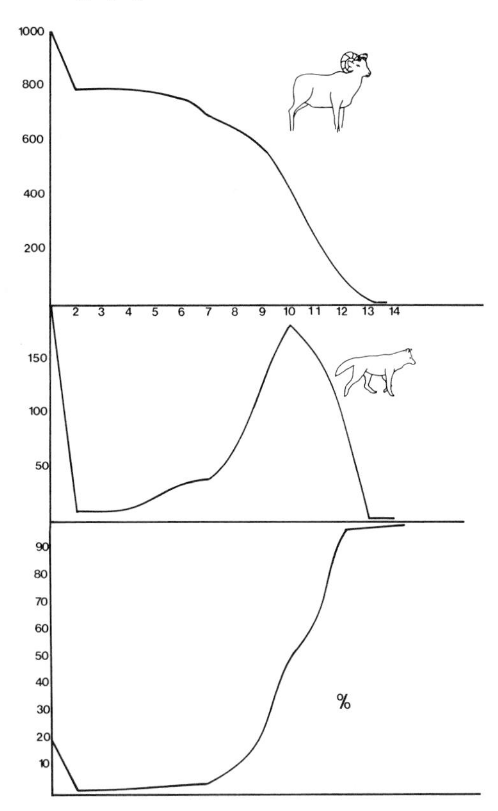

Fig. 86. Survival curve of the bighorn sheep in North America *(top)*, number of bighorn sheep in the different age classes killed by wolves *(middle)*, and fraction of sheep in each age class that are killed by wolves *(bottom)*. Almost 100% of the very old sheep fall victim to wolves, whereas the sheep between 2 and 7 years old are practically invulnerable. (From data of Geist, redrawn and partially recalculated)

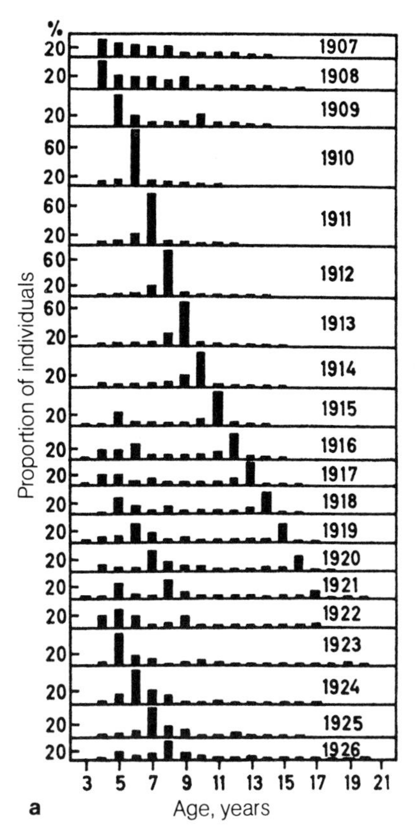

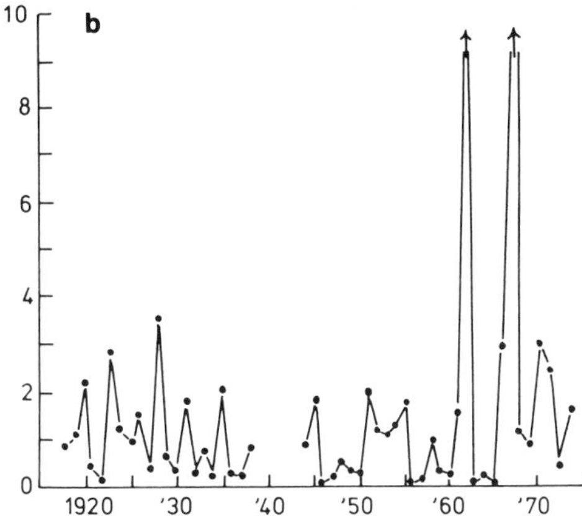

Fig. 87 a, b. Departures from the usual age distribution in a population, because of the large number of births during a particular year. **a** Age composition of the herring catches in 20 successive years. (Schwerdtfeger, 1968). **b** Fluctuation in the birth rate of North Sea haddock, measured by the number of fish (in 1,000) caught in 10 h of fishing by Scottish research vessels. (Hempel, 1977)

Germany with the June beetle and in North America with the famous 13- and 17-year cicadas. In a "June-beetle year" there may be very many more adults than the total number of larvae, for the following three years' populations tend to be very small. Extreme cases of this kind are also known in fisheries biology, where the fish born in one year can constitute the major part of the catch for several subsequent years (Fig. 87).

In interpreting such a curve, therefore, it is important to know not only the age classes but also which stages reproduce to what extent. Usually we think in terms of the familiar sequence egg-larva-sexually mature animal, but there are many departures from this rule. Some species reproduce in various stages (for example, parasitic forms like the tapeworm Echinococcus or the digenic trematodes). In many species a large number of adults never reach sexual maturity under conditions of high population density (cf. pp. 131, 141). Moreover, the simplified sequence reflects the widespread error that the significance of an individual to the population and the ecosystem is exhausted once it has reproduced. In many species, as is known for flies, lepidopterans, and mammals, the part of the life span that follows the reproductive phase is longer than the development to sexual maturity. So far we have only speculation about this phenomenon. Perhaps these individuals are of considerable importance in distracting predators from the fertile individuals; among mammals an experienced but barren animal can take over a crucial leading role in the social structure.

Further, life expectancy, growth, age, and mortality are only partially genetically determined. The genetic proclivity can often be suppressed or modified by external factors. In trees, for instance, development under favorable climatic conditions and with no pressure of competition leads to very rapid growth. But this growth stagnates relatively soon, and the trees do not become very old. Under the pressure of competition in a natural primeval forest, trees at first exhibit much slower aboveground growth, but in the long run become much larger and live a great deal longer. The situation is similar for trees in the interface zone between taiga and tundra. Here pines can reach an age of over 1,000 years, whereas in the monoculture produced by the clear-cutting that predominates in present-day forestry about 300 years is the maximum age (Backmang's Growth Law). Really massive quality trees can be obtained only from primeval forests or places where conditions are similar.

The same is true of animals. Many animal populations contain large numbers of adult individuals which do not proceed to reproduce until the number of reproducing individuals falls below some threshold; then these "reserve" animals also reproduce, in appropriate numbers. In such cases the age spectrum of a population reveals nothing about its potential for reproduction. Such relationships have been demonstrated among many birds and mammals, and are likely to prevail in a large number of other species (parasitic wasps, for example).

If there were no mortality before the attainment of sexual maturity, a species would reproduce exponentially (Fig. 88, Block 3). It is irrelevant whether mortality

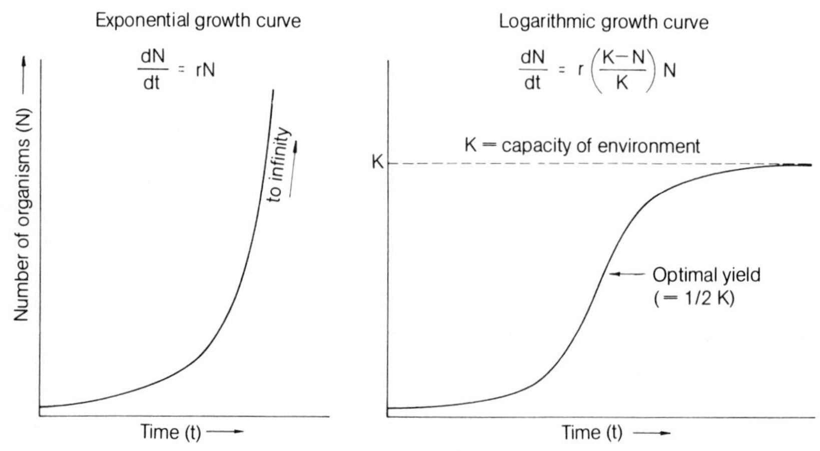

Fig. 88. The two basic curves for population growth: exponential and logarithmic. (Wilson and Bossert, 1973)

Block 3. Some data related to exponential population growth

A population stays constant if the birth and death rates are identical. If in a population of 1,000 individuals there are 25 births per unit time, we have a birth rate of 25 per thousand. If 15 people die in the same time (death rate 15 per thousand) the population grows by 10 per thousand. That is the Malthusian parameter or specific reproduction rate, usually designated by "r". In this case it is 0.01.

Having empirically determined birth and death rates, one can compute how the population will change, assuming that circumstances remain the same in future. This is generally done by the differential equation

$$\frac{dN}{dt} = rN, \qquad r = (b - d)$$

where t = the time in arbitrary units
N = the number of individuals in the population at given time
b = the birth rate
d = the death rate

The size of the population at a particular time can easily be calculated in the same way as compound interest:

$$N_t = N_0(1+r)^t \quad \text{or} \quad N_t = N_0(N_1/N_0)^t$$

where N_t = population size after time t
N_0 = initial population size
N_1 = population size after one unit of time.

In the example above, then, if we take t to be 1 year, after a year we have

$$N_t = 1{,}000 \times (1+0.01)^1 = 1{,}010,$$

and after 8 years

$$1{,}000 \times (1+0.01)^8 = 1{,}083.$$

This is a case of exponential growth. Because such a process is characterized by a constant doubling time (in our example the population doubles every 70 years), it is necessary to know this doubling time. In general it amounts to

$$\frac{70 \text{ time units}}{r \times 100}$$

These calculations are significant in estimations of the human population (cf. Ehrlich et al., 1975); moreover, they can be used to check whether and to what extent r changes under changing conditions.

occurs at the place of birth or after the organism has migrated elsewhere. This fact is the basis of Darwin's theory of selection. In nature such exponential population growth is evidently no great rarity.
Exponential growth almost without mortality is always to be suspected when an organism colonizes a habitat offering no competition, which for whatever reason has not previously been invaded. Multiplication of the arctic fulmar – pairs of which can raise at most one young bird per year – in the last 100 years in the British Isles is an example of such growth (Fig. 89). The reasons for this sudden colonization of the British Isles are not clear. A similar population explosion accompanied the conquering of Europe by the collared turtledove and occurs when there is an outbreak of insect pests. Even species as sensitive as the black grouse can undergo massive reproduction in this sense if conditions are

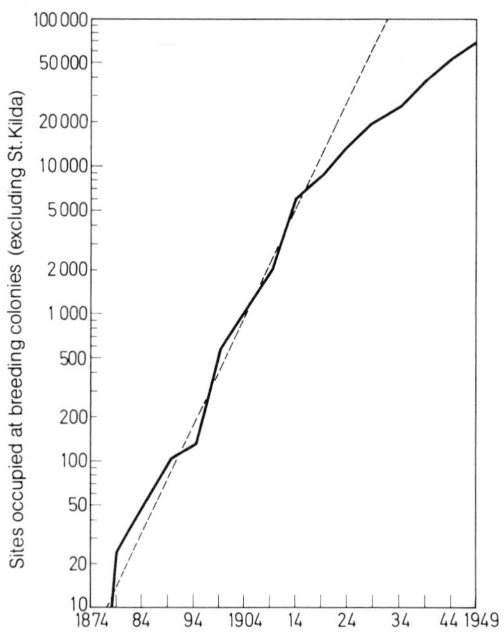

Fig. 89. The number of occupied breeding sites in the breeding colonies of the arctic fulmar in the British Isles (logarithmic plot). The straight line indicates the increase computed under the assumption that the birds never die and continue to reproduce indefinitely. (Fisher, 1952)

favorable. It evidently did so when the moors came under cultivation in northern Germany, presenting the grouse with an excellent environment, as well as in the Reichswald region near Nürnberg when the stand of pines that had been planted there fell victim to a mass outbreak of pine lappets (Sperber, 1968). In all such cases an area becomes available to the animals in which they can live without competition and without predators.

Obviously, unlimited exponential growth is an impossibility. If an organism reproduced without limit it would soon cover the earth in a uniform layer, which would eventually become thicker with the speed of light. In reality, the exponential curve flattens to give a sigmoid curve (Fig. 88). The range in which the curve flattens is the capacity of the habitat for this species.

But these curves are purely formal descriptions. What capacity is, and which factors limit it – competitors, predators, parasites, abiotic factors, food, space – are not thereby explained, nor are the mechanisms that cause saturation of the initially exponential curve. Is mortality increased? Does the rate of reproduction fall? Do the animals emigrate from the habitat? If mortality does in fact increase, is it due to starvation, parasites, predators, abiotic factors? At what stage does mortality intervene? Do birds, for example, continue to lay their eggs in nests and brood, or are the eggs "misplaced"? All these questions must be answered if we are to understand the process of saturation. Moreover, the capacity of a habitat changes in the course of a year. And finally, the capacity of a habitat can be utilized to different degrees. A species can exceed the capacity of its habitat by overuse of resources in some years, and in others be so few in number that capacity is not reached.

Such cycles of population size are not uncommon in the northern hemisphere. Formally, they can be described as a dead time following excessive growth; when capacity is exceeded, the animals' reaction is delayed. Overexploitation of the habitat leads to collapse of the population; only a few survive. Depending on the duration of this dead time and on the associated degree of habitat overexploitation, the fluctuations can vary in amplitude. Again, this is only a formal description. In its details, the process of cycle development involves a vast number and variety of causes.

Animals of the northern hemisphere that are known to go through such cycles include mice and lemmings, the alpine hare, the lynx, the rock ptarmigan, and many others. The period of a cycle is either almost exactly four years (for most Old-World species) or nine years (most species in the New World; only mice and lemmings have a four-year cycle). All species with such cycles can be categorized as r strategists, although position on the r-K continuum varies with phase of the cycle. It is remarkable that these cycles are known to occur only in the boreal and arctic regions of the northern hemisphere; suggestions that they may happen in the temperate zone, even in the high mountains, are not well documented, and no cycles have been observed in the tropics or in the southern hemisphere. Where they do occur, they are synchronous within a region. Sometimes, however, the lemming cycle reaches a maximum over the greater part of Norway while in small areas it is at a minimum (Myrberget, 1973).

Other species keep strictly to the capacity of the habitat, with a population size that fluctuates only slightly, or only as a consequence of altered environmental conditions and thus irregularly. It is thought that the majority of animals in the tropical rainforest belong to this K category. However, there have not been enough long-term investigations. Occasional massive multiplication of toucans occurs in the Central Amazonian jungle – evidence that in certain species, at least, and under certain conditions population density can be subject to relatively wide fluctuations even in the tropical rainforest.

IV. The Distribution of Organisms in Space

Normally people take the naive view that whenever the situation permits, organisms are uniformly distributed in space – that the distance between one individual and its neighbors is the same over a large area. But this kind of distribution is not the rule. When it is found, it probably always indicates territorial behavior of the species concerned. Much more often one finds two other patterns of distribution. One is the so-called "cluster." Clustering can be brought about by a social attraction among the organisms, so that newcomers settle in the vicinity of the previous inhabitants, but a great many other factors may be responsible. For example, some woodpeckers transport pine and spruce cones to "smithies," special cracks into which they can wedge the cones to get out the seeds. Large piles of picked-over cones can accumulate under these, and eventually the seeds the woodpeckers have missed give rise to a very small thicket of young conifers. Because the stand is too dense, however, it soon dies out. Ants can also accumulate certain seeds (Chelidonium) in the vicinity of their hills. The third possible distribution is a random one. Whether such distributions are really the result of chance cannot be decided at present. In some cases the selective advantage provided by a random distribution can be exploited (cf. p. 147, Fig. 90).

In studying the spatial distribution of a population, then, one must first measure the distances between the individuals and analyze these data. Uniform distances indicate territoriality. If the distance histogram has two maxima clustering is indicated, and if it is broad with no clear maximum the distribution is of the random type – but here one must bear in mind that this randomness has not yet been definitively interpreted. [The distribution patterns can be defined as follows: (1) random distribution has variance equal to mean; (2) clustered distribution has variance greater than mean; (3) uniform distribu-

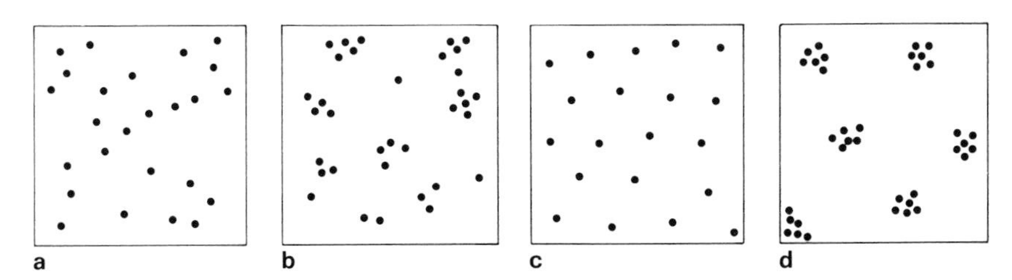

Fig. 90 a–d. Four types of distribution of a population in space. **a** Random, **b** clustered, **c** uniform, **d** clustered uniform distribution. (Whittaker, 1970)

tion has variance smaller than mean.] A good example of an animal that clusters is the barnacle (Balanus; cf. p. 95). This sessile marine crustacean is hermaphroditic, but must be cross-fertilized – which can occur only if the animals are close together. A larva that finds an attached barnacle tries to settle and metamorphose as close to it as possible, so that clusters are produced. Larvae that fail to find conspecifics already attached settle in isolation and can then be the nucleus around which a cluster develops. Among higher animals, clusters take the form of colonies or herds; the colonies of rooks, fieldfares and prairie dogs are examples, as are herds of reindeer and gnu and prides of lions.

In spite of all we know about distribution patterns, our information about the ecological significance of life as an individual, as compared with life in herds and colonies, is inadequate. For sea birds – gulls and arctic fulmars – it has been shown that rate of reproduction increases with colony size. In small colonies it is near zero; fewer nests are built and fewer eggs laid, some of those that are laid are incompletely brooded and fail to hatch, and some of the young are irregularly fed and never fledged. As the size of the colony increases all these parental skills improve and the reproduction rate rises. Moreover, all such species have a tendency to form large colonies; after years in which mortality is high small colonies are given up and the remaining population gathers at the site of a large colony. Fluctuations in population size hardly ever occur in the large colony, whereas they are evident in small marginal colonies. Two possible explanations of the increased reproduction in large colonies have been proposed. According to one, the stimulating effect of activity is decisive. The presence of animals engaged in courtship, nest-building, brooding, and care of the young incites neighboring animals to do the same thing. Many instances documenting this sort of effect have been found among birds. The second argument is that old birds, experienced in breeding, gather in the large colonies and crowd the younger animals out into the less favorable small marginal colonies. Each explanation is plausible, and both may be true. Might similar effects exist among mammals?

Recently the territorial behavior of higher animals and its significance to ecology have attracted particular interest. Mammals and birds move about within the area they inhabit in a way that is by no means irregular; they follow quite precisely established routes, and the boundaries are often just as exactly fixed. This arrangement provides for optimal utilization of resources and, at least to some extent, compensates for the increased vulnerability of uniformly distributed animals to predators, in that the inhabitant of a territory is intimately familiar with its topography. Three types of "areas of sovereignty" can be especially clearly distinguished:

1. The *home range* comprises the area normally visited by an animal during its lifetime. That of migratory birds and nomadic ungulates can be extraordinarily large, and even with relatively stationary species like the capercaillie (Fig. 91) it can be of considerable size. The home range is only partially utilized at any time (in winter rock ptarmigans form flocks that wander through the entire range which in summer is divided into territories). The home range is not defended against other animals.
2. A more important demarcation is that defining a *territory*, in the most common sense of the word. Songbirds are especially well-known territorial animals; one pair or a small family occupies a fixed region. Among songbirds it is usually the male that defends this region against intruding conspecifics not part of the family. The territory offers sufficient space and food for raising young. Its size, therefore, is determined less by (possibly inherited) space requirements of the occupant than by the amount of food it is likely to provide. In poor woodland the territories occupied by birds of a given species are larger than those in rich woods, and redshanks in the subarctic zone have territories smaller than

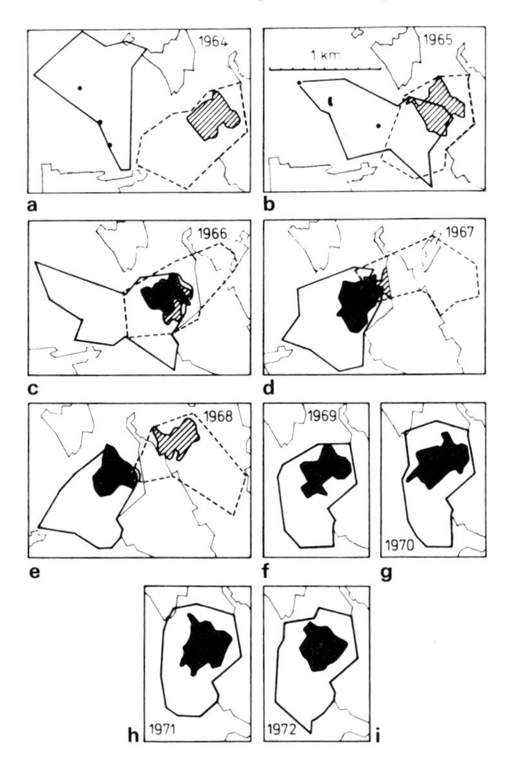

Fig. 91 a–i. Territorial relationships between two capercaillie in successive years. *Dashed line:* boundary of home range, and *cross-hatched:* territory of cock 4. *Solid line:* boundary of home range, and *black:* territory of Cock 12. Cock 12 gradually takes over the home range and territory of the initially dominant Cock 4. (Müller, 1974)

those living further north. On poor soils (granite, as is found in Scandinavia and parts of Scotland) the territories of arctic hares and rock ptarmigans are large; correspondingly, the density of these populations is low. On rich soils (volcanic soils in Iceland and Scotland) the same species have much smaller territories and they colonize the region much more densely (the primary production of the main food plant, Calluna, is about the same in the two habitats, but the plants differ in phosphorus content). If animals are provided with additional food the size of their territories can be considerably reduced and their numbers increased – a phenomenon that can be readily observed in parks and cemeteries.

The occupants of different territories know one another well, and they maintain a genuine rank order. Among capercaillies, the highest-ranking cock is possessor of the optimal territory, and lower-ranking birds are left with the surrounding territories. Müller (1974) monitored these relationships for many years (Fig. 91). The strongest of the roe deer in a region gradually come to occupy the best-structured parts of the habitat, with the best food supply, and as they do so their home ranges and territories grow smaller; deer in the marginal regions less well supplied with food have very large home ranges and territories (Ellenberg, 1974b, 1978; Figs. 81 and 82). Territories can consist of several noncontiguous parts. They need not serve only one breeding family; small family groups can "own" a single territory. This sharing is well known among field mice. When the population density is low one pair occupies a territory for itself, but as density increases the territory is occupied by a small family group and finally by an extended family. In the high Andes of South America each small family of vicuna has a feeding territory, containing a watering place and sufficient grazing land, and a separate sleeping territory a few hundred meters away. Animals with no territory have access to the water hole only early in the morning, when the occupants of that territory are still in their sleeping grounds. This situation illustrates the difficulties faced by individuals of territorial species that lack a territory of their own (Geist and Walther, 1974).

The boundaries of territories are established by vocal signals, chemical marks (such as feces), or special threat gestures. One of the best known methods of demarcation is the singing of birds. A striking feature of this behavior is that if tape-recorded calls are played they very soon become ineffective. The first time the song of a stranger is played within a territory the occupant reacts with violent aggression, but after a few repetitions it ceases to do so. Only if the song is regularly varied does it represent a real opponent; a stereotyped song repeated over and over again is not interpreted as coming from a rival. Because neighbors are very well acquainted with each other's songs, very little energy is required for the delimitation of territories; all the neighbors know the limits. Marking by feces and urine is particularly common among mammals. The vicuna mentioned previously always leave their droppings at specific places; indeed, the animals that occupy a given territory successively over many centuries use the same places as their predecessors, so that tall fecal pyramids of quite characteristic shape are built up. Threatening gestures are widespread in many classes of animal, as are subordination postures. The latter enable zebras, for example, to pass through the territories of others on their way to the water hole. This example brings out one of the problems associated with territoriality in some regions; the animals have requirements they cannot always satisfy within their own territory. From time to time, at least, they must leave their territory to drink or find special ions (cf. p. 67f.). It is relatively simple for birds to do this, by flying over the adjacent territories, but in the case of mammals it is possible only if the territorial boundaries are not too rigid.

A particular division of an area into territories, of course, applies only to the single species concerned. The boundaries established by animals of one species need not correspond at all to those of other species. The map of chaffinch territories within a wood looks quite different from one showing the territories of great tits. But in very rare cases so-called "interspecific territories" can be formed. In some parts of the Scandinavian birch-forest zone willow warblers and chaffinches behave as though they belonged to the same species, defending their territories against one another. This territory demarcation resembles that between plants. For instance, the distribution within a primeval forest of trees of many different species but approximately the same diameter is relatively uniform. By our definition, this counts as territorial behavior – and on closer examination, it proves to be so. By way of root competition (however this may operate) each large tree keeps other trees out of the region its own roots occupy. The distribution of individual species within this uniform pattern is random. Here precise allocation of the resources is correlated with the best conceivable defense against infection (cf. p. 147).

3. The *immediate vicinity* of an animal's shelter – the nest of a bird, for example – can be regarded as the third important type of "individual space." This region is defended against all invaders, not only conspecifics but any creatures of about the same size as the occupant. Colony-breeding species of bird, and mammals that live permanently in large herds, probably have territories only in this sense; with respect to feeding, the available space is shared by all. Among birds colonies are formed chiefly by species living on the seacoast or lake shores, which seek food far from the breeding biotope, on the open ocean, in periodically flooded zones, and so on.

These three categories, of course, do not exhaust all the possible forms of territoriality. The bucks in some populations of eastern African topi set up small male territories during the mating season, within which copulation occurs. These male territories are extremely stable for long periods. The territorial behavior of the European roe buck is similar. Many migratory birds establish individual territories during migration and in their wintering grounds. In such circumstances the females behave like males, singing and adopting threat postures as the male does at the nesting site. Dragonflies of the genera Aeschna, Anax, and others exhibit temporary territoriality; a male occupies a territory for a few minutes to an hour (Kaiser , 1974). We cannot dwell further on this multitude of different forms; of the three main types, the second – the territory established for maturation and reproduction – is particularly significant in ecology.

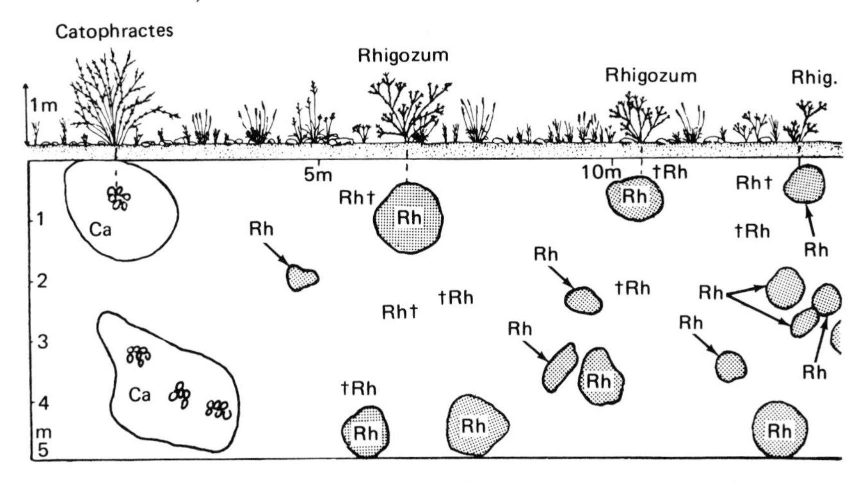

Fig. 92. Distribution of plants in the transition region between steppe and desert in southwestern Africa. *† Rh*, dead Rhigozum. (Walter, 1973b)

The random distribution presents special problems. It can convey a selective advantage, as discussed on p. 147, and this advantage will surely be relinquished only if the advantages of alternative distributions are greater (the uniform utilization of resources associated with territoriality, or the social effects in colonies). Random distributions have been found for many organisms (Figs. 92, 120, and 121). Let us examine the problems involved in one sample case.

As has been mentioned, the distribution of tree species in a mixed primeval forest is random. Various conflicting factors are at work here. Wind pollination is a disadvantage in a multi-species system, because the pollen must drift wherever the wind takes it. We find wind pollination mainly in stands where the chance that the pollen will encounter a conspecific tree is high – where the number of species is limited to one or a few. Where the number of species in an area has increased, with unpredictable distribution, there has always been a tendency to evolve a more specific method of pollination, by animals (insects, birds, bats). On p. 147 it is shown that a random distribution within a system reduces an organism's vulnerability to predators in the broadest sense. On the other hand, the distribution of insect-pollinated trees cannot be too heterogeneous if the pollinating animals are to bring about at least occasional gene flux. At the same time, it is necessary that the seeds do not simply fall straight down from the tree; they must be as widely scattered as possible, so that the heterogeneous, random distribution is maintained. In this regard, again, animals facilitate matters. Seeds that stick to an animal's body can be carried long distances, as are indigestible seeds enclosed in an edible fruit (for example, the mistletoe). The third possibility is the production of edible seeds, in which case they must be produced in large quantities so that enough are left over after the animals have eaten. Wide distribution is made still more likely if the animals can be caused to store the seeds a considerable distance away. The irregular, but then voluminous, production of seeds by forest trees is probably an adaptation in this direction. In times of abundance foragers, black-throated jays in particular, fill numerous depots and then forget about them. In the process, acorns and beechnuts are carried long distances and "sown" by the jays. Development of present-day multi-species forest ecosystems required the coevolution of plants, pollinators, and seed distributors (Regal, 1977). In Costa

Rica over 90% of the tree species are dependent on animal pollination, and just as many have fleshy fruits adapted to transport by birds. In a tropical rainforest where competition is intense, the seeds most likely to produce seedlings are the heaviest, those containing the most nutrients. On the other hand, high seed weight reduces the number of seeds that can be formed as well as their transportability. Selection probably favored the development of fruits attractive to long-distance carriers, for no seed has a chance immediately below the mother tree. Furthermore, carriers would have been favored if they deposited the seeds in places where their chances for germination and growth were good. Selection therefore tends to encourage the evolution of specific fructivores rather than generalists.

V. Maintenance of an Average Population Density

In the long term, all organisms under approximately constant environmental conditions maintain an approximately constant population density. The factors and mechanisms responsible are often classified as density-dependent and non-density-dependent. But it is most likely that on close examination all factors will be found to have some sort of density-dependent effects. In the following discussion no classification will be attempted; groups of factors that operate differently will simply be treated in sequence.

There has been much debate about the appropriate terminology. The expressions "control" and "regulation" are used quite differently. I prefer to follow Wilbert (1962) and Enright (1976) in using the neutral phrase "maintenance of an average density." "Self-regulation" has been rejected by many as too anthropomorphic; the equivalent term "autoregulation" perhaps makes clearer that no connotation of "insightful" responses is intended. Either expression simply describes a situation.

Optimal population densities can be mentioned here only in passing, and the advantages and disadvantages of constant and fluctuating populations cannot be extensively treated. The latter phenomenon can usefully be analyzed only in relation to ecosystems, and it will be taken up again in Sect. D. Some very interesting problems have had to be omitted. For example, mimicry can function only if the mimic is less common than its model. How does the mimic manage to adjust its population size to that of the model?

1. Autoregulation

If a pair of tree shrews (the primate Tupaia glis) is kept in a cage under otherwise favorable conditions, the female bears young at regular intervals. The population increases; the space available remains constant. When the first offspring reaches sexual maturity, the parents come under stress. The external sign of this stress is erection of the tail hair. If the young tree shrew is male the father is affected, and if it is female stress is evident in the mother. The tail bristles for the following sequence of reasons. Under stress the adrenal cortex becomes enlarged and corticoid hormones are secreted. This stimulates the sympathetic nervous system, which in turn activates the muscles responsible for erection of the tail hairs (a reaction corresponding to gooseflesh in humans). At the same time, the flow of blood through the kidney is reduced, and the blood is no longer properly cleaned. If the stress continues for too long, the urea produced by metabolism is not excreted and the animal dies of internal poisoning (Fig. 93; von Holst, 1969). The stress experienced by the parent when the young grow up does not go this far, however. Once a certain level is reached, puberty of the subsequent offspring is greatly delayed or is never reached at all. Growth is inhibited, and animals that have already matured may undergo regression of the sexual organs. The mother no longer marks her young with the secretion of her

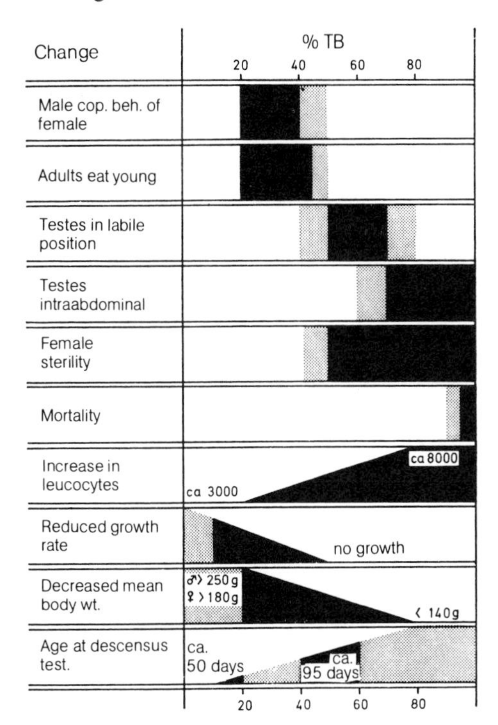

Fig. 93 a. Diagram of the ethological and physiological changes Tupaia undergoes at different population densities, indicated by the tail-bristling index (%TB; dense population associated with high %TB). Statistically significant %TB black, probable range gray

sternal gland; as a result, the young are not recognized as such, but are considered as food and eaten. In this way population size is kept at a constant high level with no external intervention, when food is abundant. Only if the stress is intensified by reducing the size of the cage or introducing another animal, so that population density is artificially increased, does the ultimate toxicity and death of some adults ensue.
This is a typical example of autoregulation of population size, and one which reveals the physiological bases of the process. Even though food is plentiful and neither predators nor parasites can attack, this species is able to keep its population at more or less the same level for long periods.
This concept of autoregulation has repeatedly been challenged, for a number of reasons. It is in fact questionable whether these laboratory findings apply to animals

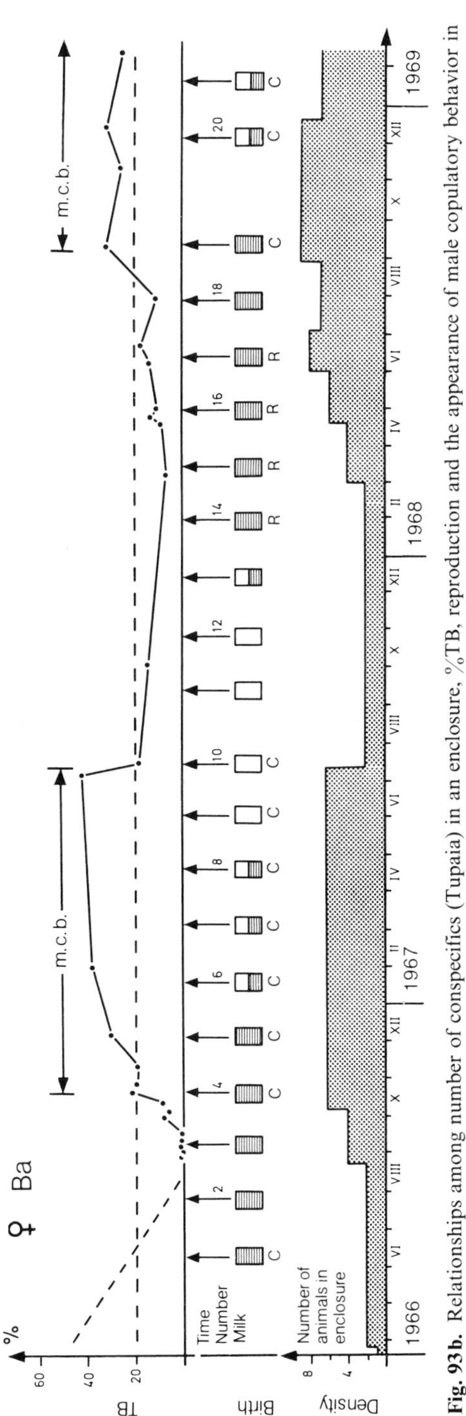

Fig. 93 b. Relationships among number of conspecifics (Tupaia) in an enclosure, %TB, reproduction and the appearance of male copulatory behavior in a female. Milk production is rated as follows: *full shading* offspring normally suckled at birth; *half shading* reduced suckling; *white* no suckling. *C* cannibalism; *R* testis retraction. (von Holst, 1969)

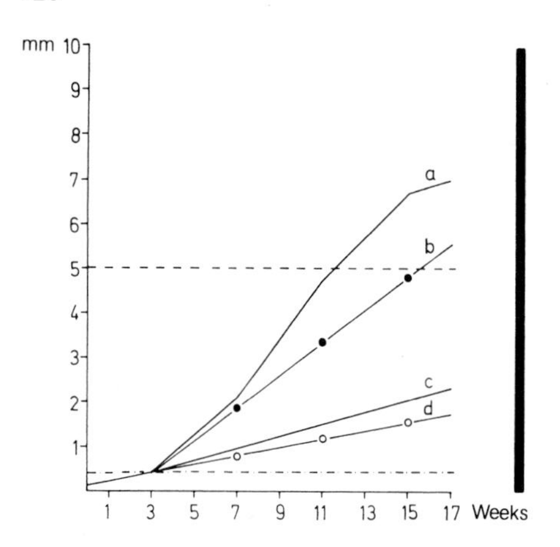

Fig. 94. Differences in rate of growth of the shore snail Ovatella myosotis (Mediterranean population). *a* animals kept in isolation; *b, c, d:* dense population; *b* largest animal; *d* smallest animal; *c* mean. (Seelemann, 1968)

in the field. The experimental population densities are much higher than would be encountered under natural conditions, where an area of one hectare is shared by at most 4 pairs of tree shrews. In the laboratory the space available to each pair is at most 5 square meters. To what extent do the results apply? A great many field observations of other animals are best interpreted in the sense of the autoregulation hypothesis. One example has already been cited: the insects that adjust the number of eggs laid to the amount of suitable substrate present (p. 48). Snails are known to grow more slowly under crowded conditions (Fig. 94).

When the density of a wapiti population is low twins are born to 25% of the pregnant females, whereas when it is high the incidence of twins is less than 1%. Among white-tailed deer the proportion of pregnant females was found to be higher at lower density (92% as compared with 78%); when density is low 33% of births are single, 60% twins and 7% triplets, whereas at high density 81% are single and 18% twins. A population of great tits in the Netherlands was low because the opportunities for nest-building were inadequate; when plenty of nest boxes were provided, the number of breeding pairs tripled. But at the same time the average clutch of eggs was smaller – in general the number of eggs was two less than previously – and fewer second broods were raised. Whereas 64% of all the breeding pairs had raised two broods, now the proportion fell to 16%.

Another population feature that can vary with density is the sex ratio of the young. In small populations of roe deer two females are born for every male, but as density increases males become more numerous; a male:female ratio of 3:1 can be reached. There is also a general inhibition of ovulation, so that the number of young in the total population is smaller; juvenile mortality is high and growth is slow, so that under unfavorable conditions the animals do not mature until they are several years old (under optimal conditions maturity can be reached in the first year). Certain insects also exhibit shifts in sex ratio (Søgaard-Anderson, 1961).

Still another element in autoregulation is mass emigration from the home range. This differs from the familiar bird migrations in that it can begin at various times of year, take unpredictable directions, and is not followed by a return to the home range. The prime example of such an emigrant is the migratory locust. These locusts exist in two forms. When an individual in an early larval stage is put into a cage with mirror walls, so that it sees conspecifics all about it even though it is actually alone, it changes to the migratory form; the same effect can be achieved by continual disturbance, or by both treatments together. The animals are normally solitary, but in the migratory form they become gregarious, gather in swarms, and leave the home range. Very similar developments occur among many birds (the sandgrouse Syrrhaptes paradoxus and many others), mammals (some of the lemming migrations), and insects (processionary caterpillars, army worms).

On the other hand, there are species that live in very uniform populations over long periods. Among them are the wolves on

Isle Royale in Lake Superior (North America) (Mech, 1966). During more than 10 years of close observation the island was inhabited by one group of 15–16 animals, a smaller group with 3–4 animals, and (perhaps) 2–4 single wolves. It was definitely established that the wolves could not leave the island in order to keep the population so constant. As happens among tree shrews, not all of the adult wolves in this population actually reproduced. The rate of reproduction was thus very small. In a wolf pack evidently only particular females, characterizable by their position in the social hierarchy, come into heat. Under definable conditions (for example, a barely adequate food supply), a wolf pack can have no progeny for years at a time. In this case autoregulation is governed by food availability, and the effect is exerted by way of social behavior. Again, however, the behavior of wolves under entirely natural conditions has not been satisfactorily analyzed. It is hardly justifiable to apply these observations of half-tame animals to wolves in the field. No analysis of the mechanisms underlying strict size constancy in a field population has as yet taken full account of social behavior. Mech (1966) suggests that such superbly functioning autoregulation is to be expected primarily among animals that have no natural enemies (these would be typical K strategists).

An array of arguments can be presented against interpreting these examples in the sense of the autoregulation hypothesis. It is very difficult to rule out food shortage as an external influence on population size. Especially in the case of the roe deer, it has been demonstrated that lack of food is actually the controlling factor. This deer is so selective in its diet that under normal conditions it never really has enough to eat. The fact that the territories of all territorial animals become smaller when food is ample lends strong support to this notion. On the other hand, conversion of locusts to the migratory form can be triggered entirely by disturbance and/or the sight of other locusts, when abundant food is available.

Accordingly, there are emphatic controversies in this field. Many scientists who have studied autoregulation claim that it is the only valid form of population control – that predators in the broadest sense, food availability, and other external factors can never guarantee a long-term limitation of population size. But the following argument has been put forth against this hypothesis. Migratory locusts multiply rapidly under favorable conditions, and this leads to appearance of the migratory phase and emigration. The animals that leave never return. Only a few remain to build up a new population. But this presents a difficulty. Because the founders of the new population are individuals that did not exhibit the autoregulatory behavior of transformation and emigration, the capacity for autoregulation should be eliminated from the population within a few generations. According to all we know about genetics, retention of this trait is impossible. This example applies to all cases of autoregulation.

The theoretical explanation of any instance of autoregulation thus faces a dilemma. Dawkins (1978) has proposed a hypothesis that is intended to do away with this problem. He departs from previous ideas of evolution in suggesting that evolution acts not on individuals or populations, but rather directly at the level of the gene. The gene, he says, is the basic machinery of an organism; it is solely to preserve itself that the gene creates the accessory structures we see as an organism. Cells, organs, physiological functions, and behavior – all are there only to aid the selfish gene in maintaining its existence. In this context one might assume that it is entirely advantageous for a gene to elaborate a program in which many of its carriers are set at risk to ensure that a few are guaranteed to preserve the gene for the future (Dawkins, 1976). Whether we agree with this hypothesis or not, we cannot refuse – simply because we have no explanation – to accept the evident fact that autoregulation does occur under laboratory conditions. Indeed, very recently some examples of the operation of autoregulation under field conditions have been found which are probably irrefutable. Chief among these is the demonstration that blood urea concentration is high in animals engaged in massive reproduction. Elevated blood urea has been documented, for example, in the lemming. The animals are thus under stress despite the fact that their population density, even at the peak of the explosion, is considerably lower than that of all the organisms that have been tested in this regard in the laboratory. At the moment, perhaps, only the following hypothesis remains: A territorial animal in a region where food is scarce will spend more time and cover longer distances in search of something to eat than in regions where food is plentiful. For a given density of this species, conditions of food scarcity will be associated with more social contacts (necessarily always involving aggressive behavior) than conditions of abundance. Because stress depends directly on social contacts, it will be greater when food is hard to find. The size of a territory, then, is not fixed according to particular length, width, or areal measures, but by the number of tolerable (non-stress-eliciting) social contacts. Because in the laboratory animals are regularly fed, the densities required to produce stress here are much higher than in the field (Fig. 95).

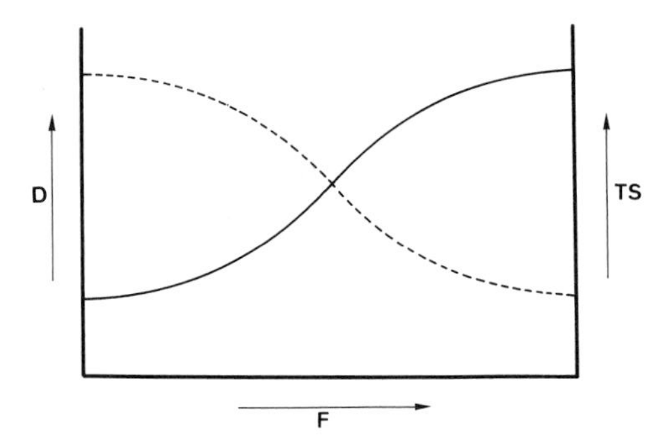

Fig. 95. As the amount of food *(F)* in the habitat rises, density (*D*) increases from a minimum to a saturation level, and territory size (*TS*) falls from a maximum to a minimal plateau

Another finding is relevant in this regard. The population cycle of the American meadow mouse Microtus pennsylvanicus begins in early summer, when the population is sparse – about 5 animals per hectare. It grows steadily, so that in the following spring its density is 125–750 animals per hectare. Then it falls off to about half that density, recovers slightly during the summer, remains almost constant over the next winter, and in spring is again dramatically reduced. In this second year it can either rise or continue to decline. During the growth phase of the population mortality is very low among both young and adult animals. At the peak, and during the collapse of the population, juvenile mortality is greatly increased, whereas the death rate of adults increases only during the declining phase. The increase in mortality is accompanied by a decreased rate of reproduction. Krebs (Myers and Krebs, 1974) studied the pattern of distribution of various enzymes in the mouse population, at different phases of the cycle, and found characteristic differences – most notably, in the protein transferrin, which transports iron (cf. Fig. 96). Myers and Krebs based the following hypothesis on these differences. During the growth phase relatively many females of reproductive age leave the home region. These emigrants can have a profound effect on the gene pool of neighboring populations. During the declining phase loss by emigration is very low. Emigration appears to remove selectively from the population those animals that are intolerant of overpopulation. But this effect is significant only during the growth phase. The genotypes remaining behind are predominantly those adapted to survival at high densities, but which have low rates of reproduction and high death rates. For them, aggressiveness is of greater selective advantage than a high rate of reproduction. That is, within a species that would be placed in the r category genotypes appear which are really K strategists. Because of

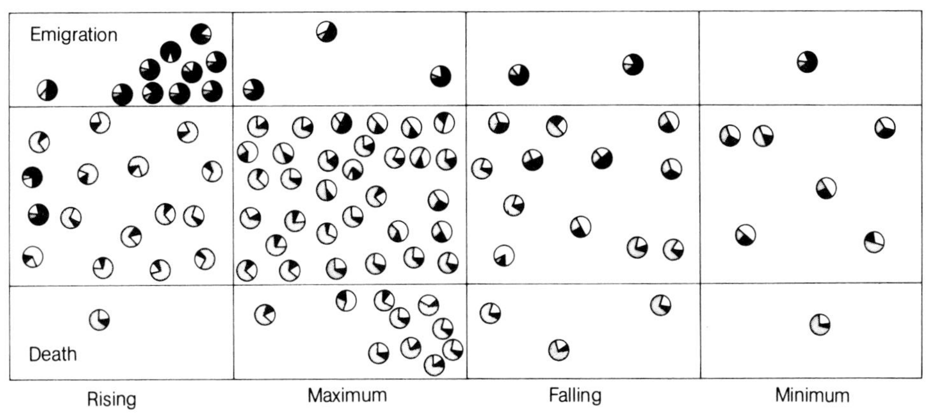

Fig. 96. An attempt to interpret the population cycles of the field mouse. The genotypes of individuals are represented as follows: *black* high reproductive rate and strong tendency to emigrate; *gray* aggressiveness, high reproductive rate, high death rate; *white* aggressiveness, low reproductive rate, low death rate. (Myers and Krebs in Sengbusch, 1977)

the large number of these K strategists, the numerically declining population declines still further. Near the minimum population size the genotypes characterized by rapid reproduction slowly begin to take over again.

Another, perhaps related, hypothesis has been developed with reference to red grouse in Scotland (Moss et al., 1974). This, too, is based on the different responses of individuals during a population cycle; the research provides evidence that the individuals really do respond differently (Fig. 97). Well-nourished hens in the spring produce a large number of chicks, which show little aggressiveness. These come through the next winter well, so that the density of the population is greater when the second year begins. Brooding occurs earlier, and less food is available. In the spring of the third year the hens are poorly nourished; they produce only a few offspring, but these are aggressive. Even in an incubator the fraction of eggs that hatch is low. Winter mortality is high among these chicks, population density falls, brooding in the fourth year is delayed until summer, the plants recover, and in the following spring well-fed hens can begin the cycle anew. The hypothesis derived from these observations, unlike that of Myers and Krebs (1974), assumes a direct effect of food supply. But so far no genetic studies have been done, and it remains quite plausible that there is a relationship between the genetic structure of the population and its nutritional state. Meanwhile, though, it has been shown that there are genetic differences between grouse populations at the maximum of the cycle and those at the minimum. In this case, one could find a way out of the cul-de-sac in which we are left with group selection.

We may be quite sure that these examples have not exhausted all the factors that can control a population cycle. It is possible that in some species feeding on flowers in the spring plays a central role. Pollen contains enormous quantities of steroids, which resemble the sexual hormones and have the same effects. Their intake could result in increased reproduction. The collapse of a population can be due to predation as well as to food scarcity. Weasels, for example, are about the same size as mice and lemmings and therefore can keep pace with their rates of reproduction. Evidence has been produced that weasels very probably hasten the collapse of these populations in Alaska and northern Canada.

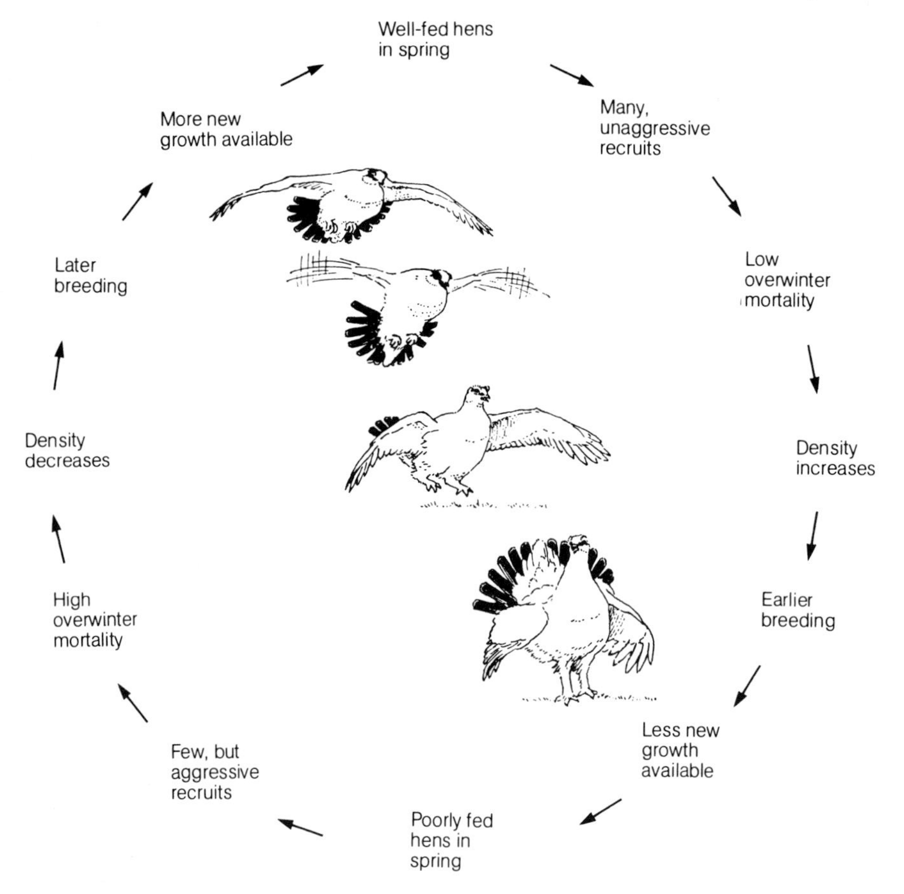

Fig. 97. Population cycle of the red grouse and rock ptarmigan (Moss, Watson and Parr, 1974)

Moreover, during the rapidly declining phase of a population cycle bacterial or parasitic infestations occur almost routinely; these, if not a major cause of it, at least accelerate the collapse.

Species with populations that fluctuate so widely can easily fall below the critical minimum if the additional factors affecting the collapse of the population become too strong. This is presumably the reason the passenger pigeon became extinct in the United States. Evidently this pigeon was of a solitary habit in normal years, gathering into the great flocks for which it is remembered only at times of population explosion. It is thought that humans took too active a part in the normal population collapse, so that the critical lower limit was passed. It is a fact that species that tend to oscillate in this way need much more room than do K strategists; if such species are to be protected, very large areas must be made available to them (Schorger 1955).

Autoregulation is also known to occur in plants. The most probable place for a seed to strike the ground is directly below the parent plant. But the more long-lived this plant, the less likely that the seedling can really thrive beneath it. The further away the seed falls from the mother plant, the greater is the probability that it will actually produce a full-grown plant. This is due in part to the shade cast by the mature plant, and it may be that allelopathic effects also play a role. There is a distinct difference, then, between the number of seeds on the ground and the number of seedlings, and the difference is still greater if the comparison is with older plants, that have really taken root and begun to grow.

The intraspecific competition that leads to autoregulation has roughly the following action in the case of plants.

1. The weights of the seedlings initially are normally distributed, but when intraspecific competition is intense, the weight distribution becomes more complicated as they grow, developing several maxima. A very few plants grow extremely quickly and are thus much heavier than average, whereas a great many plants are somewhat lighter than average.
2. Density-dependent mortality plays a large role; seedlings very close to one another have low probabilities of survival.
3. Structural changes appear. Elongation growth increases, the weight-per-area ratio of the leaf rises, and the weight of the seeds relative to the vegetative parts becomes less.
4. The result is a uniform spatial distribution of the mature plants.
5. Superior individuals grow very large.

Probably these criteria can also be applied to animals on the sea floor (clams) that continue to grow throughout their lives.

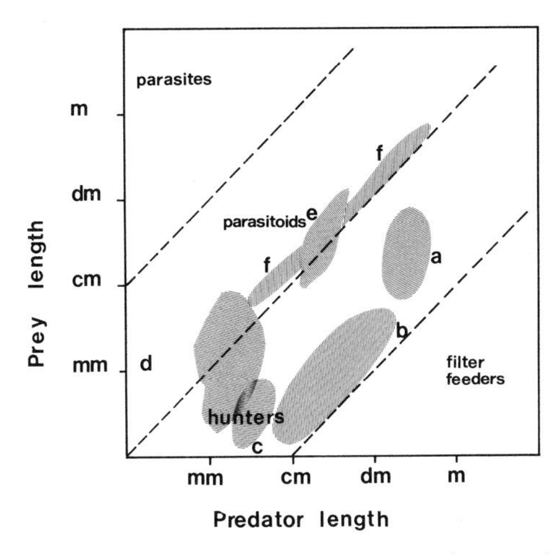

Fig. 98. Size relationships between predator and prey. Parasites are all smaller, and filter-feeders considerably larger, than their prey. *a* birds of prey and owls; *b* insectivorous songbirds; *c* hunting spiders; *d* web-building spiders; *e* orb-web spiders; *f* pack hunters (ants, wolves). (Enders, 1975; modified)

2. Predator-Prey Systems

In general predators, in the broadest sense, are regarded as extremely important in the control of population size. Presentations of this view frequently draw upon the mathematical analysis of Volterra, describing the population fluctuations of predator and prey in terms of two sine curves shifted by about 90° with respect to one another (Fig. 99). This description takes Volterra's calculations, and later individual experiments that appeared to show similar results, too literally. In reality, things are considerably more complicated. The complications begin with the question of what a predator actually is, and continue with the question of whether the (mathematically quite correct) repre-

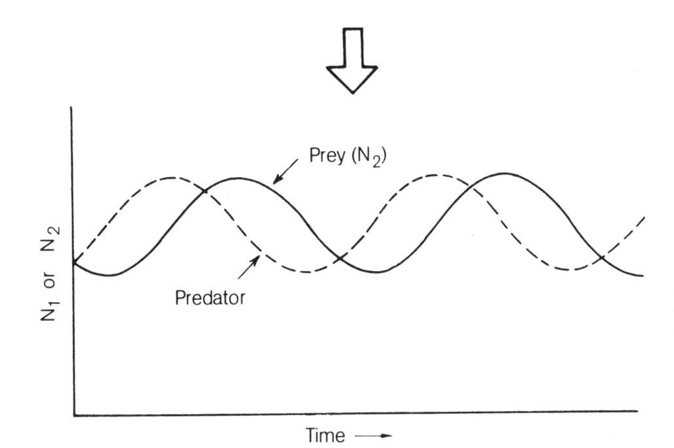

Fig. 99. Predator-prey interaction according to the Lotka-Volterra equation. This diagram is mathematically correct, but that of Fig. 100 is preferable for didactic purposes. (Wilson and Bossert, 1973)

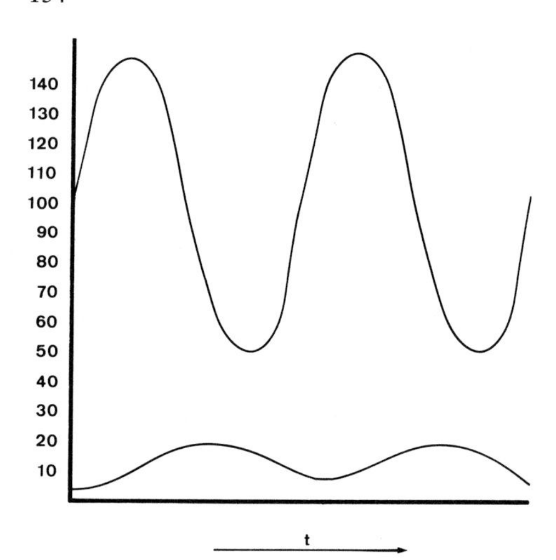

Fig. 100. Like Fig. 99, but with the numerical relationships more realistically displayed. The predator can never (unless it is a parasite or disease pathogen) be more numerous than the prey. As a result, the cycles of the predator are considerably less conspicuous than those of the prey

sentation is biologically reasonable and whether a diagram like Fig. 100 might not be less likely to cause misunderstandings. Finally the question arises whether all predators have the same effect on their prey populations.

To get around these problems we shall treat disease pathogens, parasites, parasitoids, predators, and herbivores together. Interspecific competitors can of course also affect the population of an organism as predators do (cf. Fig. 98). It is common to all of these that they at least do some damage to a living organism and can perhaps in this way control the size of its population. This simplification involves an obvious disadvantage: the "predator" in this very extended sense can be much smaller than its "prey" (in the case of a pathogen, for example), it can be about the same size (many parasitic wasps), or it can be a great deal larger (blue whale feeding on krill). The size relations of the different predator types – which, as we shall see, have quite different effects – and their prey can be schematized approximately as in Fig. 98. Research in which the plant-animal relationship is analyzed as a predator-prey system in a modern and quantitative way is extraordinarily rare. Mice are said normally to consume certain proportions of the seed output of various plant species: about 75% of Avena fatua, 44% of Hordeum leporinum, and 37% of Bromus diandrus. Of these three, Avena is preferred. The populations of these annuals are thus reduced by 30%–62%. Avena responds by enlarging both the individual plants and the production of seeds. The yearly differences in the relative frequency of different annual grasses thus presumably depend to a great extent on the mouse populations present, these being the chief feeders on grass seed (Borchert and Jain, 1977). But there are few data of this sort; we shall concentrate on zoological examples.

On pp. 43–45 we considered the fundamental relationship between body size and metabolism: smaller organisms have relatively higher metabolic rates than larger organisms. In general, this difference is also expressed in the rate of reproduction. A rabbit reproduces at a higher rate than a wolf, a mouse more rapidly than an ermine, and a fly more rapidly than a flycatcher. The relationships diagrammed in Figs. 99 and 100 are applicable (to closed systems) only if predator and prey are of about the same size, and thus have the same potential reproductive rates. If the predator is very much larger than its prey it will produce fewer offspring in a given time, and cycles of this sort cannot occur. One of the most famous examples of field observations that support Volterra's calculations is the nine-year cycle of alpine hare and lynx (Fig. 101); this does indeed occur, but was wrongly interpreted. In regions where the lynx has become extinct the hare population continues to oscillate as before. The oscillation of the hare population, whatever its cause, dictates that of the lynx population; it is not the lynx that controls the hare, but the other way round. It may well be that the situation is the same in all known vertebrate predator-prey cycles. The high

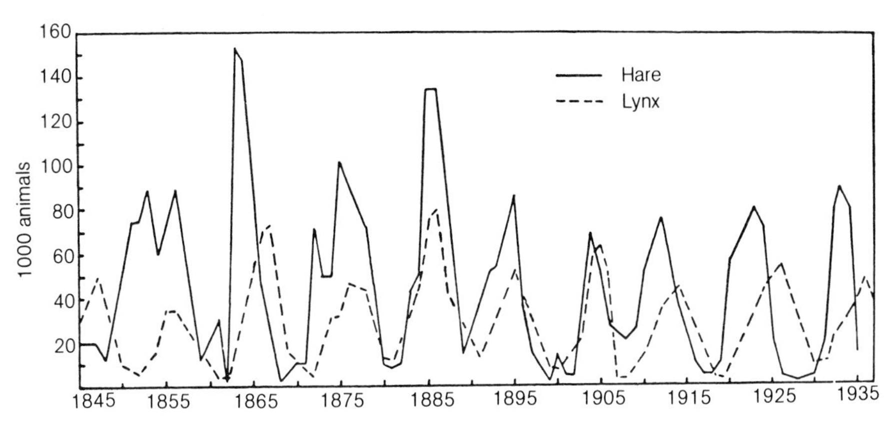

Fig. 101. Cyclic fluctuations in Canadian populations of hare and lynx. For interpretation see text. (Elton in Tschumi, 1973)

rate of reproduction of the tawny owl, long-eared owl, and buzzard is determined by the size of the mouse population, rather than controlling the mouse population.

This simple consideration of metabolic aspects suggests a general prediction of predator-prey relationships, as follows. The smaller the "predator" as compared with the "prey," the more likely it is that the former can control the latter. A predator considerably larger than its prey cannot a priori – without further evidence – be considered as controlling the prey population size.

Taking this as a point of departure, let us examine some interactions between predators and their prey that have been observed in the field. In fact, relatively few really reliable studies of this sort have been done.

Most such studies are in the area of biological pest control. In the 1880s scale insects threatened the existence of the citrus industry in California. Ladybirds were imported, and after only 10 years they had the scale-insect plague completely under control. No further problems appeared for 50 years. Then DDT began to be used on the citrus plantations and in their vicinity. DDT annihilated the ladybirds (cf. p. 49), and they had to be imported again and cultured in order to regain control over the scale-insect population (Ehrlich et al., 1975). The cacti imported to Australia to serve as fences turned out to be vigorous weeds; nothing interfered with their spreading, and they took over fertile land. In a massive research effort Australian scientists examined all the biological antagonists they could think of, and eventually several species were released. One lepidopteran proved especially useful. It brought the cacti almost completely under control, forcing them to retreat to a few small areas. Since then a state of equilibrium appears to have been reached. The common St. John's wort, Hypericum perforatum, when imported from Europe to America spread like wildfire and became a genuine pest. A leaf beetle brought from Europe to control the plant did the job well; St. John's wort has not exactly become a rare plant, but it is no longer a menace to American agriculture (Franz and Krieg, 1972). Introduction of the pathogen for myxomatosis into Australia stopped the expansion of the rabbit population in this part of the world, so that for the first time pasture land that had been overgrazed by the rabbits became available for raising sheep (but cf. p. 140). Viruses are frequently used today to combat forest pests, for they have a highly specific action and in many cases become endemic to the region in which they are applied (Bulla, 1973). Nuclear-polyhedrosis viruses have proved a useful weapon against the European spruce sawfly (Diprion hercyniae), the black arch (Lymantria monacha), and the gypsy moth (Lymantria dispar). Viruses sprayed from a helicopter can specifically halt these animals, even in the midst of a mass outbreak, and kill them off (Franz and Krieg, 1972; Zethner, 1976; for biological pest control see Franz and Krieg, 1972; for animals as plant pests see Ohnesorge, 1976). Mass mortality due to parasites and/or pathogens also occurs in natural populations. A large part of the eider duck population in 1947 succumbed to heavy infestation by acanthocephalans, associated with coccidiosis (Christiansen, 1948). Owing to a poly-parasite complex that becomes effective only under conditions of high host density, there have been wide fluctuations in the eider duck population of eastern Sweden since about 1950 (Persson, 1974).

In these cases the controller is smaller, and can reproduce more rapidly, than the organism it controls. All attempts to achieve similar success in controlling pests with familiar large predators have either failed or the circumstances were special in some way. Under what circumstances can a large predator control a smaller prey, as illustrated by the curves of Fig. 102? Such effects are explicable in regions with marked seasonal changes. Warm-blooded animals need a lot to eat even in winter and are active all winter long, whereas ectothermic animals must spend the winter inactively, in a state of diapause. Most of the former thus have a period of five to six months in which they can act to reduce the population of a prey object that is not reproducing itself. Control of krill (planktonic crustaceans of the genus Euphausia) by many specialists (baleen whales, seals, penguins) in the Antarctic Ocean can probably be interpreted in this context; at the very low temperatures prevailing there the krill needs two to four years to reach sexual maturity, whereas the warm-blooded predators are permanently highly active and – precisely because of the low

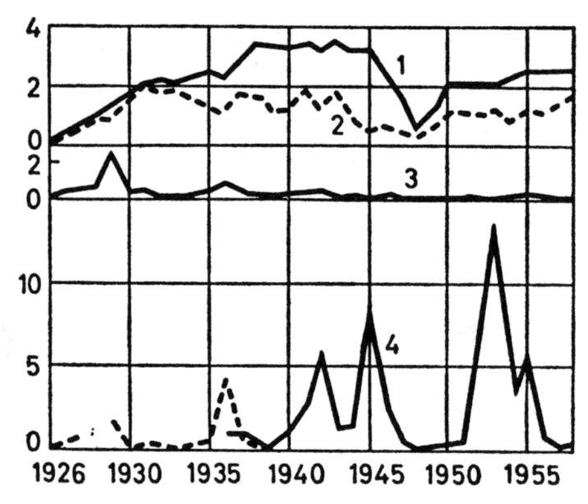

Fig. 102. Number of nest holes *(1)* and breeding pairs *(2)* of hole-breeding birds per hectare, and the number of overwintering pupae *(3)* of Bupalus piniarius per m^2, in the Steckby bird sanctuary; *(4)* number of Bupalus pupae in neighboring pine forests where birds are not protected. (Herberg in Schwerdtfeger, 1968) But the results of encouraging bird populations are equivocal; in general the number of second and third broods falls, so that the productivity of the birds is hardly any greater. Moreover, although in one case protecting the birds tripled or quadrupled their number per unit area (Stein, 1960a, b), thus reducing the insect population by one-third, not all insects were equally affected. The plant-eating insects were reduced by 33%, predators by 28%, detritus eaters by 26%, and parasites by 54%. That is, the increased bird population had the greatest impact on those insects most "deserving of protection" in the context of forest management

temperature – have extremely high food requirements. The effect of ungulates (roe and red deer) on woodland, the subject of much recent discussion, also has its roots in this situation. The animals consume only a minuscule part of the organic matter produced by the plants. Nevertheless, even at population densities so low as to appear negligible they can destroy a forest by preventing new growth – a process that can eventually convert woodland to steppe. Again, the effect of grazing is enormously increased by the fact that it continues all winter; the deer eat the buds of the plants and the bark of the young twigs of softwoods. (It used to be that red and roe deer regularly migrated from their summer range to spend the winter months in river valleys. After the spring thaws they returned to their home woods. The routes they followed are cut off today.) Ectothermic prey with warm-blooded predators in regions with distinct seasons can constitute an exceptional situation. One more factor is involved: the predators must be generalists – species not dependent on a highly specific food. An absolute specialist would not be able to exert such a controlling influence.

A similar opportunity for control exists when the predator – again a generalist – includes uncommon prey in its diet. A species that is not very numerous can be controlled in this way – but then, of course, we must ask why the prey population is so low in the first place? If the prey population should exceed this critical low level, control by a large animal is no longer possible. In view of these arguments, how should the size ratio of a warm-blooded prey animal (a mouse, perhaps) to its warm-blooded predators (fox, buzzard, owl) be evaluated? Modern research in game biology has been concerned with this question for some years. Particularly impressive studies have been done on the willow grouse in conservation areas of Scotland. This grouse has an array of predators – marten, wildcat, fox, hen harrier, and golden eagle. In winter the grouse gather in flocks that wander through the home range. In spring the cock establishes a territory and defends its boundaries against neighboring cocks. The size of the territory depends on various factors that we have discussed previously (cf. p. 104, 122). After a year in which the rate of grouse reproduction was high there are too many cocks for each to have a territory in a favorable habitat. The surplus hens and cocks move off into less desirable areas, or they remain in the habitat and are continually attacked by the cocks that have territories. They must stay near the boundary lines between territories, and even here they are disturbed and must keep moving. These persecuted birds without territories are caught by the predators, as are the individuals that were crowded out into an unsuitable environment. (This fact is highly significant with regard to conservation. A species that has become rare as a result of a change in its biotope, which forces it to live in suboptimal habitats, is exposed to greater predator pressure than it is in its optimal habitat. The very high losses of capercaillies to predators in forests under commercial management, where conditions are less than ideal for the birds, can probably be explained in this way.) Essentially none of the birds with territories fall victim; they know their territories in every detail, with all the possible hiding places. Moreover, they are not constantly involved in aggressive encounters, and can spend enough time watching out for approaching predators. These advantages lend them virtual immunity to attack. Not every individual in a prey population, then, is equally accessible to the predator. This fact greatly complicates the situation. If the predators are eliminated, during the time following a year favorable to prey reproduction the territory defenders are incessantly occupied with fighting intruders, so that even the birds with territories suffer from continual disturbance. Under these conditions, far fewer young are raised. The predators, in decimating the "surplus" animals, raise the production of the prey population.

It appears that this example can be generally applied. Corresponding data have been obtained for various mice and for the muskrat. Further evidence of its generality is provided by the following study. For a long time there have been claims that birds of prey respect a "protected zone" around their nests, within which animals ordinarily hunted are not attacked. Many observers have reported birds nesting in the immediate vicinity of the nests of predators. Now the experiments of Wyrwoll (1977) have shown that this finding requires another interpretation. Hawks strike any prey they can reach, even very close to the nest. However, the birds nesting nearby are very well acquainted with the hawk and its flight patterns; they are not taken by surprise, and are therefore unlikely to be caught. Moreover, they also enjoy a certain amount of protection from other hawks; strangers entering a territory do not know it very well, and are attacked by the hawk in residence.

If the songbirds that occupy territories within a wood are caught and taken away, their place is quickly taken by others of the same species. It seems to be quite generally true of songbirds that a great number of birds are without territories, living between the territories or in less favorable habitats. Among bacteria, too, the possibility cannot be ruled out that a large fraction of the population remains relatively inactive in the soil, whereas only a few show full activity. Perhaps we are faced here with a general principle, which as a regulator of mortality has evolved at the wrong – for the species – time and place.

In laboratory studies of prey and predators of about the same size, phase-locked sinusoidal oscillations have often been found. They have been demonstrated for protozoans (Paramecium under the influence of the predatory Didinium) and for insects and their parasites (beetles and other insects under the influence of parasitic wasps). Under natural conditions, such oscillations are to be expected in a predator-prey system only if either the predator and its prey are the only elements in the system, or if the predator is a specialist. For specialists, when the density of the host population increases the prey is easier to find (or to infect). High densities favor the spread of parasites and pathogens. For this reason, collapse of a massive population is almost always associated with disease and parasite infestation. In any given case it is hard to say whether the collapse was brought about by the disease, the parasites or factors of autoregulation (cf. p. 126). It may well be that all of these are ordinarily involved, to different relative degrees on different occasions (see, e.g., Hörnfeld, 1978). The organisms at the peak of a mass outbreak are particularly vulnerable to pathogens and parasites, owing to factors of autoregulation and perhaps to nutritional deficiencies and to their genetic constitution. Whereas generalists can affect the massive reproduction of an organism only in the initial stages, when the population is beginning to increase (p. 118), specialists as a rule act at the peak of the cycle. But if the predator can fall back on other kinds of prey to any appreciable degree, phase-shifted oscillations can be excluded at the outset.

The nonspecialized predator can respond in two ways to an increase in the prey population (Holling, 1966). First, there is the so-called numerical response (Fig. 103); the number of predators per unit area rises. This can happen in nature by immigration or by reproduction of the predator. But the number of predators cannot increase indefinitely. There is almost always a certain degree of territoriality, which prevents too-dense colonization by predators. Then a "functional" response occurs (Fig. 103). The predator takes more than it normally would of the prey that is becoming more numerous, as compared with other species. Suppose that 10 prey species are available; if each is equally numerous, the predator will consume equal numbers of each. Now, if one species becomes much more populous it constitutes a larger proportion of the captured prey. Up to a certain level this effect is even more pronounced than would be expected on purely mathematical considerations. Evidently the predator learns to become more skilful and quick at finding this common prey, and makes use of its knowledge.

Again, however, we are oversimplifying matters. On one hand, the predator preferentially hunts down the prey it has learned to recognize and neglects other possibilities; on the other, predators are known to concentrate on conspicuous organisms that deviate from the general pattern. These two findings are contradictory, but there are well-documented examples of each. And of course, if it is preferentially consumed a particular prey will change from a common to a rare form. A chicken that has learned to peck at yellow kernels picks these out of a pile and leaves those of

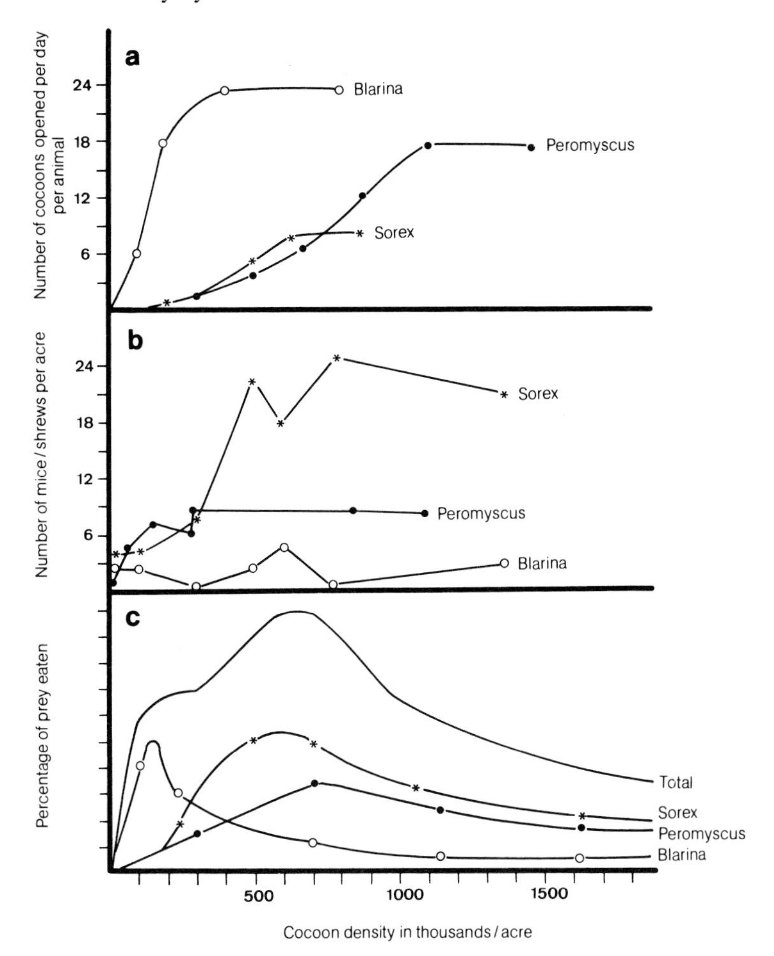

Fig. 103 a–c. Numerical and functional responses in predator-prey systems. **a** When the cocoons of a forest insect are present in great numbers, mice (Peromyscus) and shrews (Blarina and Sorex) open more of them than would otherwise be expected. **b** The number of shrews, as well as that of mice, increases as the population density of a lepidopteran (measured by the pupal cocoons) increases. **c** The result of **a** and **b**: with increasing density of cocoons there is at first a distinct increase in the percentage of cocoons destroyed by predators, a maximum is reached, and then the curves fall off rapidly. (Holling in Schröder, 1974; modified)

other colors. If the yellow kernels are far more abundant, this is a typical functional reaction. But after a while the yellow ones will be nearly all eaten; at this point, the chicken is no longer seeking out the most common "normal" color, but the few remaining "oddities." This simple experiment teaches us that it is not at all a simple matter to infer from field observations whether an animal prefers the "normal" or the "abnormal" prey; the recent history of the predator must be known (cf. Curio, 1976). Here, in the region where ethology and ecology meet, there is a great deal of work still to be done.

At an intermediate level, then, predator pressure is particularly high because of the addition of numerical and functional reactions of the predator. If the prey population exceeds this critical intermediate level, the density of the predator can no longer be increased; simultaneously, the number of individuals of the numerous prey animal that are eaten decreases to a more than proportional degree. This is a reaction entirely comprehensible from a human point of view – we also prefer not to eat exactly the same thing every day. When the prey population is very high, then, the effect of predatory generalists is severely restricted. In this case, only the specialist can intervene effectively. In considering the general interactions between predator and prey, one must distinguish between models according to whether they involve specialists or generalists.

Let us summarize these interactions. If the "predator" is considerably smaller than the "prey," we can observe oscillations in population size, but these as a rule fade out relatively soon. A new stabilization of the "prey" usually occurs at a density level appreciably lower than that at the beginning of the experiment. Successful cases of biological pest control exemplify this behavior. The damping out of the oscillation probably results from a genetic change in the populations of predator and prey. Predators that exterminate their prey die with it. Prey organisms that find even low-level predation lethal are also eliminated from the population.

Examples of such coevolution include the cycles of myxomatosis and rabbits (Fig. 104) in Australia. When the cuscus was transplanted from Australia to New Zealand, a similar situation resulted. These marsupials feed on the leaves of the eucalyptus tree. In their Australian homeland they account for a negligible fraction

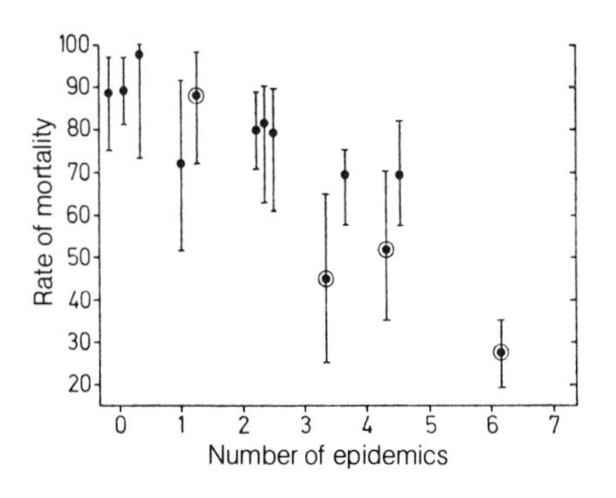

Fig. 104. Altered lethality due to myxomatosis in wild Australian rabbits, injected with an unaltered virus, after a series of epidemics have run their courses. The data indicate the average mortality with a confidence range of 95%. (Stern and Tigerstedt, 1974)

of the herbivore activity. But in New Zealand a mass reproduction of the cuscus occurred, and very many eucalyptus were destroyed. Since then the relationships have changed, and now the role of the cuscus in New Zealand is as insignificant as in Australia. Presumably this shift is, again, the result of coevolution. The trees in Australia contain enough secondary plant substances to prevent mass reproduction of the cuscus. At first those in New Zealand did not, but as the numbers of cuscus multiplied the trees were subjected to heavy selection pressure. Only those individuals with secondary plant substances survived. When these toxic substances are present, the cuscus cannot live exclusively on eucalyptus leaves, but must supplement their diet with other plants (Freeland and Winter, 1975).

In this sense, then, long-term oscillations involving predators and prey of similar size, such as are observed in the laboratory, are probably not realized in nature. The correlated oscillations of warm-blooded prey animals and their warm-blooded predators have another implication (cf. p. 144). If we try to put all these findings together in a single picture, we shall perhaps arrive at the following hypothesis. There are predators that are especially effective regulators when the prey population density is low, but which have little or no effect at intermediate and high density. Other species become more effective when prey density is rising through the intermediate range, and still others really affect the prey population only when its density is very high. Chief among the first group are species distinctly larger than their prey; the intermediate group comprises predators of about the same size as the prey, and the predators that act at very high densities are primarily very small pathogens and parasites. Large predators find their prey even when it is relatively sparsely distributed, so that their activity is influential even under such conditions. Small organisms, microorganisms in particular, must have large numbers of potential hosts available. Taking this – admittedly poorly documented – notion as a point of departure, we see that the joint action of a number of predators can keep the prey population at a relatively low density.

In regions where the number of species is large, therefore, an animal cannot exploit to the full its ecological potential; it never reaches the population size that would otherwise be permitted by the prevailing conditions. In such a case, it becomes possible for several ecologically similar species to coexist in the same habitat. The simultaneous presence of several species of Daphnia in the same lake is interpretable in this way (Seitz, 1977), as is the abundance of species, each represented by relatively few individuals, that characterizes tropical coral reefs and rainforests.

Low density of prey populations is accompanied by a very high rate of reproduction – very high productivity – of the species; this was exemplified by the red grouse. These organisms are almost always in the exponential phase of reproduction. This fact has profound implications for the turnover of materials in the ecosystem (cf. p. 192). Not only does primary production rise (because old trees with lower vitality are damaged more severely than young trees, and thus eliminated from competition), but the rate of decomposition is higher (the soil bacteria population is kept small and thus in a state of intensified reproduction; cf. p. 228).

Finally, then, we can put together a general hypothesis of predator-prey relationships. The complexes of predator and prey have coevolved in such a way that the prey population is kept at the level of highest productivity, and that there is rapid turnover in the ecosystem. If adaptation to predator complexes takes the form of regularly recurring population increases, in which the capacity of the habitat is exceeded (cf. p. 144), it becomes evident that large population and high productivity are mutually exclusive. At the peak of a population explosion the species essentially ceases to reproduce and production stagnates (or becomes negative). The highest productivity can therefore be attained only when the population is relatively small. This conclusion is of central importance whenever the question of obtaining the highest possible yields is at issue.

One would expect, therefore, that when all predators are excluded the populations of animals in a given area would show a distinct increase. This postulate has been supported by a number of recent studies. Reise (1976) set off circumscribed areas of sand flat on the island of Sylt by putting gauze cages over them, having first removed most of the predators there. The result of this experiment is shown in Table 8. It is clear that under the protection of the cage the numbers of animals increased, even though not all the predators could be eliminated (Carcinus occupied the area). Similarly, it has been shown that by keeping fish out of lakes and ponds the mass of plankton could be increased – whether Daphnia (Seitz, 1977; Confer et al., 1978), Chaoborus (Stenson, 1978), or other aquatic insects (Henrikson and Oscarson, 1978). Thus when predators are present the

Table 8. The effect of excluding predators by inverting a cage over an area of sand flat. The number of individuals can increase by a factor of more than 4. (Reise, 1976)

Sand-flat inhabitants	Control	Cage
Hydrobia	21	30
Young clams:		
Cardium	8	747
Other (3 species)	6	25
Adult clams (4 species)	3	1
Pygospio	183	764
Spio	15	21
Tharyx	8	15
Microphthalmus	3	27
Capitella	4	16
Scoloplos < 1.0 mm wide	115	113
≧ 1.0 mm wide	3	8
Other polychaetes (9 species)	20	27
Peloscolex	22	35
Amphipods (2 species)	5	2
Carcinus (juv.)	1	26
Species density	21	22
Individual density	417	1,827

prey population density falls below the theoretical maximum. As a result, it becomes possible for ecologically very similar species to live side by side in the same area; competition among the individuals is very low (cf. p. 180).

The possible results of predator activity, in the broadest sense, can be summarized as follows.

1. The predator brings about genetic changes in the prey population, and the predator population can undergo genetic changes in the process.
2. The predator reduces the prey population.
3. The predator increases the productivity of its prey. These effects may appear separately in some cases, but usually they occur simultaneously.

In the course of evolution, prey animals have developed numerous strategies by which to escape predators. The swift running of the hare and rapid flight of the pigeon are examples, which also demonstrate that such developments do not guarantee immunity to coevolved specialists. Coevolution of predator and prey leads to an array of adaptations by the prey to counteract the steadily improving attack technique of the predator, and by the predator to overcome the increasingly effective flight technique of the prey (see, e. g., Roeder, 1968; Curio, 1976). The formation of packs and flocks is a widespread strategy by which animals ward off their enemies. A closely packed flock or swarm has an effect similar to a single larger animal. The predator finds it more difficult to take aim at an individual, for the targets continually intermingle, disappear from view, reappear and vanish again. In addition, it is more difficult for a predator to approach a flock unobserved. Finally, a flock can defend particular individuals with relative ease; consider, for example, how herds of musk oxen draw up defensive rings, with the young in the middle. When confronted with such behavior, the predator's strategy is to try to break single individuals away from the flock. Dobler (1977) demonstrated this principle in experiments on a sunbleak (Leucaspius delineatus) that swims in schools. When the carp are alone in an aquarium the school is ill-defined, but when a pike (Esox lucius) is introduced, the carp gather in a dense mass (Rüppell and Gößwein, 1972). This closing up of the school is under visual control, and at lower light intensities the school is more open. The pike, which evidently has more sensitive eyes, can hunt under these conditions. Therefore there is a brief period, in the evening twilight, during which these carp are the preferred prey of the pike. Similarly, aquatic insects mostly emerge from the pupa late in the evening, when it is too dark for the fish to see very well. Of the pupae that ascend through the water during the daylight hours, most are eaten by fish. Land insects, by contrast, predominantly emerge in the morning (Remmert, 1963, 1976).

The silver fir (Picea abies) and the gall-forming aphid Adelges piceae are a classical example of an evolved system. The aphids settle on the bark and multiply rapidly. Under these conditions the fir develops necroses in the cortical parenchyma. These block the flow of sap so that it is inaccessible to the aphids, which can survive only in small numbers at a few sites on the tree (cracks in the bark). After a few more years of growth the necroses have disappeared, and the aphids can multiply again. Other fir species (in North America, for example) have not developed this defense mechanism. Fir aphids introduced here become dangerous pests (Zwölfer, 1977).

Usually coevolution occurs in small, unspectacular steps. But evidently "technological breakthroughs" can modify entire systems. The return of the teleosts from fresh water to the ocean – the invention of a means of coping with the ocean salinity – and the return to the sea of terrestrial predators introduced to the oceanic environment great innovations in the way of sense organs. The lateral-line system, high-performance eyes, and the large central nervous system are examples. At first the original marine animals had no defense against these new developments. The disappearance of the ammonites (extinct fossil cephalopods) concurrently with the appearance of fishes can probably be so explained. Another such breakthrough probably occurred when even small animals became warm-blooded; these were far superior to ectothermic reptiles of the same size when the weather was cool. As we now know, the large dinosaurs were likely to have been warm-blooded simply in view of their enormous dimensions; but the extent to which they were capable of genuine regulation is uncertain.

Many plants have developed secondary plant substances which today are generally regarded as protective mechanisms; they prevented herbivores from eating the plants (and have an allelopathic effect; cf. p. 72). In this case, too, specialists developed. Tobacco hawk-moths are not disturbed by nicotine; koala and cuscus eat the poisonous leaves of the eucalyptus, as do many specialized insects. All these means of defense work only against generalists, not against the specialists that have continued to evolve in parallel with the evolution of the prey. The most important strategy of avoiding predation, therefore, is the strategy of unpredictability. This mechanism is evident even in the normal movements of flight from an attacker. The hare doubles back in a way and at a time that the predator giving chase cannot predict. Many animals alternate between habitats. When the familiar migratory fishes move from fresh water to ocean or in the reverse direction, their parasites are all killed; the physiological shock of this change is not predictable by the parasites. The alternation of insects between fresh water and land gives rise to a similar situation. Unpredictability can also be programed into chemical defense mechanisms. If the secondary plant substances in many cases are inherited polygenically, the specialists cannot tell in advance which of them will be present in a particular individual; evolution to deal with an unpredictable combination of secondary plant substances is impossible. However, in such a system some individuals will lack secondary substances altogether, and these are vulnerable to attack by nearly all herbivores. This is the price exacted for this sort of adaptation (Dolinger et al., 1973). Innate resistance to parasites and pathogens is also evidently based on such a polygenic system. Certain mutants of some Drosophila species are capable of enclosing the eggs that parasitic wasps lay in their bodies with a special tissue, so that the eggs die. Not every penetration by a wasp, then, is successful. For the wasp, the question whether its egg-laying will be successful is unanswerable. Unpredictability as a strategy for the avoidance of enemies is also represented by the well-known cyclic changes in population size of many animals (Fig. 105).

The effectiveness of such cycles is particularly well illustrated by the 13-year and 17-year cicadas in North America (Alexander, 1962). In a given region there is usually only one brood, so that cicadas appear above ground only every 13 or 17 (respectively) years. The larvae, like those of most cicadas, live in the ground. At the time of metamorphosis, animals that would feed on the numerous larvae are suddenly deprived of all food and must starve. On the other hand, vast numbers of imagines suddenly appear in the crowns of the trees – far too many to elicit a numerical or functional response of the predators. Furthermore, the singing of the cicadas drowns out the songs of the birds, so that territory marking becomes practically impossible; the density of breeding songbirds in woodland swarming with cicadas is distinctly lower then in a normal wood. From the human point of view this occurrence is part of a regular 13- or 17-year cycle, but for other animals it is an unpredictable population explosion or collapse. Evolution to accommodate such a situation is impossible.

In the last analysis, the same argument holds with regard to outbreaks of migra-

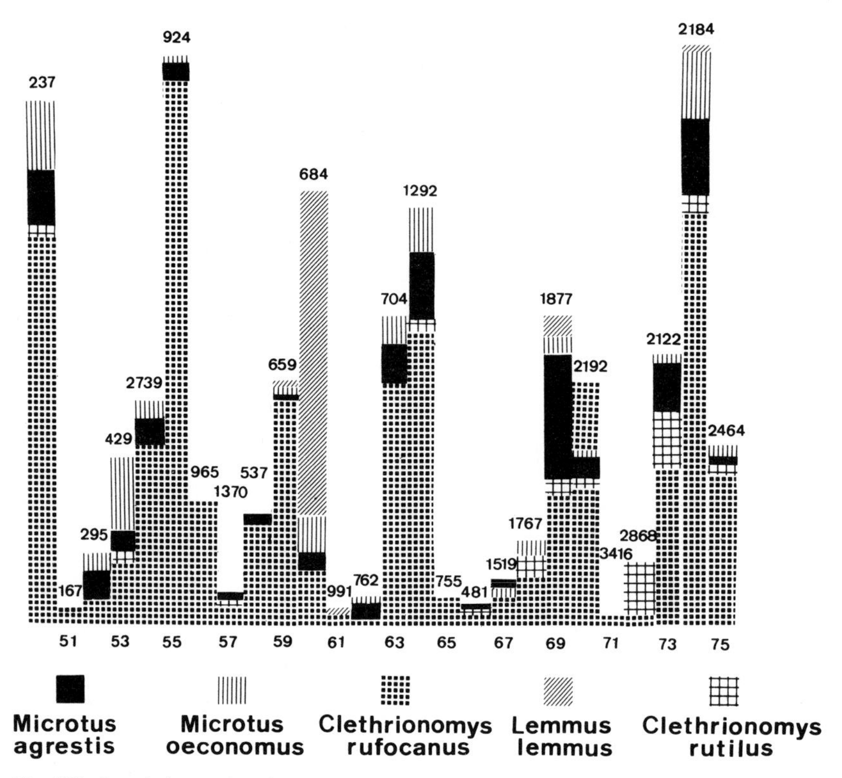

Fig. 105. Population cycles of small mammals in northern Finland, as measured by trapping from 1950 to 1975 (Lahti et al., 1976). (The numbers represent nights × traps.)

tory locusts and to the well-known cycles of lemmings, alpine hares, and mice. The animals are either so numerous that they are beyond the numerical and functional response range of the predator, or so rare that it is hardly worthwhile to search for them. The number of predators feeding on these animals is therefore lower than one would calculate, on the basis of the calculated average size of the prey population over many years.

Under these conditions, synchronous oscillation of different species is favored by selection. In central Europe the imagines of different species of June beetle feed simultaneously on a single tree – in apparent violation of the principle of exclusion discussed above. But in this case the enormous number of individuals acts to eliminate predator pressure, so that their simultaneous appearance is a selective advantage. By the time these short-lived beetles have reduced the food supply so much that competition among them could become effective, they are about to die anyway.

To recapitulate: unpredictability of the time of appearance of a prey species, with all the associated adaptations – in cicadas, precisely programed (in an unknown way) development spanning many years, mechanisms for controlling the oscillations in populations of alpine hares, ptarmigans, mice, and lemmings, and a reproductive potential adequate to permit occasional crossing of the density limit set by the predator – is a strategy for the avoidance of predators.

Comparable "population fluctuations" as an adaptation to predators in the broadest sense are found among forest trees. These, in a normal stand, produce seeds only at more or less long intervals. In central Europe a beech forest produces beechnuts every two to four years (Fig. 106), and the

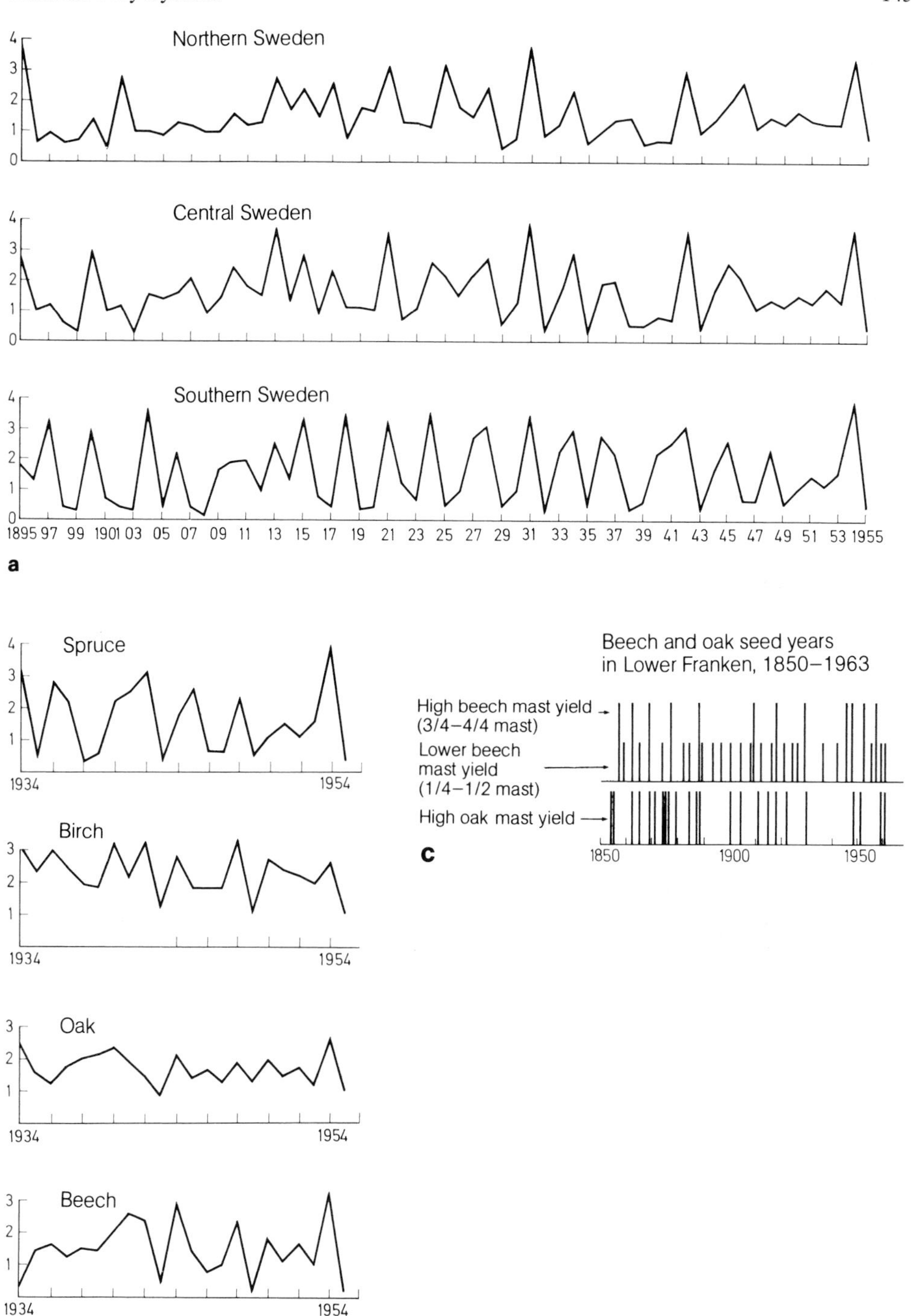

Fig. 106 a–c. Fluctuations in the production of seed by trees. **a** Seed production by spruce in Sweden between 1895 and 1955. (Svärdsson, 1957). **b** Seed production by (top to bottom) red fir, birch, oak, beech, and pine in southern Sweden, from 1934 to 1954. (Svärdsson, 1957). **c** Seed production by beech and oak in southern Germany during the last 100 years. (Maurer, 1964)

acorns in an oak forest appear every three to 15 years. These intervals are the most likely statistically. But in extreme cases two acorn years can occur in succession, only to be followed by 15–20 years with no acorn production. After a good seed year the pith rays of the trees, the storage organs, are nearly empty, so that it is very rarely possible for even favorable climatic conditions to trigger seed formation in the following year. As the years go by a supply of energy is gradually accumulated in the pith rays, and it becomes increasingly more likely that a good climatic situation will give rise to an acorn year. When it is not a seed year, one can find essentially no fruit at all in square kilometers of beech or oak forest. In beechnut or acorn years an increased population of roe deer can survive the winter entirely by reliance on this food supply. As a consequence of these fluctuations, which cannot be predicted by the animals, only a relatively low percentage of the readily accessible seeds, with their high energy content, is eaten by animals. Indeed, many animals store the seeds up and then forget about some of the depots, thus assisting wide distribution of the "prey." In the Middle Ages acorn and beechnut years were crucial to the fattening of pigs and thus to human nutrition. In a very open stand, unlike that of a primeval forest, oaks and beeches produce large quantities of seed each year. Some of the so-called "primeval forests" in central Europe were formed in this way. Such a situation is anthropogenic in origin; forests so structured were never created under natural conditions.

The point to remember about all these cycles is that they represent a selective advantage because they are unpredictable by the predator populations. As a consequence, losses are lower than they would be if the population density remained constant at the average level. A further selective advantage accrues when the cycles are synchronous over the largest possible area, and are synchronized with those of other species vulnerable to about the same predator spectrum (Fig. 105; cf. also Lahti et al., 1976; Hörnfeld, 1978).

There is an immense literature concerned with the causes of such cycles. The authors approach the question from various directions. Some take the view that the cycles are based on endogenous programs – that is, programs intrinsic to the species themselves. In discussing the internal clock we have seen that endogenous cycles can be demonstrated to be so only if, when all possible external controlling factors are ruled out, the cycle persists with a slightly different period length. It is not possible to do such experiments on mammals and birds, with their extensive cycles. However, there are a number of indications that in some species this kind of endogenous control is likely. Very probably, genetic modification of the population in the course of a cycle is involved here. In such a case external factors would serve solely as timing signals; they could affect details of schedule without themselves being responsible for the phenomenon. Among these external factors are weather conditions, food availability, predator pressure, disease, and many others – as in the case of the timing signals for the physiological clock. The fact that the results of comparable research on a given species in a given region but in different years can be contradictory supports this view. For example, one group (Tast and Kalela) found no relationship between lemming cycles and the nutritional situation in Finland, whereas others (Lahti et al., 1976) described a correlation between the two.

A second possibility that has been considered is direct control by ecological factors such as food, climate, and pathogens. The fact that many rhythms are imprecise is regarded as an argument for this view. But in many cases this "explanation" simply shifts the problem. How is it that food is available at rhythmic intervals? Särdsson has pursued this question, and come to the conclusion that the trees (as well as the Ericaceae) in Finland and Scandinavia must necessarily bear fruit at more or less uniform intervals. In so doing, they use up the accumulated reserve materials, and these can be replenished only over several years. When adequate reserves are available, the temperature threshold for the formation of female flowers is low; if there are no reserves, the threshold is high. In a fairly uniform climate, this situation must result in an endogenous rhythm of seed production. But – as has been said – many investigations have revealed no relationship between food supply and cycle. Given the complexities involved, even this observation does not refute the hypothesis. The occurrence of an acorn or beechnut year, after all, is not based on an increased production of organic matter (and is thus not predictable by methods for the measurement of production), but rather on channelling of the materials that are formed in metabolism and on the mobilization of reserves. Kalela found a correlation between lemming cycles and the number of flowers with highly visible structures for attracting insects (such flowers are a favorite food of the lemmings). Because pollen contains substances that can act as sexual hormones, a causal relationship cannot be ruled out; but this would involve differences in the quality of the food and not in amount.

A third group of authors has tried to establish a relationship among sun spots, climate, and population cycles. But these experiments have not been widely accepted. Because the direct relationship between climate and cycle is not clearly discernible, and the synchronization of the different species can be explained as a strategy for the avoidance of predators, this approach most probably no longer has a real chance.

In all these considerations, one must bear in mind that the controlling factors can be quite different for different species and at different times. Extrapolation should be avoided.

The lemming cycles, for example, follow a very abrupt course, rising sharply and "collapsing" just as suddenly. The cycles of Nordic voles (Microtus oeconomus, M. arvalis) living in open terrain may well match this description, whereas their relatives in brushy regions (M. agrestis, M. terrestris) have population cycles in the form of a smooth undulation that damps out. Wendland (1975) showed that the latter curve also describes the fluctuations of Apodemus flavicollis populations. Mass emigrations such as have been observed to occur among Pallas' sandgrouse and some forest-dwelling grouse (Bonasa umbellus), and are to be interpreted as population collapses, probably are very rare among most grouse of the forest (Tetraonidae). In the latter case, fluctuations tend to have the character of damped waves. The same can be said of the various grasshopper species; only a few exhibit extreme mass outbreaks with emigration (very abrupt population fluctuations), while the majority follow simple undulating curves. It is clear that the mechanisms everywhere are different.

Spatial unpredictability is a similar strategy. Many organisms are nonuniformly distributed within their habitat; to us, their distribution appears random. This is true of individual plant species within a plant community. The location of a particular tree in a tropical rainforest is a matter of chance (Fig. 92). A nonrandom uniform distribution always indicates specific territorial behavior of the organism concerned. An organism that feeds on a randomly distributed other organism must develop highly specific mechanisms for finding its food source. Many organisms have failed to do so.

In actuality, then, a random spatial distribution within the habitat is not truly due to chance, for it is favored by selection. Burdon and Shilvers showed that this factor is involved in the infection of grain by pathogenic fungi. When the grain is sown in the classical manner, the fungus attacks the roots in the soil very rapidly, over the entire area. The spread of infection can be greatly slowed if the seed is sown in such a way that the plants are randomly distributed over the field. By sowing other plant species between the grain plants, the

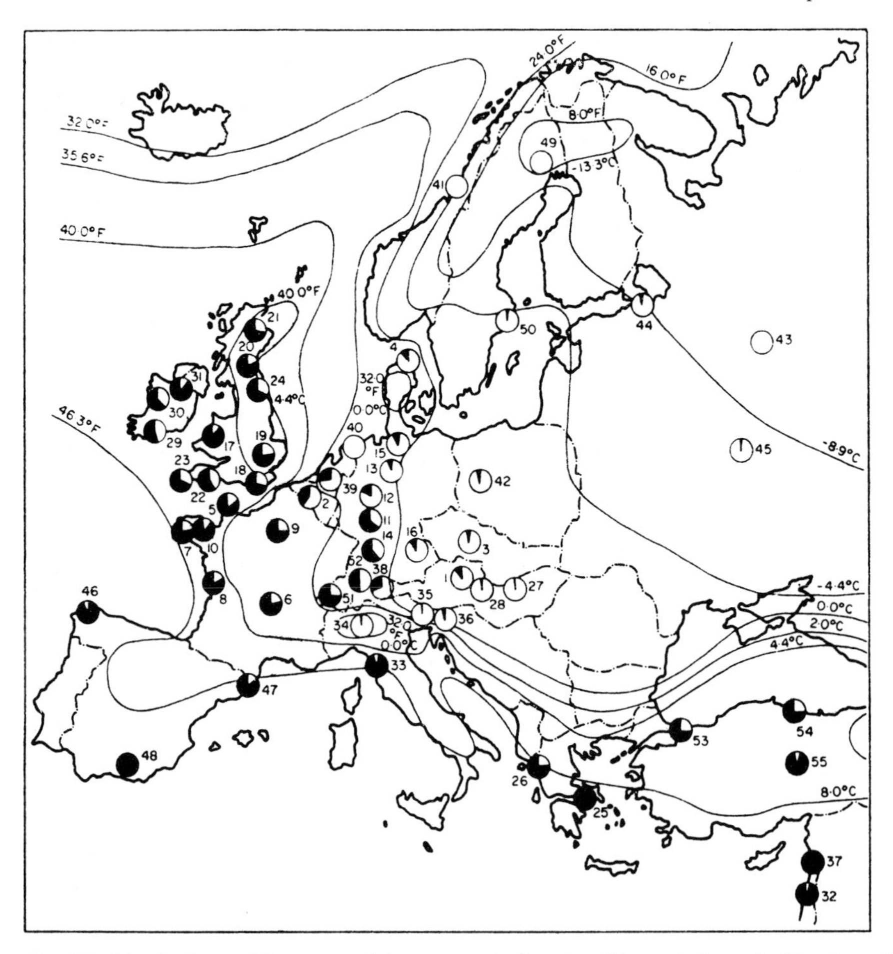

Fig. 107. Distribution and frequency of the cyanogenic form in wild populations of white clover (Trifolium repens). The black part of each symbol indicates the fraction of the total population that is of the cyanogenic form. The January isotherms are also shown. (Daday in Jones, 1973)

rate of spread and the proportion of plants infected can be still further reduced. In experiments on grasshopper mice (Onychomys torridus), Taylor (1977) observed the same principle in operation. Within an enclosure the mice had to expend much more time and effort to find mealworms distributed in "clusters" than when the larvae were distributed more or less uniformly. The price a species pays for this spatial and temporal unpredictability is the loss of certain individuals that happen to be found. But on the average the strategy of random distribution presents any enemy with a particularly awkward situation, for it cannot evolve to cope with chance.

Evolution by which an animal adapts to features of its prey is possible only if these features are predictable. If a plant elaborates a secondary substance (an alkaloid, for instance), it at first becomes unpalatable to generalist herbivores (Fig. 107). A number of herbivores can no longer eat it, but others continue to do so. If the plant forms other alkaloids as well, eventually almost all herbivores will avoid it. But certain species will adjust to the regular presence of alkaloids by coevolution. They will either store these alkaloids, excrete them, or decompose them. In general, storage seems to be the normal way in which animals render such substances harmless. In the course of a few steps of evolution, a plant with very many enemies has become a plant with only one, a specialist. Does this development put the plant at an advantage? A generalist can exterminate its prey and then turn to other prey objects. But the specialist cannot; if it reduces the numbers of its prey below a critical level, its own population is decimated as a result, and the plant has an opportunity to recover. To this extent, the evolution of secondary plant substances does convey a selective advantage.

A remarkable aspect of coevolution is a phenomenon that has occasionally been described as a biological "wage-price spiral." Other evolutionary processes can become associated with the coevolution of prey and predator. The predator stores the alkaloids as described, and as a result can no longer be eaten by the animals that would otherwise prey on it. The caterpillar of the tobacco hawk-moth is unpalatable to jays. An inexperienced jay that has swallowed such a caterpillar vomits it up and in future leaves that species of caterpillar strictly alone. The upshot is that the plant is protecting its own specialist from its enemies – unless, of course, a specialist again evolves that can tolerate the alkaloids in the first specialist. And this does happen; almost all these specialists are parasitized by wasps immune to the particular alkaloids concerned. Furthermore, such a specialist is a suitable subject for mimicry. Because it is unpalatable and avoided by the generalists among the predators, it is a selective advantage for other species to adopt the same appearance. The best-known example of such mimicry is found among the danaid butterflies; some of these have caterpillars that feed on toxic plants, species of spurge and Asclepidaceae, and these produce poisonous imagines. Throughout the world a great number of lepidopterans have evolved to resemble these poisonous danaids (in America, for example, the model monarch butterfly, Danais archhippus, has been copied by the viceroy, Limenitus plexippus).

Hydrocyanic acid is also used by some plants as a protection against being eaten. Among these are species of clover (Trifolium repens) and related plants (Lotus). But it is only in optimal habitats that these plants are capable of coping with their own poison; the two species mentioned can manage this only in a maritime climate (Fig. 107).

Our present-day knowledge about secondary plant substances is of considerable economic importance. The alkaloids that plants contain are appreciated as seasonings, but secondary plant substances are not considered a desirable part of the everyday diet. Modern breeding methods therefore aim at removing the secondary substances from all the major crops (rice, maize, wheat, millet, potato, soybean, fodder grasses). As this is achieved, the crops become more vulnerable to attack by generalist herbivores. More insecticides are required to protect the crops – and these in turn are an insidious threat to the quality of the food.

Animals have not gone so far as plants in developing chemical means of defense (cf. Figs. 108 and 109). Most of the animal toxins are acquired, as described, from plants. However, hydrocyanic acid is found in diplopods (Julus), and the defensive strategy of the American skunk is well known. Recently particular interest has attached to the central European bombardier beetle (Brachyinus), which when endangered expels hot quinones with an explosive sound. The temperature of the gas bursting out of the glands is 100 °C or more. Hydroquinones and hydrogen peroxide are produced by gland cells and stored in a reservoir. When danger threatens there is a catalytic reaction between the two components; the hydrogen peroxide is converted to oxygen and water and the hydroquinones are oxidized to the corresponding quinones. Other carabid beetles defend themselves by means of formalin (Fig. 110).

We have been discussing the fundamental relationships between predators in the

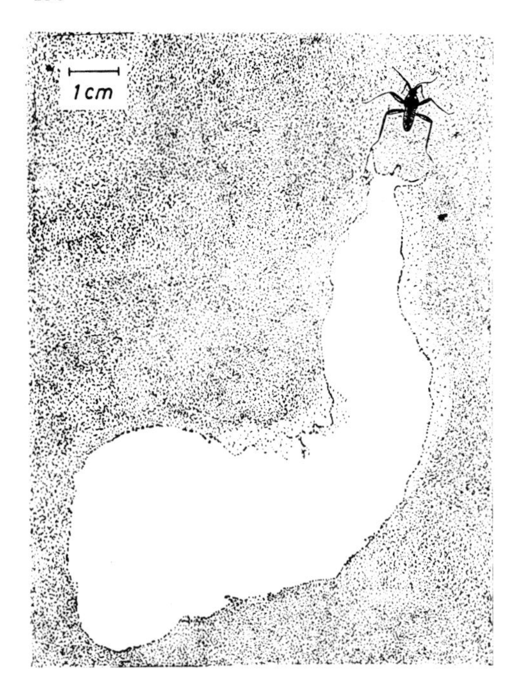

Fig. 108. A surprising flight strategy: the ripple bug Velia secretes on to the water a substance that lowers its surface tension, and the rapidly expanding film abruptly pushes it out of range of the predator. (Linsenmair and Jander, 1963)

broadest sense and their prey in the broadest sense. Before considering how these notions apply under field conditions, we must remind ourselves that they are basic concepts which may be modified in any number of ways. A small predator that spends long periods in a dormant state (perhaps controlled by photoperiod) naturally has a lower rate of reproduction than a prey object – perhaps even somewhat larger – with several generations per year. Moreover, in nature we are rarely dealing with two-species systems; almost always many species participate in a system. Interactions among them can occur in quite surprising ways. The parasitic wasp Orgilus obscurator is a specific parasite of the European pine-shoot moth Rhyacionia buoliana, as are 4 other wasp species. All these species together, however, have less effect on the host than the first would if it acted alone; the latter four can find the host only by following the trail of the first species, so that the host they find has already been parasitized. The new arrival adds its egg, which develops more rapidly than

Fig. 109. If the sea star Astropecten comes into the vicinity of Spisula, the clam leaps out of the sand. (After Thorson)

that of the first wasp and thus has a competitive advantage (Zwölfer et al., 1976). So far, then, we have been concerned with obtaining some idea of what may be possible. Each case must be analyzed separately. Particularly in view of the emphasis now placed on biological pest control, it is impossible to be too cautious in extrapolating from fundamental considerations. When such attempts at pest control backfire, it is usually because of unjustified extrapolation.

Under the rubric "The Problem of Optimal Yield," predator-prey systems have acquired great economic significance. Humans, to meet their own needs, would like to withdraw as many individuals, or as much biomass, as possible from artificial and natural populations without damaging these populations or the overall system, so that the greatest possible yield will be obtained in the long term. Here man is a predator in the broadest sense, as has been discussed in this chapter. In populations under human control – domestic animals and agricultural crops – the situation is relatively simple and in itself presents no particular problem. There are a number of marginal problems which, though no less severe, are not relevant in our context. For example, large monocultures are exceedingly vulnerable to pests of all sorts. The great famine at the beginning of the nineteenth century in Ireland, to which a third of the Irish population succumbed, was due to such a pest; the Irish at the time re-

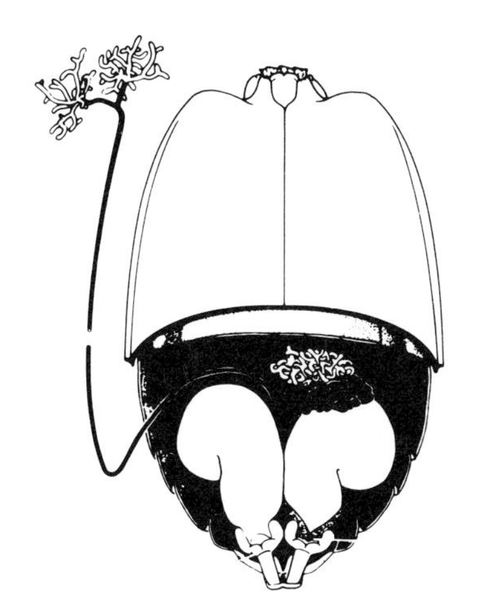

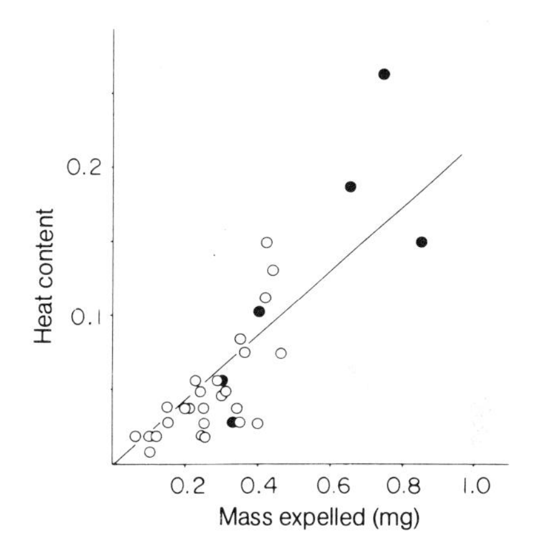

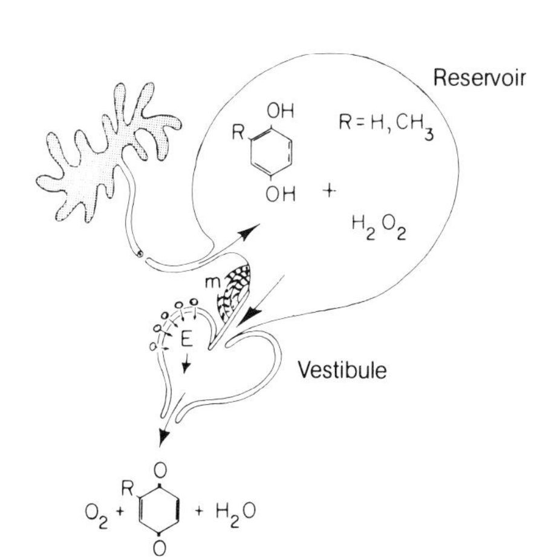

Fig. 110. The mechanism by which the bombardier beetle keeps its enemies away. (Schildknecht in Eisner; see Sondheimer and Simeone, 1970)

lied heavily on potatoes, which were destroyed by a fungus. The possible recurrence of such a catastrophe is the nightmare of modern nutrition scientists. Indeed, the present-day breeding and planting of extremely uniform varieties makes it seem that this disaster could easily come to pass. A second problem arises in that the high-performance grasses, as mentioned previously, are particularly sensitive to nutrient and water deficiencies (cf. p. 38). And a third problem is the danger of erosion inherent in modern methods of cultivation, and that of lateritization of tropical soils.

The situation is different in the ocean, where humans have remained at the hunting and gathering level, and are not yet engaged in any sort of management. It is remarkable that the ocean – an unfertilized habitat in which no means of pest control are applied – produces about 15% of the protein available to mankind. It is the goal of modern fisheries biology to ensure that this supply is maintained indefinitely. Researchers in this area have developed models and methods which could to a great extent achieve that goal (though the crucial political difficulties that stand in the way are beyond the ability of the scientists to resolve).

In recent times we have been increasingly confronted by a corresponding problem on land. It may be that some large steppe regions are best exploited not by traditional agriculture, the sowing of grain or the raising of domestic animals, but by management of the naturally occurring large mammals. The tundra of northern Europe, with its reindeer, offers a classical example. The steppes of southern Russia, grazed by saiga antilopes, demonstrate the same approach, and utilization of the large steppe mammals of eastern and southern Africa in a similar way is currently under discussion. Such animals evidently produce considerably more protein than domestic animals per unit area and time, and unlike cattle they do not destroy the steppe. The chief obstacles at the moment are the tendency of people to reject game meat because they are unaccustomed to it, and the tendency of native populations to prefer beef because it is the preferred meat of the whites. But such changes in attitude to food have occurred many times in Europe when the traditional foods are scarce – think of potatoes, for example – and should be surmountable. Another current difficulty is the transport and storage of slaughtered animals in the warm climate.

Of course, strategies for obtaining an optimal yield can be studied, in principle, more simply and equally well by using small laboratory animals. One such study in particular deserves mention here. Slobodkin and Richman (1956) investigated the results of fishing out fixed percentages of newborn Daphnia from laboratory cultures. The water fleas were first established in aquaria and given a constant diet of unicellular algae, such that a certain initial mean population size was reached. Every four days thereafter 0, 25%, 33%, 50%, 66%, 75%, or 90% of the young animals were removed from the different tanks. As time passed, the sizes of all the populations varied widely, but the fluctuations in population size were distinctly smaller when the percentage of animals removed was larger. A correlation was also observed between increased harvest and reduced average population size (Fig. 111 a). But the small population provided far higher yields than the large population (Fig. 111 b). The curve of yield vs. harvest percentage is just the opposite of the curve of population size vs. harvest percentage.

All our considerations of predator-prey systems suggest that this relationship is to be expected in general. A large population excludes high production, whereas in a small population high production is guaranteed. The danger in keeping a population low is that when unfavorable external factors are acting the normal degree of utilization may amount to extreme overexploitation, which could annihilate the organism being harvested. The problem, then, is to discover for each species under

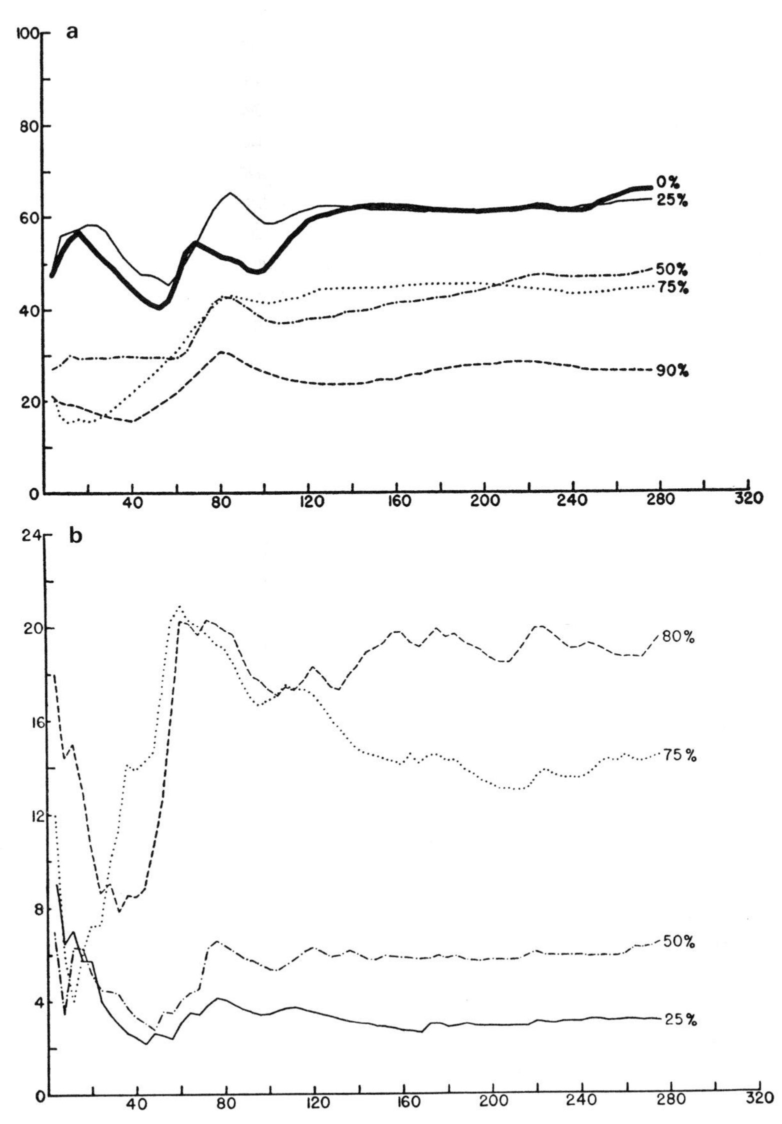

Fig. 111 a, b. The relationship between population size and optimal yield in model populations of Daphnia (Slobodkin and Richman, 1956). **a** Mean population size (ordinate) during the course of an experiment (abscissa, days) in which a constant percentage (indicated at the right of each curve) was repeatedly harvested. **b** Size of the harvest (number of animals) when different constant percentages of the total population are harvested

the conditions of its normal environment how high the harvest must be so that production is optimal, and how high the harvest may be while still ensuring that a high rate of reproduction is maintained. No general prescription can be written. A given species behaves differently under different conditions (and thus in different parts of its range), and different species are not directly comparable.

Another question is the extent to which reduction of population size – so that the growth rate of the progeny is high because they face no intraspecific competition (Fig. 112) – can achieve maximal but not optimal yield. When there is no intraspecific competition selection favors not the most resistant genotypes, but those that grow most rapidly. This problem is the same as that encountered in forestry, where really

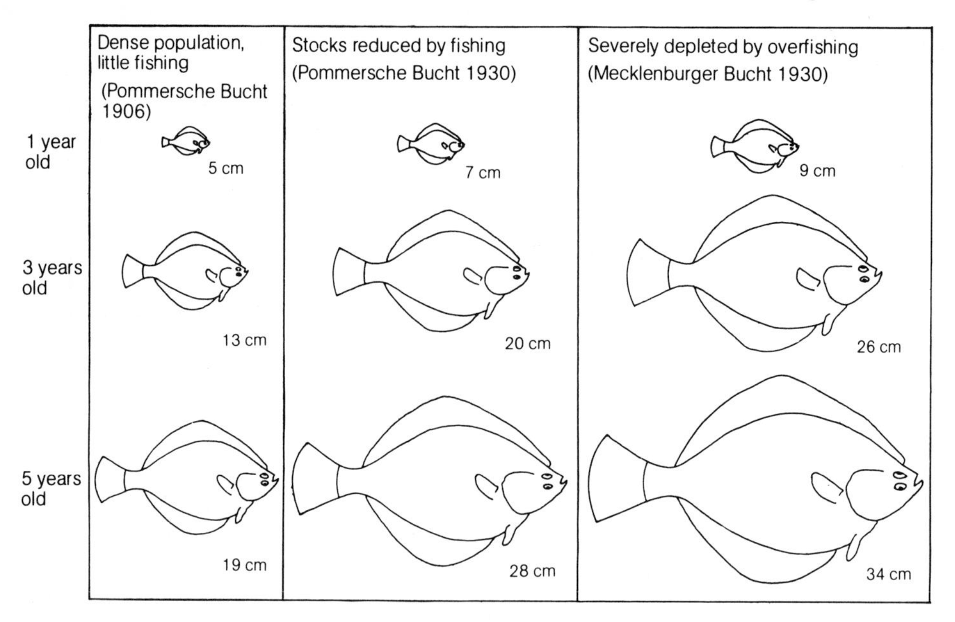

Fig. 112. Growth relationships of flounder in the Baltic Sea. (Kändler in Bückmann, 1963)

high-quality wood is obtained only from very old trees that have grown in a mixed stand under intense competition (cf. Backmang's Growth Law, p. 118). It is extraordinarily difficult to find a good compromise. This is the chief task of modern fisheries biology (cf. Bückmann, 1963; Hempel, 1977); from many field and laboratory studies a diagram has been derived that represents very well the individual parameters that must be taken into account (Fig. 113). This gives the general picture; to find the production maximum for a particular animal species in a particular region, a separate analysis of that species and region is required.

As an illustration of the difficulties encountered in transferring laboratory results to the field, the work of the fisheries biologist is quite appropriate. Three types of overfishing are distinguished. Growth overfishing is the best known of these; the stand of old individuals past their prime is thinned out, and favorable conditions are created for the young fish with a high growth rate and low natural mortality. A population so reduced and rejuvenated is extremely productive. This sort of overfishing, if it is not carried too far, leads to higher production and larger yields. The situation is different with recruitment overfishing. Some species are short-lived, do not grow appreciably after reaching sexual maturity (and then are soon caught), or are not very fertile. With these

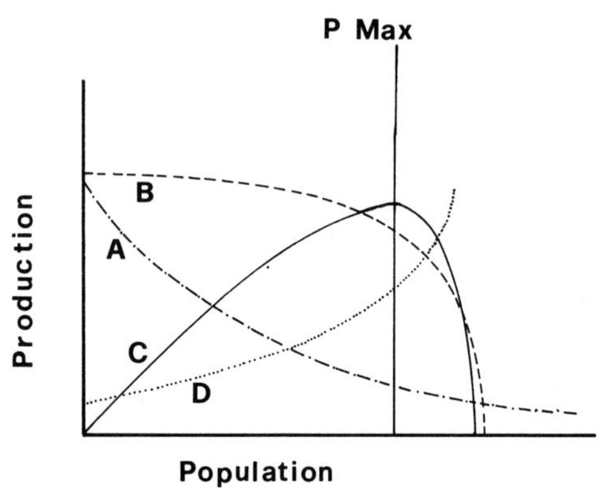

Fig. 113. Variation in quantities of interest in production biology, as population density of fishes increases (Schäperclaus in Nellen, 1978). *Abscissa:* population density; *ordinate:* production. P_{max}, highest attainable long-term production; *A* Nonutilized potential food, *B* Individual growth, *C* Total fish production, *D* Energy consumed by an individual fish in the search for food

species, it is not possible to ensure long-term constant yield entirely by regulating the size of the fish caught (by using nets of a particular mesh size). In such cases protected zones are generally required. Otherwise all the sexually mature animals are fished out, the rate of reproduction falls, and so does the size of the population. The third category is ecosystem overfishing. Here a population is overexploited by the fishery industry to the extent that it loses its importance as a resource in the ecosystem. It can then be irreversibly replaced by other species, as has evidently occurred in the case of the Californian sardine (Sardinops caerulea). The ecosystem has been permanently altered; the overfished sardine is very uncommon today, and its place has been taken by other species of no economic importance (Nellen, 1978).

Studies in marine fisheries biology, then, are complicated by the fact that the systems comprise several interdependent species, and that reciprocal predation occurs – herrings eat large numers of cod eggs and larvae, whereas full-grown cod decimate the herring stocks. It is because of such a situation that when excessive fishing of herring and mackerel in the North Sea caused collapse of these populations, there was a simultaneous massive increase in the stocks of the cod-like fish (the eggs and larvae of which were no longer being eaten by herring and mackerel). The fishermen's catch remained the same, and the overall stocks of economically important fish in the North Sea were about as large as before. One way of protecting the herring population could be to reduce the cod population drastically by fishing and thus encourage reproduction of herring and mackerel; the consequence, however, would be that herring and mackerel in turn would further reduce the cod stocks. Things become even more complicated as a result of the cannibalism which is common among these fish; cod themselves eat many juvenile cod, and herring also eat herring larvae. At present there is hardly any cannibalism among cod in the North Sea, for the large cod populations came into being at the same time, in years when the herring stocks were very low, and the individuals are therefore roughly the same age and size. All these complicated interactions have been treated by Anderson and Ursin (1977) in an extremely elaborate mathematical "North Sea Model," probably the first comprehensive large-scale model for predator-prey systems of such complexity.

One must also consider the extent to which disturbance associated with man's harvesting activities can influence yield. Reichholf (1975) has given an excellent example of this possibility, in his study of the effect of hunting on the water-bird communities of the Inn River. The number of water birds shot during the hunting season is low and could be tolerated by the populations without difficulty. But the birds are repeatedly disturbed by the hunters. When flocks of ducks are driven away again and again from the best feeding places, and kept in constant motion just when they should be finding food, the population is seriously affected. The problem is exacerbated by the fact that all the favorable bodies of water in this region are invaded by hunters at the same time, so that the flocks have nowhere at all to rest and feed. Thus the main point of conservation areas is to give the animals a refuge in which disturbances are few; the population reduction due to the actual harvest by hunters would be negligible. So far no one knows how important it is to take such effects into account when dealing with other organisms.

3. Food Supply and Population Density

No population can maintain itself unless adequate food is available. In regions where food is scarce the density of animal populations is lower than where it is more abundant. This correlation is quite apparent when the density of birds in different forests, with different food supply, is compared (Table 9). We have seen (p. 48) that "a normal animal is a hungry animal."

Table 9. Number of breeding birds per unit area in different habitats. The numbers should not be taken literally, for pronounced deviations can occur (for example, cf. Fig. 171); the data merely indicate order of magnitude. (Tischler, 1955)

	Biotope	Breeding pairs/km²
High density (1,000–3,000 pairs/km²)	Farm/garden complex in agricultural region (Rhine region)	2,700–2,000
	Old stand in oak-hornbeam forest near a village (northwestern Germany)	2,386
	Marshy alder woodland on the Samland coast (eastern Prussia)	2,080
	Cemetery in Berlin	1,630
	Frankfurt Zoo	1,459
	Marshy alder woodland in northwestern Germany	1,159
	Old stand in oak-hornbeam forest (northwestern Germany)	1,100
Medium density (250–1,000 pairs/km²)	Oak-hornbeam forest of intermediate age (northwestern Germany)	926–315
	Plateau (wood/arable/lakes) in eastern Holstein	759–682
	North American beech-maple mixed forest	570
	North American deciduous mixed forest	550
	Grovelike deciduous mixed woodland in Finland	530
	Spinneys in the Rhine region	442–400
	Various forests around Hannover	379
	Deciduous forests of the Oxalis-Myrtillus type (Finland)	360–340
	Orchards in the Rhine region	384–334
	Industrial region of a large West German city	338
	The more open parts of a forest fringing the River Spree (Brandenburg)	284
Low density (1–250 pairs/km²)	Coniferous mixed forests dominated by spruce (Finland)	240–180
	125–200-year-old pine forest in Brandenburg	236
	Deciduous mixed forest (pole timber) in northwestern Germany	197
	Field with hedges (Rhine region)	168–112
	Pine forest (Vaccinium type) in Finland	145
	The more dense parts of a forest fringing the River Spree (Brandenburg)	119
	Pine forest (Brandenburg)	96
	Dry forests of the Lüneburger Heide (northwestern Germany)	73
	Agricultural region near Berlin (large fields)	69
	Fields without hedges (Rhine region)	96–40
	Pine forests (Calluna type) in Finland	40

Nevertheless, it is very difficult to establish a relationship between food supply and population density. One cannot base the analysis on the amount of food present in the habitat, but must consider only that available to the animal concerned. It is not the entire population of mice, lemmings, or grouse that their respective predators can exploit, but ultimately only the "surplus" animals – as was explained in the discussion of predator-prey systems. It has also been pointed out that a roe deer cannot draw upon all of the plants in its habitat for nutrients, but only the extremely rare, tender buds of particular plants. To quantify food supply under such circumstances is a formidable task, and it is even more so when the supply is to be related to the energy that a roe deer, for example, must expend to find these buds. Feeding experiments have shown that food does play a decisive role in establishing the density of animal populations. By supplementing the diets of roe deer with a specially developed high-energy food, Ellenberg (1978) was able to increase the density of the deer in an enclosure to many times the ordinarily supportable level, and encountered none of the problems otherwise associated with high density. Mice, too, if sufficiently well nourished, can be brought to population densities more than 20 times that of a field population at its peak. Even under these crowded conditions, reproduction continues.

In view of these findings, it would seem that the critical factor in nature is not autoregulation but the specific nutritional situation. However, this conclusion is false. For one thing, most of the mice in the experimental enclosure would leave it despite their adequate nutritional state, if they were not fenced in. Moreover, the territory size (and thus the critical density) of a population appears to depend on, among other things, the available food; it is not species- or population-specific, but can vary within fairly wide limits as a function of food supply. The effects of the food supply – which naturally is density-dependent – and of autoregulation (also density-dependent) in the field are as a rule hardly separable. In general the two seem to act conjointly, together determining the size of the population. Whether in a particular case one or the other is decisive can be determined only by study of that situation, and the results apply only to that situation and not to the same species under other conditions.

The classical examples of exponential reproduction flattening out to form a sigmoid curve when capacity of the system is reached are cultures of bacteria and algae. Curiously, there have been very few studies of such cultures in the context of population ecology. As happens among higher animals, many factors are responsible for the flattening of the curve at system capacity. These can be grouped into three complexes, as follows.

1. At least one crucial nutrient in the medium is used up; according to Liebig's Law of the Minimum, reproduction then comes to a halt.

2. The algae release into the medium substances that inhibit reproduction once a certain concentration is reached.

3. Simple problems of space prevent further development of the culture. This phenomenon has been demonstrated only recently, by means of scanning electron micrographs of the surfaces of algae (diatoms in particular). Many algae send out exceedingly fine cytoplasmic processes that extend far into the surrounding medium. When these processes are regularly disturbed the algae appear to be affected in some way such that reproduction ceases. This effect occurs when the actual mass of algae is as little as 5% of that of the medium.

As happens among animals, then, microorganism populations are influenced by a great variety of mechanisms tending to a single end result, which ultimately is describable by the same mathematical formulation in both cases. In the process, however, the organisms themselves seem to be modified. In zoological culture experiments done some time ago, it was found that algae in the stationary phase of development are a considerably poorer food for planktonic animals such as rotifers and water fleas than are algae in the exponential phase. Pigments are difficult to extract from stationary cultures, for the cell walls are evidently much thickened.

The metamorphosis by which the planktonic larvae of marine invertebrates enter the adult stage appears to be more readily induced by bacteria in the stationary phase than by those in the exponential phase.

There is a remarkable lack of systematic investigations into the differences between algae or bacteria in the exponential phase of development and those in the stationary phase, which might help to explain these phenomena. In general the prevailing opinion seems to be that cells enter the same stationary phase as a result of various factors – an insufficiency of space, of light, or of important elements – and that regardless of the inducing factor or factors, there are no systematic differences among cell populations in this phase.

Cultures of planktonic algae ought to offer particularly good opportunities for testing the influence of different food components. It is apparent that light, oxygen, nitrogen, phosphorus and other minerals are all potential limiting factors, the lack of which produces cultures with fewer cells per unit volume than are obtained with optimal medium composition and lighting conditions. But there seem to be no really quantitative data. It has been observed that in bacteria under anaerobic conditions the ATP concentration is higher than under aerobic conditions (Ibrahim, 1973). When pantothenic acid is lacking, yeast cells contain less serine (Tokuyama et al., 1973), and a number of diatoms also appear to undergo reorganization of their chemical composition in the absence of important nutrients. It would be of great interest to establish in detail the influence of individual nutrients on rate of reproduction and population size of these organisms.

4. Abiotic Factors and Population Density

Any field ecologist – indeed, any observer of animals – knows that the more conditions diverge from the optimum for a species, the less frequently the species is encountered. The animals do not vanish all at once, but the species gradually becomes more "diluted" until it can no longer be found at all. On the island of Bornholm, for example, the beach isopod Ligia oceanica is present, together with the amphipod Orchestia platensis and the periwinkle Littorina saxatilis. But the numbers of each are far fewer than in the western part of the Baltic Sea or in the North Sea; a concentrated search is required to locate one of these animals. The beach flies Coelopa frigida and Fucellia are more sparsely distributed in the Baltic Sea, the lower the salinity of the water. We can observe a gradual change in population density as we travel through the marine supralittoral of the Norwegian coast, as far north as Spitsbergen; here, too, the same species occur but gradually become less common. In this case the change is related not to salinity, but rather entirely to the climate (Remmert, 1965b). The extraordinary differences in the density of animal populations in the Icelandic and Scandinavian tundra are correlated with the quality of the soil; the high moors of Scotland, on granite and basalt, are a model case of the effect of soil quality. The same factor determines the density of animals in the chalky parts of the Alps as compared with the central Alps.

The general rule is that as abiotic conditions – especially climatic conditions – become less favorable, the animals are more strictly limited to the more favorable sites. The term "regionally stenoecious" has been applied to such restricted animals. Crickets (Gryllus campestris) are common in the dry meadows of most parts of southern Germany, whereas in northern Germany they are increasingly restricted to south-facing slopes with specific vegetation, protection against wind, soil that is easily warmed by the sun, and favorable (neither too dense nor too open) plant cover.

But quite apart from the restriction to optimal local sites within the bounds of an animal's range, as every field biologist knows, the animals simply become rarer; the total number of individuals per unit area falls off as the limits are approached until there are no more at all. In Burgundy the populations of cicadas are quite dense, whereas in Germany they can be found only in favorable locations and then only in very small numbers per unit area; the large tettigoniid grasshoppers are extremely numerous in southern Germany and as far north as the northern boundary of the central mountain range, but in Schleswig-Holstein and northwestern Niedersachsen there are very few per unit area. The gradual disappearance of tree species at increasing altitudes or more northern latitudes is another example; yet another is given by the trees in the transition zone from forest to tundra or savanna. Of course, these considerations apply to warm-blooded animals as well as to ectotherms and plants. In unfavorable climates the hare population fluctuates about a lower average than in favorable climates. The same is true of the capercaillie; in this case the chicks are especially sensitive to rain. All these findings appear to contradict ecological theory. Enright (1976) referred to this fact as the biogeographer's dilemma. A population can be kept at a particular level only if certain density-dependent mechanisms intervene. But the only such mechanisms appear to be autoregulation, predators in the broadest sense, and quantity (but not quality) of food. Without such density-dependent regulation an organism ought, in the long run, to reproduce until it is equally common wherever it can exist at all. Is this the case?

Let us briefly summarize the implications of Enright's (1976) hypothesis. It is self-evident that we have a constant population only if, on the average over time, death and birth rates are equal. Figure 114 illustrates this relationship. The point of intersection of the curves for death and birth rates determines the population density, regardless of whether one or both of these functions are density-dependent. For instance, choose mortality as the density-dependent parameter, and take the birth rate as independent of density (Fig. 115). As Fig. 114 indicates, the picture would not differ in principle if the density dependence were reversed, or if both birth and death rates were density-dependent. At an optimal temperature T_0 we would obtain the mortality curve mT_0, which intersects the straight line representing birth rate at the point N_0. This point represents the equilibrium population at this optimal temperature. If we change the temperature – raise it or lower it, but in any case move it out of the optimal range – the result is an increase in mortality that is independent of density, dictated entirely by the temperature. The density independence is reflected in a shift, without change in slope, of the mortality curve along the abscissa. If the temperature step is too large (T_2) the mortality line intersects the ordinate to the left of zero density; that is, at this temperature the organism can no longer exist. At an intermediate temperature the mortality line intersects the birth-rate line at an intermediate value. Because of these effects, the density of a population at unfavorable temperatures must be at an intermediate level.

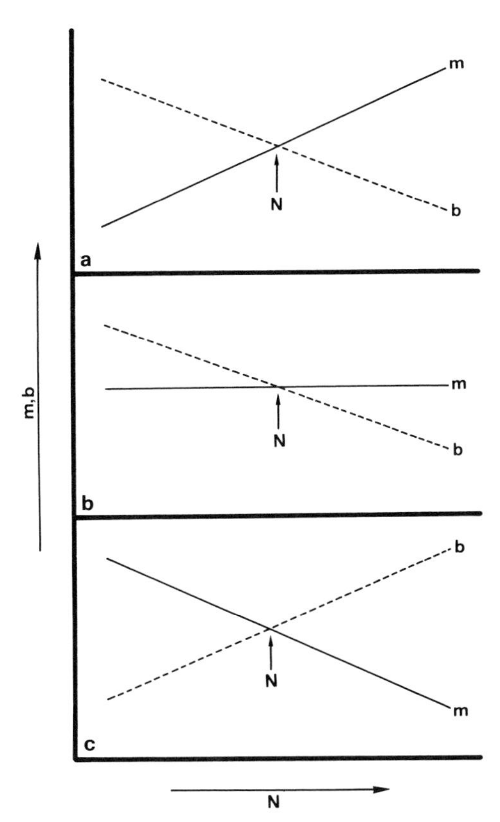

Fig. 114a–c. The density of a population (*N*, the number of organisms per unit area) is determined by the point at which the line for birth rate *(b)* intersects that for death rate *(m)*. It does not matter which of these two parameters is correlated with density. (Enright, 1976)

As mentioned, it is immaterial whether we take mortality, natality, or both as density-dependent; and the effects are similar when factors other than temperature are considered – salinity, humidity, and so on.

Of course, this is an extremely simplified model. But it applies in more complicated situations, when birth and/or death rate are not linear but curvilinear functions of

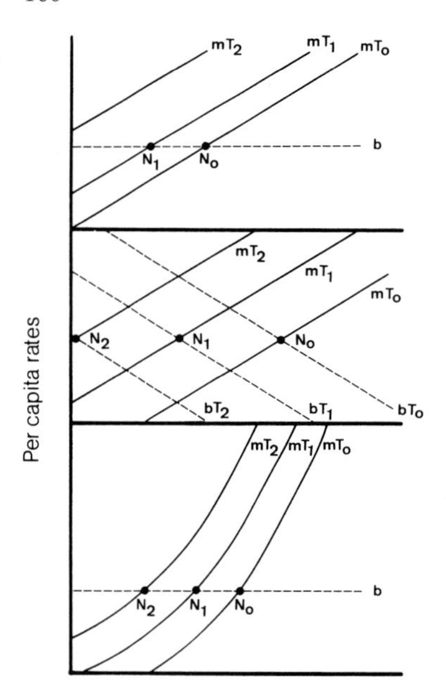

Fig. 115. Under optimal conditions (mT_0, bT_0) the point of intersection of birth rate and death rate coincides with maximum population density; under less favorable conditions $(mT_1, bT_1; mT_2, bT_2)$ it falls at a lower density. For further explanation see text. (Enright, 1976)

density (Fig. 115). This model resolves the biogeographer's dilemma; any species becomes less common, the further one moves from its optimal habitat toward the limits of its range. The observations of the field biologist are supported by theory.

As a well-documented example of such a change in population density as a function of an abiotic factor, recall the distribution of alpine hares and red grouse on Scottish moors of different soil composition (Watson et al., 1973). On moors where the ground is basaltic and rich in minerals there are usually 50–100 hares per square kilometer, whereas only about 5 (1–10) were counted where the ground was granitic (though the level of primary production was the same). The differences between the grouse populations are not quite so great, but the general finding is that about twice as many are present on rich as on poor moors. The difference seems to result entirely from the mineral content of the plants, and to be unrelated to the amount of food available (Fig. 116). The extreme

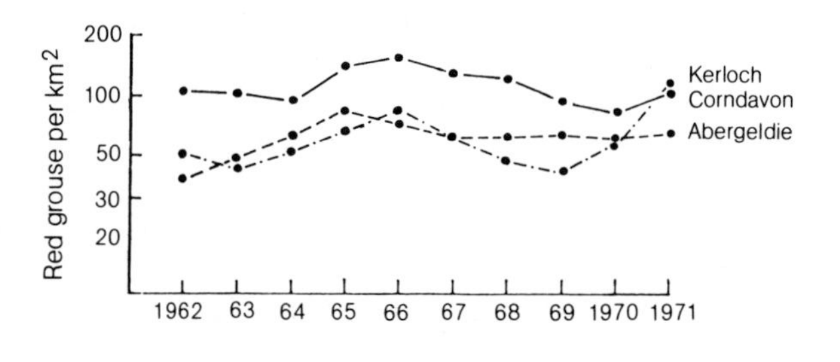

Fig. 116. Population densities of alpine hares and red grouse on immediately adjacent moors differing in geological substratum, in Scotland. Density is highest on rich volcanic soils, and lowest on poor granitic soils. (Watson et al., 1973)

differences in bird density between Iceland (volcanic rock) and Scandinavia (granitic soil) may well also be associated with the quality of the food. Although no quantitative studies have been done, the situation appears similar in the Alps, when regions of primitive rock and limestone are compared. And other phenomena probably also belong in this category – the fact that under NaCl deficiency mice maintain a low population density and avoid mass reproduction (cf. p. 68f.; Aumann and Emlen, 1965) and the observation that in beech woods growing on poor soils there are only about 1,000 litter-eating animals per square meter, whereas in the same type of forest on rich soils there are over 3,000 per m^2 (Thiele, 1968). A famous related comparison is that between central Amazonia, with its extremely poor soils and bodies of water, and the much richer outlying regions (see e.g., Fittkau, 1973, 1974; Fittkau and Klinge, 1972; Fittkau et al., 1975 a, b). Central Amazonia (most easily reached from the Amazon River itself and the capitol of Amazonia, Manaos) has particularly poor soil and thus a very small number of animals; the plants grow very slowly, and agriculture is quite impossible. In the peripheral regions to the north, south, and west, where the soils are richer, the trees grow more rapidly, and the density of the animals is considerably higher, it appears that agriculture will be practicable to a certain extent.

A Special Problem. Terns breed in colonies, a familiar sight on seacoasts. After favorable breeding years when many young are produced, the general result is not an increase in the number of animals per colony but rather the foundation of new colonies by the young birds. After poor years these colonies are given up, and most of their members return to the "mother colony." We have, then, a large colony with only slight fluctuation of the number of breeding pairs, and many other colonies in which the breeding animals can vary in number from zero to several hundred or even several thousand (evidently this is related to the Frazer-Darling effect – that in large colonies the rate of reproduction per head is higher than in small colonies). There is evidence that similar relationships prevail among other animals, such as insects. Solid research is lacking, but it seems a likely assumption that after years very favorable to reproduction many animals emigrate to suboptimal regions, from which they (or their offspring) return in or following unfavorable years. For this reason, the population in an "optimal biotope" appears to be constant, though in reality the constant level can be maintained only if periods of low birth rate are associated with a concentration of those animals that remain. Detailed long-term studies of this phenomenon are urgently required, particularly since the information would help in establishing the size of conservation areas.

VI. Case Studies in Population Ecology

1. Euphydryas: the Splitting of a Species into Separate Populations

Paul Ehrlich and his co-workers (Ehrlich et al., 1975) for many years observed butterflies of the species Euphydryas editha on the grounds of Stanford University in California. The species had been known to occur there since 1934. The first year of study (1960) revealed that the "population" actually consisted of three populations, even though the butterflies inhabited an apparently uniform grassland surrounded by chaparral (thorny shrubs). None of the three populations moved very far within this area, so that even though there were no barriers there was essentially no contact between them. In the four generations of 1960–1963, 97.4% of recaptures were made within the region of the butterfly's own population. This separation was maintained throughout the 15 years of the study – even though one of the populations twice appeared to have died out, only to

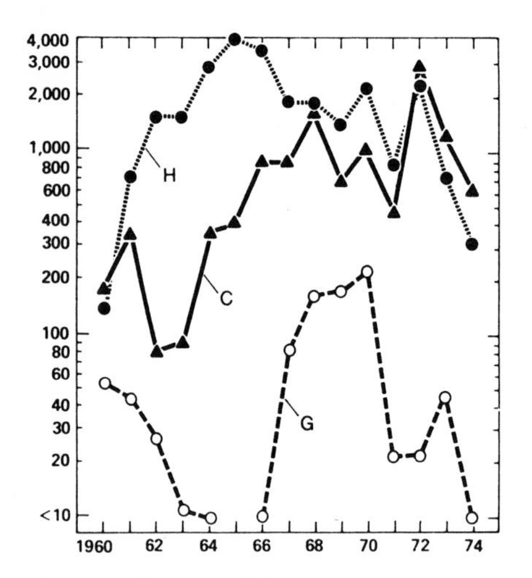

Fig. 117. The sizes of the three populations of Euphydryas studied by Paul Ehrlich's group

reappear in a subsequent year. All three populations fluctuated widely in size (Fig. 117), but these fluctuations were not synchronized. Even if an animal were to leave its own population and enter another, it would have hardly any opportunity to pass on its genetic information. Most females are impregnated immediately after emergence from the pupa; once the sperm has been transferred, the sexual aperture of the female is closed by a plug so that no more can be added. A male wandering into a new population after the beginning of the flight period will find that all the females have mated and are therefore not available for further mating; a wandering female will have already mated. Therefore one would expect that genetic differences among the populations could exist.

More detailed analysis was called for. How did the large fluctuations originate? The question was particularly interesting in view of the fact that they were not synchronous. A priori, adult mortality was likely to be relatively unimportant in establishing population size. Production of an artificial female mortality, by catching and removing the females, had no effect at all. Of course, not all the females were removed; the researchers were unable to catch more than 5%–25% of the females obviously present. Their interest next turned to the young stages. These proved to have parasites, but there were never more than 3%–24% infested, and this percentage was too low to account for the large fluctuations. The main food plant of the caterpillars is Plantago erecta, a ubiquitous plantain found on both serpentine and sandy soils. Caterpillars, however, are found on it only where the soil is serpentine. There were no chemical differences between plants that could be correlated with soil type. But then a new aspect was revealed. Meticulous analysis showed that more than 90% of the larvae raised on this one and only food plant died. A higher survival rate could be achieved if the eggs were laid very early in the year, when the plantain is still green and fresh. The result was also better when the eggs were laid on P. erecta growing near burrows of the pocket gopher Thomomys bottae. Such plants had deeper roots and stayed green longer. Furthermore, when the caterpillars had access to the semiparasitic Orthocarpus densiflorus, which they could eat in addition to Plantago, mortality was low. The availability of Orthocarpus seems to be the critical factor in larval mortality. Good years for Orthocarpus result in large populations of Euphydryas editha, whereas in bad years the populations are small. This relationship also explains the correlation between the occurrence of Euphydryas editha and the serpentine nature of the soil; in the area studied, Orthocarpus grows only on serpentine soils. There is another point to note here: all reports in the literature stated that the food plant of E. editha was P. erecta. Having accepted this, the researchers took years to realize that the distribution of the butterfly is determined by an altogether different plant. Such surprises can be expected to turn up anywhere in ecology, and time and again call for the revision of apparently well-established hypotheses.

But there was still another aspect to the problem. In at least one area occupied by the three small populations, species of

Table 10. The ecological situation of various *Euphydryas editha[e]* populations in California. (Ehrlich et al., 1975)

Characteristic	Jasper Ridge	Del Puerto	Arroyo Bayon	Agua Fria	Mud Creek	Ebbet's Pass	Sulfur Springs	Lower Otay
Altitude (m)	170	450	720	610	610	2,730	150	180
Flight period	March–April	May–June	May–June	April–May	May–June	June–July	April–May	February–March
Plant on which eggs are laid	Plantago erecta	Pedicularis densiflora	Pedicularis densiflora	Collinsia tinctoria	Collinsia tinctoria	Castilleja nana	Plantago lanceolata	Plantago insularis P. hookeriana
Secondary food plants	Orthocarpus (oblig.)	Some, less important	Unknown	Sometimes Lonicera	Some, often necessary	Post-diapause caterpillars feed on Penstemon heterodoxus	None known	Not in most years
Flight habits	Sedentary, 1%–5% move about 600 m between captures	Up to 1,200 m in search of nectar	Unknown	Short flights along river	Some urge to travel	Stationary	Unknown	Long travels in dry years, stationary in wet
Population size	A few hundred to a few thousand	Over 1,000	At most a few hundred	200–600	More than 1,000 in most years	A few hundred	At most a few hundred	1,000 to many thousand
Factors in population control	Rain in spring; density of main food plants in spring	Intraspecific food competition	Interspecific competition with A. chalcedona	40% mortality due to parasites	Rain in May and June; density of food plant; intraspecific competition	Combination of predators, parasites, and intraspecific competition	Unknown	Late winter rains, density of food plants, clear food competition
Length of fore-wing (♂) (mm)	22.5 ± 0.2	20.9 ± 0.2	21.6 ± 0.2	24.4 ± 0.1	22.7 ± 0.2	18.2 ± 0.8	20.7 ± 0.1	19.7 ± 0.1 to 21.1 ± 0.2
Egg weight (mg)	0.227 ± 0.005	0.229 ± 0.004	0.226 ± 0.004	0.251 ± 0.004	0.209 ± 0.006	0.231 ± 0.006	0.277 ± 0.05	0.181 ± 0.005
Number of eggs per clutch	113 (laboratory)	52.3 ± 4.1	85.1 ± 12.6 (laboratory)	39.1 ± 4.8	17.9 ± 1.6	14.0 ± 1.5	70 (laboratory)	39.2
Flowers providing nectar	Localized, very dense	Common along paths and ditches	Common but not dense groups	Common and dense	Dense locally	Dense locally	Not dense	Dense to very rare depending on rainfall

Lomatium flowered very sparsely in some years. These flowers are the chief food source of the adult butterflies. All the evidence indicates that scarcity of food for the adults of this population was another factor in its dynamics.

The question is whether these results apply to all Californian populations of Euphydryas editha. As Table 10 shows, this is by no means the case. Within a very restricted space we have widely differing mechanisms of population control and quite different food plants; to make predictions from these findings is enormously difficult.

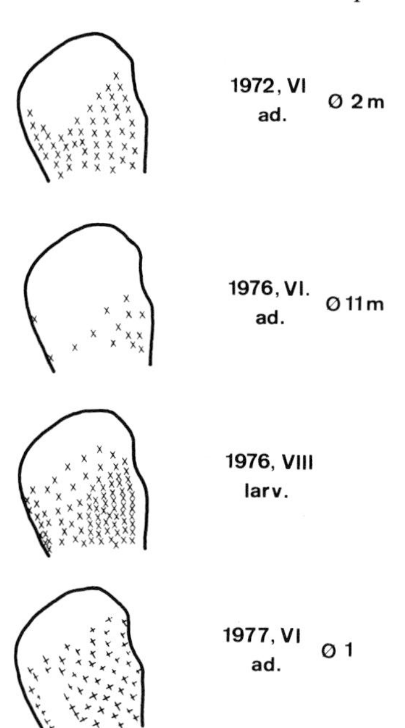

Fig. 118. Distribution of crickets on the Walberla plateau. The part of the steep slope still occupied by crickets in 1976 lies at the left of the plateau. ∅, mean distance between singing males

2. Field Crickets: the Mechanisms Underlying Population Dynamics

The Walberla is a mountain north of Nürnberg which has certainly been unforested since prehistoric times. The plant community on its steep grassy and rocky slopes and its plateau, partly cultivated and partly used as an occasional sheep pasture, is a typical mesobrometum. This uniform system in some years unexpectedly departs widely from the norm, as is discussed on p. 237 f. and in Figs. 167–169. Here we are concerned with the field crickets which are typical inhabitants of the grassy slopes.

In June, 1972, all the upper area of the Walberla and the grass-covered slopes with southern exposure were densely occupied by crickets. The mean distance between singing males was barely 2 m. In the years that followed, the size of the population was much reduced. In June of 1976 the only crickets that could be found were sparsely scattered over the slopes and only a small part of the plateau. Here the singing males were separated by about 10 m; at most 100 singing males occupied the plateau, and about twice as many were on the slopes. Assuming a sex ratio of 1:1, the total population comprised only about 600 animals (Fig. 118).

This decline resulted from the high mortality during the winters of 1972/1973, 1973/1974, and 1974/1975, and from conditions (cool and damp) unfavorable to reproduction and growth during the summers of 1973, 1974, and 1975. In the summer of 1975 those crickets that remained experienced two violent cloudbursts that killed more than 50% of the population. The same storms had similar effects in other locations where crickets were living. The winter of 1975/1976 was cooler, and the residual population apparently survived to a great extent. The following summer was extremely hot and dry, so that the few remaining crickets could entirely recolonize the Walberla plateau; more larvae were counted than ever before. Control counts in September which covered the whole plateau showed that there was at least one larva per square meter, and the density was considerably higher in the regions where the expansion originated. If we take as a starting assumption that about 100 females reproduced during June of 1976, with an average of 250 eggs per female, a maximum of 25,000 young crickets would have developed during the summer. But the measured mean density of 2.5 half-grown larvae per square meter on the plateau, which covers more than 3 ha, indicates that there were 75,000 crickets in the area. According to the above estimate, not all of these could have come from the original plateau population; some must have come from the residual populations on the slopes. But even these were not very large; the implication is that the summer-1976 generation suffered essentially no mortality. In one year the population expanded from 200 to 25,000 (or from 600 to 75,000). Most of these crickets survived the following winter, which again was cool, and in summer of 1977 there were 2 singing males per square meter. On the basis of the observations over these five years, the factors controlling the population dynamics of the crickets can be summarized as follows.

1. Predators are routinely present. In spring jackdaws patrol the countryside in search of crickets. Mice, voles, and shrews have been seen with crickets they have caught. But no quantitatively significant effect of their activities could be established. Parasites were not found on the Walberla during the period of observation.

2. In damp, cool summers the crickets display hardly any activity. Even where there are many crickets, one rarely hears them sing. Courtship and egg-laying are protracted over a long time, as is the emergence of the young. The larvae grow very slowly (Fig. 119), and many animals fail to reach the size necessary for overwintering.

3. Because of the effects summarized in (2), the sizes of the larvae in cool, wet summers vary over a wide range (2.5–5 mm at the beginning of September). If there are large numbers of larvae, because the adult population and the numbers of eggs laid were large, cannibalism now sets in, with the larger animals seizing and eating the smaller. The result is a further reduction in population density. Perhaps this autoregulatory mechanism has a positive effect, however, in that the smaller larvae, with their high nutritive value, are a good food

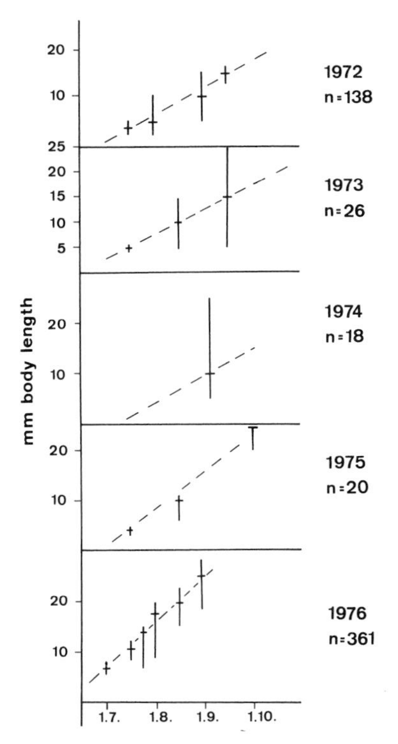

Fig. 119. Growth of cricket larvae during the summers (July–October, abscissa) of 1972–1976 on the Walberla plateau; overall variation among the larvae found is indicated along with the mean

for the larger ones and enable them to survive the winter. Gryllus bimaculatus has been found to have a lower null point for development (the temperature at which no development takes place) when given protein-rich food (cf. p. 90 and Fig. 65). In years when the population density is very low there is no possibility of cannibalism.

4. In warm winters, when the temperature rarely falls below the freezing point, the overwintering crickets may be active even in December and January, and thus be caught in the traps. After such mild winters the number of adults in May is greatly reduced. Evidently the overwintering larvae expend a good deal of energy under these conditions – perhaps more than they can consume in the winter. The mortality is high. Certain experimental results also support this explanation; apparently neither feeding nor digestion of food can occur at temperatures below 10 °C. Too little

attention has been paid in the ecological literature to the possibility that warm winters can be a factor. This factor is probably responsible for the relative paucity of insects in boreal-maritime regions (Schleswig-Holstein; Emeis, 1950; Fig. 26).

5. The literature contains descriptions of populations in which the mean separation of cricket burrows is only about 50 cm. Under such conditions Klopffleisch (unpublished observations) found that mass migrations occur (Figs. 120 and 121). The population he studied exhibited a marked change on June 5, 1972. At about noon on this day, probably due to the density of the population, two or more males encountered one another directly above the area under study. Unfortunately, this encounter could be followed only by sound; very vigorous and shrill rivalry songs were produced. When this happened, many crickets that had been inside their burrows or on the platforms in front of them suddenly fled down the slope. Only two minutes afterward, the mass emigration was over. The entire event occurred so quickly, and the number of moving crickets was so great, that it was not possible to determine where they had gone. However, because all of the animals within the area had been marked, by examining those that remained the number and original positions of the missing crickets could be precisely established. A total of 42 males (out of 59 in the 35-m² test area) had fled, but not a single female. During the study of the crickets on the Walberla, such a high population density was never observed, nor were any such migrations seen to occur.

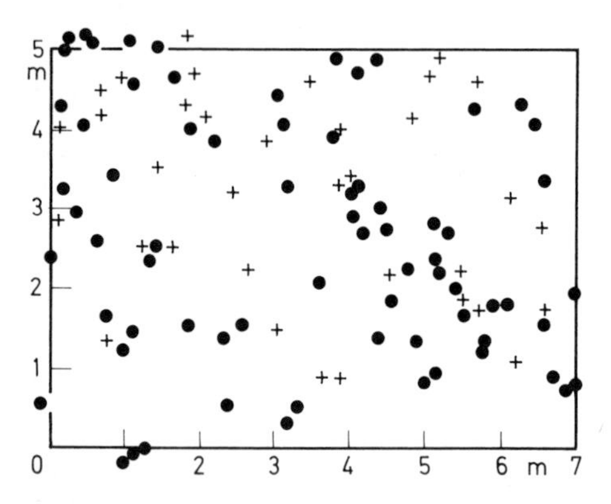

Fig. 120. Locations of male (●) and female (+) crickets in a very dense population in the Schwäbische Alb. (Klopffleisch, unpubl.)

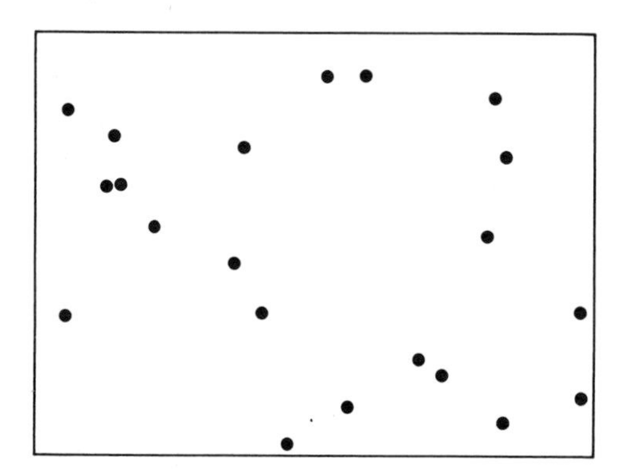

Fig. 121. Locations of male crickets remaining after most of the males shown in Fig. 120 had disappeared in a mass emigration (the females were not affected). The mass emigration occurred on June 5, 1972. (Klopffleisch, unpubl.)

6. In very favorable summers the population can be dramatically reduced by sudden heavy rainfall. Frequently, on such occasions, less than one-tenth of the crickets survive. Evidently the animals drown in their burrows. Although the burrows are dug in well-drained soils, the drainage seems inadequate for a downpour. It is entirely a matter of chance which of the crickets are affected. Burrows particularly resistant to predators prove hazardous in a sudden rainstorm, because it is less easy for the water to run out. The direction of the wind, and thus of the rain, plays an important role.

7. Another cause of mortality can be found in the genetic heterogeneity of the population. Ordinarily the final or penultimate larval stage enters obligatory diapause. But in the warm summer of 1976 a few crickets metamorphosed to the imago stage; in September there were five singing – that is, mature – males in the entire area. We can take it as certain that, as Danilewsky has shown for lepidopterans and Sauer for Panorpa, the population comprises a number of genotypes with different responses to the photoperiod, and that these

are selected differently in different years, depending on the weather conditions. But selection involves mortality due to external factors. In our case we have calculated that there were 75,000 individuals present in the area studied at the beginning of September; of these, 5 males developed into adults. Assuming that the same number of females metamorphosed, we derive that 0.013% of the population can develop directly, without diapause, and thus appear as adults (under apparently favorable conditions like those of 1976) in autumn – too early. It may well be that in every year, depending on the prevailing environmental conditions, different genotypes are removed by selection.

8. A warm, dry summer favors adult animals and the larvae they produce. Masses of eggs are laid simultaneously, and the larvae grow rapidly and synchronously.

9. Synchronous growth of the larvae, under the conditions of (8), prevents larval cannibalism and thus increases population density.

10. Cold winters with snow cover reduce winter mortality. Catches in the winter of 1972/1973 indicate that mortality then was negligible, whereas in the following winters it was extremely high.

11. Dense vegetation impedes predators (Factor 1) in search of the crickets. Crickets regularly seek out regions with a dense plant cover 15–30 cm high. When such an area is mown or grazed most of the crickets leave it. The situation is similar when, under certain conditions (a warm winter), the vegetation opens up so that the ground is only 60%–80% covered. The crickets avoid such regions.

It is impossible to arrange these factors in order of importance. The only factor that can be disregarded is predation. Others must be taken in combination (warm and dry summers with synchronous growth of the larvae; cool and wet summers with differential larval growth). Favorable years can suddenly become unfavorable if there is a rainstorm at the wrong time. Regulation of cricket population density thus seems to be brought about almost entirely by climate. Mortality, because of the climatic conditions alone, is almost always very high. The optimal temperature established in the laboratory, 27°–34 °C, is almost never reached in the field. In regions where the temperature does go so high, Gryllus campestris has already been displaced by Gryllus bimaculatus, which is a superior competitor under such conditions. As far as its temperature range goes, then, Gryllus campestris inhabits suboptimal regions – even though the Walberla must be counted among the typical cricket biotopes. Density-dependent emigration probably occurs extremely rarely, only when there is coincidence of a number of favorable factors several years in a row. There must be a cold winter which is survived by a high larval population, followed by a warm dry summer with no sudden heavy rainfall, during which many eggs were laid simultaneously, to hatch into large numbers of larvae growing in synchrony. A second harsh winter is necessary, to allow this dense population to survive with essentially no mortality. Then there must be a second warm summer, so that such a large population can reach the main courtship stage. This sequence of events is a rare exception. As a rule, regions with such a climate are inhabited by Gryllus bimaculatus rather than Gryllus campestris. This example illustrates the inordinate complexity of weather factors and other mechanisms that influence the size of populations. The surprising fact that cannibalism – in general a typical autoregulatory factor at excessive population density – occurs here when density is low, as a mechanism to permit survival of a few animals, again underlines the precariousness of generalization.

It also shows how huge fluctuations in population size can be under quite normal conditions. To repeat once more: over three years conditions were such that the population declined fairly steadily, only to shoot up in a single generation from 200 to 25,000 animals per hectare, because of a favorable summer. Regions too small for such fluctuations – where the populations cannot survive several years of drastically suboptimal conditions – cannot be inhabited by these crickets (in fact, field crickets did die out during these years in many parts of Oberhessen). If conditions match closely the long-term average for several successive years, the eventual result is extermination of the crickets. Because of the exponential temperature dependence of insect growth processes, a "too warm" summer causes the enhanced reproduction described, and this suffices for the species to survive several "normal" or "cold" summers.

3. Bat and Moth: the Coevolution of a Predator-Prey System

Few animals have presented zoologists with as many puzzles as the bats. The most intriguing question was: how can they orient at night, and even catch prey, with their poorly developed visual system? Finally, 40 years ago, the principle was revealed when Griffin demonstrated the presence of an active system for localization by ultrasound, similar to radar or sonar. As research on this system progressed, such astonishing details were brought to light that the bat is now one of the best-analyzed and most fascinating experimental animals of neurobiology.

A bat hunting normally, as one can easily observe, follows a certain route precisely time and again. It does this "blind," so that it is caught with little difficulty by nets set up in its way; when travelling a familiar path a bat normally does not send out echolocation signals. These vocalizations, which could reveal the presence of an obstacle, are normally given only when the bat is actively in search of a prey object or is in an unknown area. The vocalizations of each species have a particular, specific pattern (Fig. 122). In some cases the tones are sent out through the mouth and in others through the nose. Nasal tone production is always correlated with the elaboration of a special nose structure that concentrates the sound "beam." This structure is a familiar sight; horseshoe-nosed bats are named for it. Bats in this category have the mouth free to seize the prey, whereas the others must make the catch in some other way. The bats called pipistrelles, for example, "shovel" their prey into the expanded skin of the tail and later remove it from this pocket with the mouth.

The following discussion will be limited to the large horseshoe-nosed bat (Rhinolophus ferrumequinum), which is well known owing to the work of the groups of Neuweiler and Schnitzler (Neuweiler, 1978; Schnitzler, 1978). The bat sends out an ultrasonic pulse and orients by the echo.

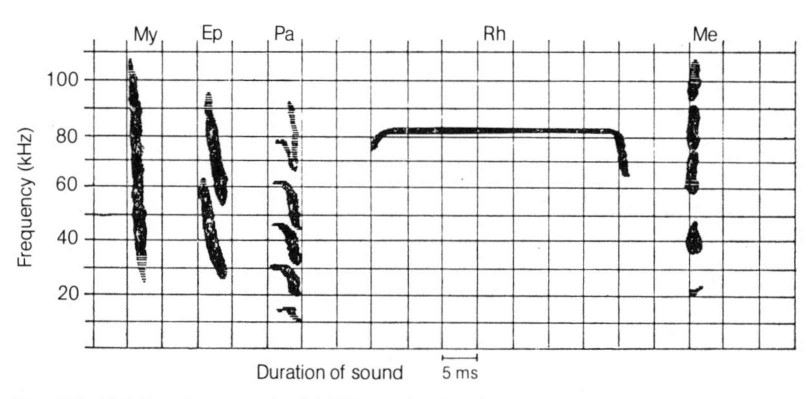

Fig. 122. Echolocation sounds of 5 different bat species. *My* Myotis myotis (FM type); *Ep* Eptesicus fuscus; *Pa* Paphozous melanopogon; *Rh* Rhinolophus ferrumequinum (CF-FM type); *Me* Megaderma lyra (HF type). (Neuweiler, 1978)

Ultrasound is favorable because the sound waves spread out in a relatively straight line, as compared with the sound audible to us, which spreads fairly uniformly in all directions. The disadvantage of ultrasound is that one cannot hear it. That is to say, we cannot; the bats can. They hear well in the normal range, which actually contains information important to all animals; between 40 and 80 kHz they hear practically nothing, but just above 80 kHz lies a narrow band of frequencies to which the bat ear is extremely sensitive. Frequencies of about 86 kHz or higher elicit no further response. That is, the bats are quite specifically sensitive to tones in a range corresponding to their own echolocation sound (Fig. 123). From the time it takes for the sound to return to the ear the bat can infer its own flight speed with respect to another object. It does this in a particularly clever way. When we hear the whistle of a train approaching us it seems very high, whereas when the train is going away the pitch of the whistle falls. This phenomenon is called the Doppler effect, after the man who discovered it. When the bat flies toward a stationary object it hears tones higher than those it sends out. It then lowers its sending frequency from about 83.4 kHz to whatever lower level causes the echo to have a frequency of 83.4 kHz (Fig. 124). The shift in pitch contains the information required to calculate flight speed, and at the same time permits the echo carrier frequency to remain constant. Constancy of the carrier frequency is necessary, because there is an additional change in the echo which the bat neither can nor wishes to compensate for. Depending on the object reflecting the sound, the carrier frequency is modulated in particular ways – especially when the object itself is in motion. Every flying insect has a characteristic wingbeat, by which it can be identified. Insect wings are of different sizes and shapes, and those of the various species move up and down at different speeds. All of these features affect the way the carrier frequency is modulated (Fig. 125), so that the bat should be in a position to tell what sort of prey it is approaching,

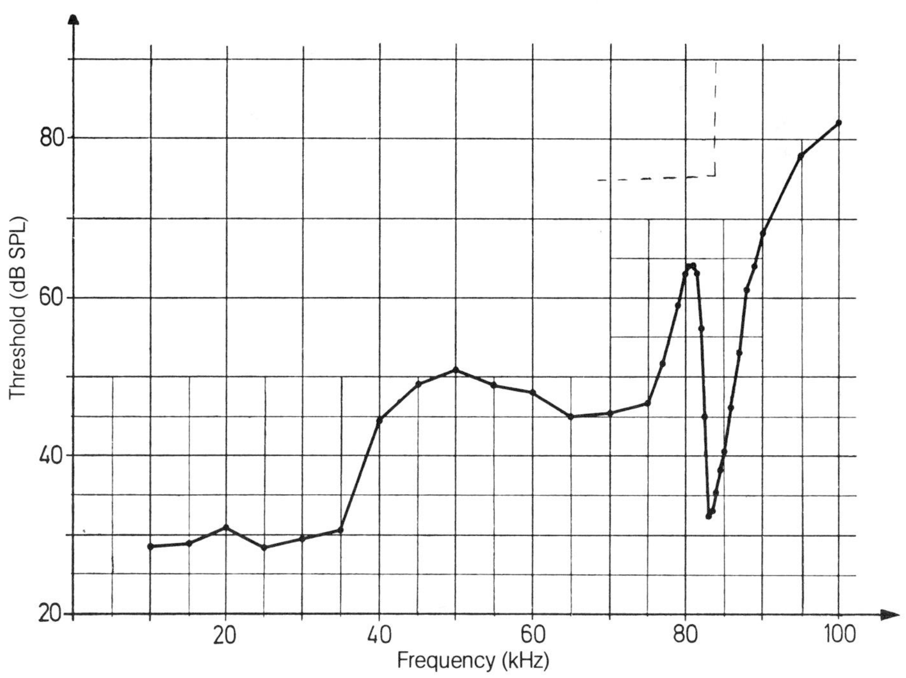

Fig. 123. The auditory threshold curve of Rhinolophus ferrumequinum. Each point represents the average of the responses recorded from the inferior colliculus of 10 animals. (Neuweiler, 1978)

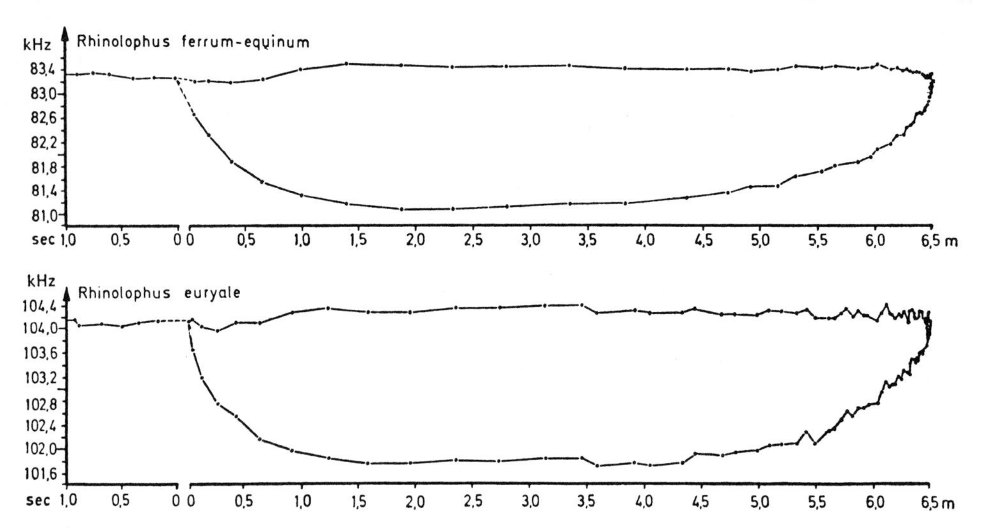

Fig. 124. Frequencies in the CF part of the echo (upper curve) and in the sound emitted (lower curve) by Rhinolophus ferrumequinum, while flying over a distance of 6 m and in the last second before the start (o—o). The Doppler shifts of the echo frequency by the bat's own flight speed are compensated by lowering the emitted frequency, so that the echo frequency is kept constant at ca. 83,4 kHz. (Schnitzler, 1978)

on the basis of this modulation. In fact, bats are quite capable of discriminating between falling leaves and insects; a brief sounding suffices. The precision of such discrimination, and the kinds of information contained in the returning echo, are among the focal points of present-day ecological neurobiology.

We all know that bats are skilled fliers. Against an animal with so many adaptations to the nocturnal life, no moth or beetle flying by night should have a chance. But they do, for during evolution these animals have become modified along with the bats. What possibilities are there? One is simply to taste bad – a feature that

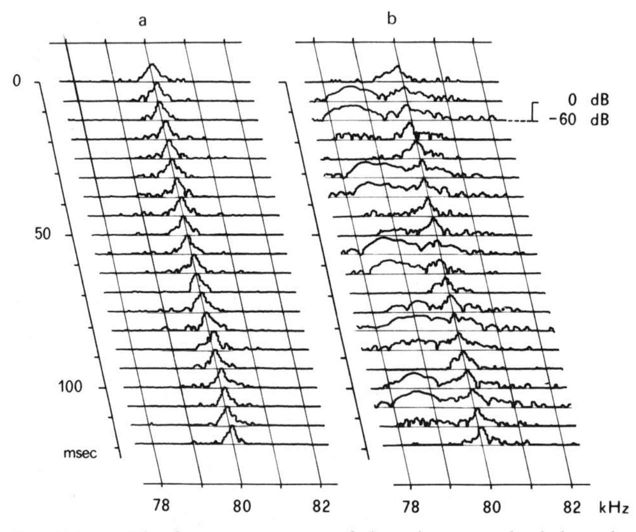

Fig. 125 a, b. The frequency spectra of the echoes sent back by a hawk-moth (Daphnis nerei) depend on whether the moth is stationary **a** or in flight **b**. The echoes of sounds with a carrier frequency of 80 kHz, presented in front of the insect at an angle of 45° to the long axis, are plotted at 6.25-ms intervals. (Schnitzler, 1978)

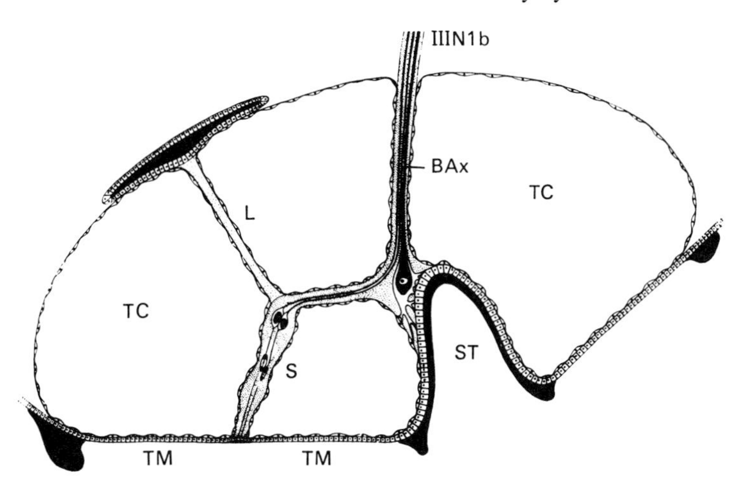

Fig. 126. Diagram of the tympanal organ of an owlet moth (Noctuidae). The sensillum *S* contains a pair of auditory receptors, or A cells. It is attached at one end to the tympanal membrane *TM*, and within the air-filled tympanal cavity *TC* is supported by the ligament *L* and the nerve from the sensillum. At the skeletal strut *ST* the A fibers pass close to the B cell; then they run together with the B axon *BAx*. These three fibers together form the tympanal nerve III N lb. (Roeder, 1968)

we know protects many insects. But that is of use to the moth only if it can let the bat know in time. The sounds that many moths make during flight have been the subject of much discussion; it may well be that they serve as an auditory warning signal. There are even instances of acoustic mimicry; some species that are evidently highly palatable also call as they fly. A second possibility, in theory, is to distort the bat's calls. This method appears not to have evolved; it would require very high and specific sound energy. A third possibility is to reflect the signals so poorly that the echo is useless for localization. Many have interpreted the very dense coats of hair covering some bombycid moths in this way, but they have not been shown to have such a function. A particularly interesting possibility is that of hearing and responding to the sounds of the bats. For many years zoologists investigating the hearing of insects have faced problems as great as those studying bat orientation. Moths in the families Noctuidae and Geometridae were found to possess organs so structured that they could only be auditory organs, and they were located in a place where many insects do have auditory organs – at the waist, the junction between thorax and abdomen. But these animals gave no responses to tones, and there was no evidence that moths in these groups produce sounds. Now we know that these tympanal organs are specific ultrasound receivers (Fig. 126). This can readily be demonstrated. It is quite common in summer to see a few noctuid or geometrid moths fluttering about street lights. If one rattles a bunch of keys, or turns a cork in a wet bottle (so that it squeaks), ultrasound is produced along with the audible tones; the animals around the lamp panic. In the noctuid moths the tympanal organ has only three sensory cells, one of which (the B cell) probably functions chiefly in monitoring the state of the tympanal organ. The two others, the A cells, respond to ultrasound. Their range of sensitivity is not restricted to the frequencies in the calls of, for example, the large horseshoe-nosed bat. The moths must be able to detect the calls of a variety of bats, and these contain different frequencies. In general, maximal sensitivity extends to frequencies as low as 50 kHz.

But the evolution of the ability to detect such frequencies is not enough to allow a moth to escape a present-day bat. In par-

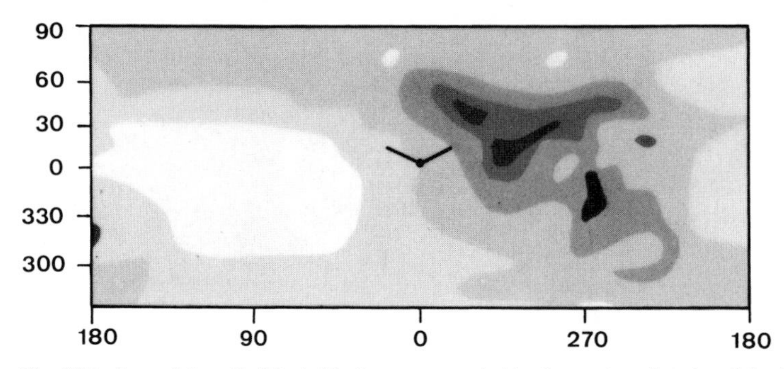

Fig. 127. An owlet moth (Noctuidae) was suspended in the center of a wire globe in such a way that it "flew" as though free (*middle:* flight toward the observer). Sound was then directed toward the right tympanal organ from various positions on the surface of the globe, and the intensity of the response was measured. The result is shown here in the form of a map, using the Mercator projection. The darker the area, the smaller the response of the tympanal organ to the stimulus (Roeder, 1968). The map, of course, applies only to the wing position shown here

ticular, the moth must discern the direction from which the calls come – an extremely difficult task when it is beating its own wings. Imagine that we fix a moth in such a way that it flies in the center of a globe; now we send sound waves toward its tympanal organ from various points on the surface of the globe, at various phases of the wingbeat. Sometimes the wing will completely cover the tympanal organ, and sometimes it will completely expose it. From the responses of the A cells to these sounds we can construct a "world map," using the Mercator projection of the globe (as is normally done in maps of the world), and so obtain a picture like that of Fig. 127. The picture varies with wing position, even though the sound source has not changed its position. Therefore calculation of the direction of the sound source must involve more than just comparison of the A-cell responses on the two sides of the body; the central nervous system must also take account of the position of the wings at the time the signal was received. Evidently the B-cell participates in this aspect of the analysis, along with specific sensory cells at the bases of the wings.

Having developed this apparatus, the owlet moth can detect the sound and localize its source. Now two possibilities are open to it, and the choice between them apparently depends on the intensity of the bat's signal – that is, on the distance between bat and moth. Either the moth flies away from the sound, or it lets itself fall in spinning flight almost to the ground. It is the latter response that we see when we rattle keys under a street lamp.

An echolocation system was developed to permit nocturnal activity, and has been applied to the task of finding food. This echolocation system is so highly specialized that it is outside the gamut of sensations we know. The animals affected by this system have developed means of defense – a bad taste, poor echo reflection, auditory organs in animals that produce no sound themselves. Here again we have a process that can be described as a wage-price spiral in evolution. The two groups have kept pace with one another.

In view of the large number of bat species that coexist in a given habitat, we can expect their food spectrum to be wide; it seems impossible that all these species would take the same food. If so, it is likely that each species has developed specific echolocation mechanisms – specific sounds and ways of processing these sounds, adapted to its particular prey spectrum – and, conversely, that the particular prey species have coevolved to deal with precisely these sounds. But so far this is all speculation; we do not even know from field observations what food is taken by the different bat species living together in any habitat.

4. Larus: the Fusion of Species

In many textbooks and handbooks the great gull (genus Larus) of Europe is taken as an example of speciation. Here we turn to this example once again, but from the point of view of ecology. The generally accepted view is that following the Ice Age there were two separate populations of Larus in the herring-gull line, forms with yellow feet in the Asian interior, and forms with pink feet in both inland and coastal regions of North America. The pink-footed gulls expanded their range within North America and across the Atlantic to Europe. Their descendants are our present-day herring gulls, Larus argentatus, with the subspecies argenteus in Ireland, the British Isles, Iceland, western Norway, France, Spain, and the Bay of Helgoland, and argentatus in the Baltic Sea. The same region was also invaded from the east, from inland Asia, by a group of yellow-footed forms. These proceeded by way of the Mediterranean (michahellis), the Atlantic islands (atlanticus), the North Sea including the Norwegian coast, the North Atlantic islands and Iceland (britannicus = graellsii), and Denmark (intermedius), as far as the Baltic Sea (fuscus). The forms fuscus, intermedius, and britannicus have relatively dark plumage, and are called black-backed gulls; sometimes atlanticus is included in this group. The black-backed gulls are given the status of a species, Larus fuscus, because in many areas they live side by side with herring

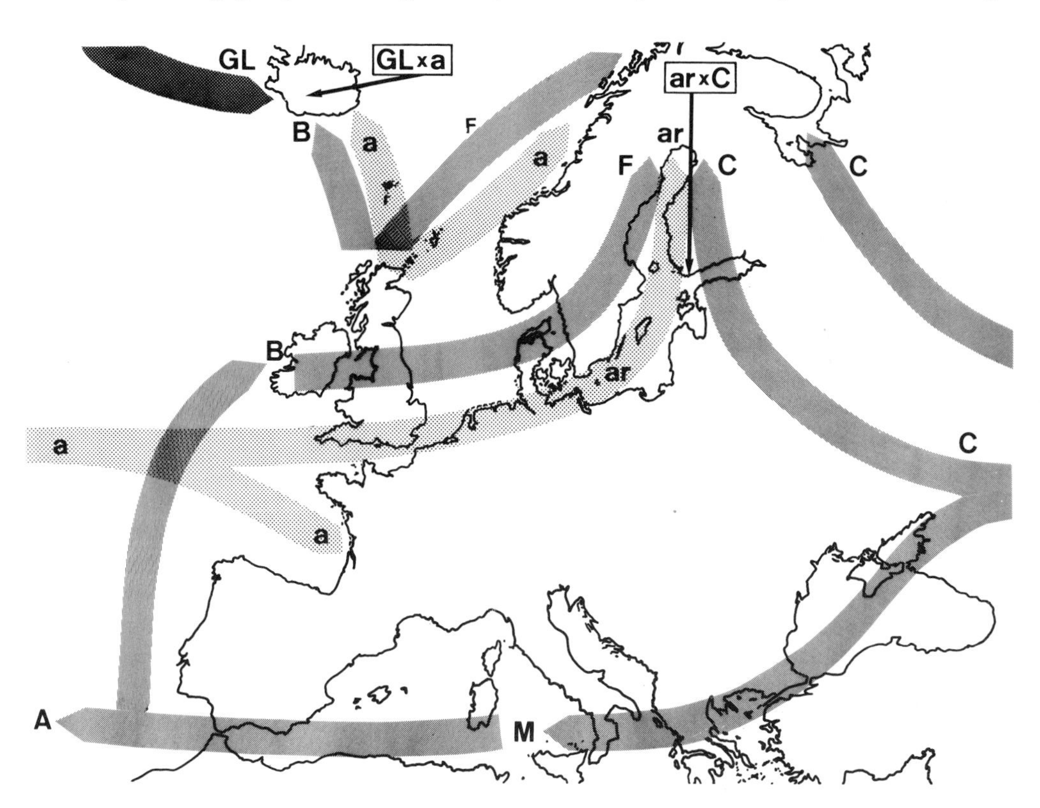

Fig. 128. The forms of black-backed and herring gulls that breed in Europe, showing the history of their ranges. The breeding grounds of Larus glaucus are also shown. Regions of hybridization are indicated by □ →. *Mid-gray, capital letters:* yellow-footed forms; *C* = cachinnans, *B* = brittanicus, *M* = michahellis, *I* = intermedius, *A* = atlanticus, *F* = fuscus. *Light gray, small letters:* pink-footed forms; *a* = argenteus, *ar* = argentatus. *Dark gray, 2 capital letters (GL):* L. glaucus (= L. hyperboreus). Combined from various authors

gulls with essentially no hybridization. The two are also distinct ecologically. Black-backed gulls predominantly inhabit rocky coasts and small rocky offshore islands, whereas herring gulls live chiefly on dunes. The eastern black-backed gull (fuscus) is a strictly migratory bird, flying overland as far as Lake Victoria, and the western forms (intermedius and britannicus) also migrate as far as equatorial Africa, along the coast. The herring gulls mostly stay in Europe.

None of this is particularly exciting in itself; in several cases species that were split into two groups during the Ice Age intermingled again in some regions and then proved to have become truly separate species. And this is just what has happened here; but most descriptions of the situation obscure the fact that there is an awkward problem. The forms moving westward through Asia also spread along the coast of the Polar Sea, to the White Sea, and along the rivers, moors, and lakes of the inland region south of the Gulf of Finland (Fig. 128). These were light-colored gulls, resembling herring gulls, but they had yellow feet and thus clearly belong to the group of black-backed gulls. There is no consensus as to the names of the subspecies here in the European region; many authors designate the whole group by the name cachinnans, and for simplicity this will be adopted here. About 100 years ago these cachinnans gulls advanced as far as Finland. Since then, the region around the Gulf of Finland and the Gulf of Bothnia has been inhabited by 3 forms of Larus – cachinnans mainly on moors and moor lakes, argentatus on the coast (particularly in dune regions), and fuscus on the coast and primarily on rocky islands. Argentatus and fuscus were also to be found, in small numbers, on the large Finnish lakes where there were suitable breeding sites. Hartet pointed out this fact some time ago, and in recent times it has repeatedly been emphasized by Voipio (1968, 1972).

It is a familiar fact that herring-gull populations expand where there is contact with human civilization, and this has occurred in the Bothnia and Finnish Gulf region. In the process fuscus was to some extent displaced by argentatus; argentatus tended to spread to the inland lakes and cachinnans to stay on the coast. Today the Finnish "herring-gull" population is a mixed population of argentatus and cachinnans. The mixing began about 40 years ago and is now nearly complete. There is no preferential pairing; panmixis between the two groups is absolute. Once more it must be emphasized that though the cachinnans group is most closely related to fuscus, mixing occurred not between these two close relatives but between cachinnans and argentatus – a more distant relative, but one similar in appearance and ecology. The ecological resemblance consists in the fact that cachinnans and argentatus have a stronger tendency than fuscus to feed in the coastal region and on land, and thus are more closely associated with human civilization. Fuscus primarily seeks its food on the open sea. Here, then, we have a fusion of two different populations while a third population, phylogenetically closely related to one of the two, behaves like an entirely separate species.

As the numbers of herring gulls multiplied, the species expanded from the southeast to colonize Iceland. The area was also colonized, at about the same time, by britannicus. The two are ecologically distinct; argentatus tends more to be associated with humans, while britannicus is an ocean-feeding bird. Two Larus species were already in residence in Iceland – Larus marinus and Larus glaucus (=hyperboreus). Both of these are a good deal larger than either argentatus or fuscus, and many systematists place them in a subgenus of their own. Marinus is usually nongregarious and not a follower of civilization; it seeks its food on the ocean or the coastal rocks. Glaucus, on the other hand, has become associated with man throughout the Arctic and, as in Europe, can be found on trash heaps everywhere. Glaucus and the much smaller new arrival argentatus had very similar ecological requirements.

Ingolfson (1970) has shown that the Icelandic population now is a hybrid of the two, neither glaucus nor argentatus remaining as a separate species. Only in the far northwest, not far from Greenland, are there perhaps still pure glaucus colonies.

Here, again, different populations – in this case actually quite different species (or even subgenera) – with barely distinguishable habitat requirements have fused to form a single hybrid population.

Examples of this sort in ornithology have now accumulated in appreciable numbers (e.g., Immelmann, 1962). There is also evidence of such hybrid populations among insects (Remane and Koch, 1977). The closely related cicada species Muellerianella fairmairei and M. brevipennis tend to hybridize in the locations (Netherlands) which both inhabit. The result is a triploid form comprising only females and resembling fairmairei. To reproduce, these triploid females must mate with males of one of the two original species. The sperm does not actually fertilize the egg, but simply initiates embryonic development of the triploid progeny. In the Netherlands, then, there coexist three forms which are entirely separate genetically and can pass on their distinct genomes indefinitely, although one is a product of crossing of the other two and permanently requires males of at least one of the original species (Drosopoulos, 1977).

Many fish are known to be similarly related. The Amazon molly, Poecilia formosa, which lives in coastal and inland waters from northeastern Mexico to Texas, is an all-female species (hence the name). The animals are hybrid in nature, with characteristics exactly intermediate between P. sphenops and P. latipinna. Formosa females must copulate to reproduce, but fertilization does not occur; the egg is merely stimulated to develop. From eggs that have developed in this way only females hatch. The gibel, a European variety of goldfish, is similar; at the periphery of the species' range one no longer finds both males and females, but females alone. Their eggs must be triggered into development by the sperm of related fish – carp or other Carassius species – but are not fertilized.

The theoretical significance of such relationships is considerable. In these cases the females of one species can be considered parasites of the males of another, and at least occasionally they must occupy the same habitat.

When species are spatially separated, mechanisms for sexual isolation offer no selective advantage. Such mechanisms as have been evolved for this purpose can now disappear. Once this has occurred, if the isolated species should reencounter the original species or a similar one intermediate forms and hybrid populations can come into being.

In zoology, one views this situation with mixed feelings; in botany it is an everyday matter. I have intentionally chosen a zoological example in order to spotlight a phenomenon that has received too little attention. It represents a method by which the genomes of two species each can be preserved for the future despite a broad overlap of ecological requirements, by fusion of the species. As such, it deserves further study.

The relationships among the Icelandic gulls can be summarized as follows. We find as breeding residents two closely related forms within a group of races (britannicus and argenteus), which ecologically and systematically behave as different species. A third species from the same group, not previously mentioned, appears as a winter guest. This is Larus glaucoides. It differs ecologically from the first two; with its long, narrow wings it is an open-ocean form that even from a distance appears conspicuously divergent. Finally, the island is inhabited by two species of a distant subgenus, marinus and glaucus, which are quite distinct systematically from the others. But two of the systematically separated forms, argenteus and glaucus, which are ecologically similar, have fused to form a hybrid population.

D. Ecosystems

D. Ecosystems

I. Theory of Ecosystems

The earth continually receives light energy, which is changed into chemical energy and then, in several steps, into heat energy. This flow of energy drives a circulation of many substances from the abiotic environment through the organisms and back into the abiotic domain. Within the brief time span that humans can survey, a rough equilibrium seems to prevail in the ecosystem – between uptake and release of materials, and between input and output of energy.

Because the dose-response relationship of organisms is nonlinear, their distribution over the earth tends not to be smoothly graded. The relatively abrupt transitions between one set of organisms and another can be used to demarcate subsystems within the ecosystem Earth, and the state of equilibrium within these subdivisions can be studied separately. The definition of such a lower-level ecosystem is like that of a population – the boundaries are not distinct. The system must be specified to suit the particular purposes of the study.

In theory, the minimal requirement for a system is one organism that produces chemical energy from light energy and another that reconverts this chemical energy to heat. The situation in nature is never that simple; a large number of species are involved. From the general requirements of organisms discussed in the theoretical introduction to autecology (pp. 5–6), we can infer that the process of speciation occurs more rapidly and extensively in regions generally favorable to life than at the biological frontiers. The number of species on tropical coral reefs and in tropical rainforests – habitats with about the optimal temperature – would be expected to be higher than in habitats where conditions are inconstant or at the limits of the biological range. In the former regions, therefore, we would expect intraspecific and interspecific competition in the broadest sense to be the most important ecological factor operating at present, whereas in the limiting regions abiotic factors would currently play the greatest role.

During evolution, organisms have progressively improved their adjustment to their environment – both the abiotic and the biological environment. In the latter regard, we find that coevolution has occurred, as discussed in the section on population ecology. We must assume that the organisms within an ecosystem have evolved together and have affected each other's evolution – that there is a mutual great coevolution of the members of an ecosystem. This evolution must necessarily have progressed toward optimization of the ecosystem. It must proceed in this direction, but need not have reached the optimum point.

The ecosystems on earth vary greatly in age. The oldest still in existence are probably the tropical coral reef and the tropical rainforest. Both are characterized by large numbers of species and extreme constancy, by rapid turnover of matter and energy, by the binding of almost all substances necessary for life in the organisms themselves, and by a complete recycling of the biologically essential materials. Younger systems contain fewer species, and often (in high moors, for example) recycling is incomplete.

In the course of optimization, which we can now describe more precisely, evolution proceeded from systems with few species and relatively pronounced oscillations in both the quality and the quantity of the elements, to systems in which many species are represented and the qualitative and quantitative fluctuations are relatively slight. Each species in the latter systems is represented by but a few individuals; this implies high turnover, as we have seen in considering predator-prey systems. The coevolution of various species was required to achieve this state. Forests with many species, for example, are possible only if instead of wind pollination there is a better-targeted pollination by animals, and if the seeds are dispersed by animals rather than drifting on the wind (Regal, 1977).

But basic physiological and biochemical facts like those presented in the first chapter could not be overcome by the coevolution of plants, ectothermic organisms, and warm-blooded animals. For example, it is a priori impossible for plants and ectothermic animals to respond in parallel to a particular change in the weather; cooling affects ectothermic animals much more than the plants on which they feed. Because physiological and biochemical responses to changing environmental factors are superimposed on the mutual adjustments of plant and animal, parallel oscillations in response to a cold or warm summer – if they occur at all – must be very rare. This fact must always be kept in mind when analyzing ecosystems (or populations). Moreover, every discussion of plant sociology and vegetation science (Willmanns, 1973, Ellenberg, 1978) teaches us that the composition of plant communities in central Europe can vary greatly, depending on the climate. The coevolution of the members of a system, at least in the young systems of central Europe, is completely suppressed by autecological and physiological phenomena. This applies even more to the animals in these systems. Our ecosystem theory, based on coevolution and optimization, is to a great extent postulated for very old systems – it is a postulate for the direction in which ecosystems evolve.

Another aspect of optimization is the complete exploitation of all resources by these systems. In old systems with many species all the nutrients are fixed in the organisms; there are essentially none in the abiotic parts of the system. For this reason, disruption of the system causes irreversible damage. Old systems have not evolved to cope with such offences. In young systems the nutrients are not so fully exploited; as a result, young systems are better buffered against total destruction. They are more resilient, and can regenerate.

The coevolution of organisms does not necessarily imply a permanent relationship. "New technologies" that arise during evolution can fundamentally alter systems in a relatively short time. The return of the teleosts from fresh water to the sea, the development of warm-bloodedness, the migration of mammals into the oceans, and the evolution of seed-bearing plants probably all fall into this category. In each case an ecosystem that had long been in existence and was quite stable was suddenly unbalanced and slowly restabilized as a new system.

Many ecosystems evolved together with the human race (in central Europe, for example, where after the Ice Age humans immediately began to affect the systems and their development). Here man is an element in the ecosystems, and the other elements have coevolved with him. Regions that have been under human cultivation for a very long time are therefore subject to the same considerations as so-called natural regions in the same geographical area.

In all these discussions we must not forget what was pointed out in the first section of this book – that organisms must be physiologically adapted to their habitats. This is the primary requirement. In relatively young systems physiological adaptation can predominate over all other factors. For example, the Baltic Sea acquired its present form and its present salinity very recently, only about 2,500 years ago. Before that time it had been a basin of nearly fresh water, with a fresh-water fauna. The brackish water of the present day could be invaded only by certain species from the North Sea, those capable of adequate osmoregulation. These animals, selected practically at random on the basis of a single physiological parameter, make up the fauna that characterizes the Baltic Sea today. It is self-evident that in this habitat, which we can regard without reservation as being in equilibrium, no coevolution of the animals could have occurred. That is, it is possible for a habitat to be in balance even without appreciable coevolution of its elements.

Moreover, the idea that in old tropical regions all the conceivable niches are occupied by specially adapted forms is certainly wrong. For example, in the tropics there is hardly any bird that has evolved in the same direction as the sandpipers of the subarctic and arctic. These cold-adapted species are thus the only ones that can colonize the tropical sandy beaches and coral reefs off the tropical American or African coasts. The fact that no indigenous tropical bird species utilizes this habitat contradicts several theoretical constructs of ecologists.

And finally, ecological systems are not related to one another as organisms are. Each ecological system is unique and cannot be repeated. The processes we have analyzed in great detail in one system may be quite different in another, even in one that is superficially very similar. Here is a real difficulty in ecosystem research, and a severe obstacle to the establishment of general rules.

II. "Natural" Ecosystems

The naïve term "natural ecosystems" is often used for those that are or were not subject to human influence. Today no such natural systems exist. The early settlers were certainly responsible for the extinction of the giant armadillo and giant sloth in South America (which followed a way of life similar to the ungulates of the African steppes) (Martin, 1973). They were responsible for the death of the large prosimians and giant ostriches on Madagascar, and for the disappearance of the giant ostriches and many other animals in New Zealand. They transported the dingo to Australia, where it returned to the wild and without doubt fundamentally altered the Australian fauna. The advanced civilizations in man's early history became established in regions relatively easy to colonize – not in forested or rainy regions, but in steppe regions with a relatively dry climate but sufficient water for agriculture and domestic animals. These were delicately balanced habitats, and proved highly susceptible to irreparable damage. As a consequence of their destruction there was less precipitation, and the land became a desert – in India, North Africa, Mesopotamia, Asia Minor, and parts of China. It is said that the decline of these civilizations was due to climatic changes; this is correct, but the climatic changes were at least in part due to disruption of the habitat by man. The Vikings moved out of Scandinavia to Russia, the Black Sea, England, France, the Mediterranean, Iceland, and Greenland, when they had exhausted the supply of wood in Scandinavia. Today one must look long and hard to find trees in Norway of the size they used to build their churches and ships. As early as the Bronze Age the forest in northern Germany had been destroyed, and the formation of the Lüneburger Heide began (cf. p. 261). At the end of the Middle Ages man turned his attention to the sea; the stocks of birds, seals, whales, and fish were decimated, first near Europe, then in the North Sea, and finally in the Antarctic. Only very recently have European marine birds begun to win back territories where they were exterminated at that time. Today we can but speculate about the significance of the whales in the high seas.

Nowhere, then, can we speak of a "natural" ecosystem. The problem must be rephrased. In categorizing ecosystems, there are a number of points to consider.

1. By autecological and population-ecological criteria, regions can be designated which at least in terms of vegetation are "close to nature." Studies of such regions can provide information about the way the components of a natural ecosystem can interact.
2. By comparing natural systems with systems long ago created by humans (the north German heaths, arid grasslands in Germany, the Mesopotamian desert, and many islands that have long been inhabited by man and the animals and plants he introduced), we can learn something about the extent to which a genuine "ecosystem" evolved with humans.
3. If pretechnological man, with relatively primitive methods, could turn fertile steppes into uninhabitable deserts, how long will it take for modern man with his elaborate techniques to destroy the earth as a habitat? Can he preserve this habitat? Can he correct the mistakes of earlier generations?

III. The Climax Concept: Microcosmic and Seral Successions

Natural and near-natural ecosystems are not the same thing as the climax community. The latter concept has been especially important in botanical ecology. Its definition involves the notion that under given climatic conditions, regardless of the nature of the soil, in the long term the same ecosystem must develop; the system ultimately established is the climax. The climax vegetation in central Europe would thus be a beech forest. Many botanists still put forth this argument, and claim that every community other than this climax has resulted from human intervention. In many cases nonclimax communities are so derived, but not always. There has simply not been enough time since the glaciers receded from central Europe for a climax vegetation, as defined above, to have come into being. Even without human influence lowland forests and swamps would not today be beech woods, but would bear vegetation including alders, willows, and poplars. Regions flooded for long periods each spring would be the same fens (and there are plenty of these in central Europe) today if there had been no humans. Furthermore, some animals exert an influence – the beaver, for example. As studies in Canada have shown, in regions where the vegetation is very near climax the dam-building activities of beavers can create extensive shallow lakes; when the dam breaks these become large meadows which are only very slowly reclaimed by the forest. Burrow-digging animals such as fox, badger, wolf, and bear have repeatedly provided habitats for "pioneer plants" (r strategists). Fires have certainly played a considerable role. Finally, the climax concept is a fiction to the extent that a forest under some circumstances can endure a drastic change in climate lasting 100 years or perhaps more, with no fundamental modification. A sequoia forest in which all the trees are between 1,000 and 3,000 years old can in theory withstand a too-warm or too-cold climate for 500 years; it can survive such a long period without producing any seeds, and is competitive enough to tolerate invasion by other trees to which this climate is more favorable. In this purely hypothetical situation one would expect a rapid flux of subcommunities to occur – a great variety of animals and short-lived plants – without changing the sequoia stand itself in the slightest.

Perhaps this argument is not as hypothetical as it sounds. We know that in historical times central Europe has been subjected to pronounced changes in climate. Between 1600 and 1700 there was a drastic deterioration in the vineyards, the cultivation of almonds in the Rhine region came to an end and saffron, which had been an important crop around the cities of the Rhine plateau, could no longer be grown. Mountain meadows that had served as pasture disappeared under ice. We do not know whether these events were associated with a general cooling or whether the climate simply became more maritime; either could have the same effects. Climax in this sense, then, is a fiction. A better approach, and the one that is now prevalent when speaking of such vegetation and the associated fauna, is to use the terms "site-adapted," "natural," or "close to nature." A factor that has been underestimated until relatively recently is fire (cf. p. 60f); it can play a major role in altering the potentially natural vegetation in a region. From what is known to occur elsewhere we must conclude that the European landscape was also strongly influenced by fire in earlier times, but that humans have always tried to suppress this influence. Neither the "climax" nor the "potential natural vegetation " gives a realistic picture of the landscape as it was before the large-scale intervention of man.

An ecological discussion in which only the "potential natural vegetation" is considered is therefore incomplete. Even in regions not affected by humans there are regular successions which must be analyzed in detail. In many presentations, small evanescent substructures are used as relatively uncomplicated models of such successions – carrion, dung, or the detritus thrown up on the shores of oceans and lakes. The decay of a tree trunk is such a microcosmic succession. Although the stages through which such systems pass are interesting, they tell us little about the actual sequence of events or the underlying principles in the process of succession on a large scale – after a forest fire or, a more important situation currently, after disruption by humans (the recolonization of fallow fields and the like). And one difficulty in comparing large-scale communities (seres) with microcosms is often overlooked. The sequence of events in a microcosm differs from a seral succession in that the latter can be arrested at a particular stage by abiotic factors or by the actions of man or animals. The decay of an animal body, a dung heap, or a tree trunk is simply that – a process of decomposition in which a certain amount of organic energy is available and is relatively soon used up. By contrast, the vegetation that first colonizes the soil after a forest fire, the first seral stage in the succession, can continue indefinitely as a self-sustaining habitat under certain conditions – if the shoot tips of the newly growing trees are grazed off, or as a result of regular mowing. Many and perhaps all of the dry grass communities in central Europe fall into this category, as probably do large parts of the Hungarian pusta, the Spanish steppes, and many regions in the zone where steppe and forest compete. If it had not been for the bison, the North American oak savanna might have been a dense oak forest; without the large land mammals that range the African steppes, larger areas there would be forested.

The process of succession has often been described. One such example is a shallow lake which first fills up with reeds (Phragmites), then becomes a fen with moisture-loving trees such as willow (Salix) and alder (Alnus), and eventually is a forest. In central and northern Europe the first colonists of previously forested land, following fire or clear-cutting, are grasses and birch trees, with seeds easily dispersed by the wind. Tree pipits (Anthus trivialis), woodlarks (Lullua arborea), and yellowhammers (Emberiza citrinella) take up residence. The birches grow rapidly and the land is soon covered by brush; warblers (Sylvia and Locustella) appear. When this brush stage has reached a height of about 2–3 m, consisting predominantly of conifers, the crested tit (Parus cristatus) becomes common. All these bird species disappear later, when the tall forest, with sturdy trees, becomes established; now others move into holes in the trees, while large birds such as the black stork (Ciconia nigra) and birds of prey (Aquila) breed in the tree crowns. In water firm substrate material, to which animals and plants can attach, is always a factor in minimum. When a solid object is newly introduced it is colonized in a similar way. The first to appear are unicellular algae; because of their highly motile zoöspores these spread rapidly. Where algae are established animals settle – primarily chironomid larvae. But occasionally, depending on the particular situation, a solid object in fresh water can all at once become completely covered by larvae of the zebra mussel (Dreissena). Larger species then make a place for themselves between these first colonists; in fresh water these may be caddis-fly larvae and filamentous algae.

The list of examples could be continued indefinitely. Are there general considerations that would help us to understand these regular, characteristic sequences? There are indeed. If we compare only the first stage and the late phase (equivalent to the climax), we see that the situation corresponds to Table 7. The number of species in the early phase is small and they are in the r-selection category; a number of associated features are also found. But this distinction, in itself, is too simple. Evidently another phase is fairly regularly interposed between the early colonization and the climax; this stage is extremely interesting, although it has only recently found the status it deserves in discussions of succession. It corresponds roughly to the stage represented by a central European dry grassland, or by the steppes created by man in Hungary or Spain. The communities here are extraordinarily constant and resistant, with a vast number of plant and animal species (i.e., a high diversity; cf. p. 190).

When the climax stage is reached the diversity is no longer so great. The diversity of the coral reef in the Gulf of Aqaba (northeastern Red Sea) is distinctly higher than that of the Great Barrier Reef, which is much larger and in a much more favorable region. This phenomenon has been explained by the fact that in the Gulf of Aqaba the water level is relatively low at irregular but not too long intervals; then, because the sunlight is so strong, part of the reef dies. The coral reef here therefore never reaches the stage of maturity, in which the number of species would be less. Though this explanation is at first surprising, it appears to be generally valid. It also applies to the introduction of solid objects to ocean or fresh water; as described above, animals and plants settle on such objects. There is a characteristic curve of low diversity at the outset, very high diversity at an intermediate stage, and lower diversity in the terminal, henceforth more or less constant stage.

Our theoretical ideas about succession are thus built on a fairly sound foundation, but as far as practical applications are concerned enormous gaps remain. Agricultural areas that were given up around 1945 present a most confusing picture. In the absence of any human activity quite different communities have developed on what was originally cultivated ground – in some cases typical dry grassland, in others dense forest with trees that have grown to a height of almost 20 m, and in still others low brush communities. And these regions are intermingled like a mosaic; each is several hectares in area, but in view of the soil and the climate one would expect the successions on all of them to be uniform and synchronous. This example from central Europe gives us something to consider whenever we try to use our theoretical knowledge for some practical purpose – as we must do in the course of rural planning and the creation of conservation areas. We are confronted with problems we had not anticipated, which can be solved only by well-designed and prolonged research. Some of the intermediate stages, with a great variety of species, seem very well able to resist supplantation by the actual climax species, with no assistance from man or the large animals; here the succession can be halted for decades. Such is probably the case in many dry-grassland areas of central Europe. When the normal complement of animals – from protozoans to insects and water birds and mammals – is present, the conversion of a lake to dry land proceeds considerably more slowly than is usually supposed. Moreover, the beech forest that takes over arid grassland and represents the climax stage in such areas is much less diverse and less differentiated than the arid grassland it has replaced.

The length of time over which a change in a habitat can remain noticeable is illustrated by the following example. The Aavasaksa Mountains (Fig. 129), north of the Gulf of Bothnia, on their upper slopes bear a conspicuously rich and dense taiga vegetation – spruce, pines, birches and poplars, with abundant undergrowth and ground cover. The lower slopes are almost bare by comparison; there are extremely sparse stands of pine, with here and there a stunted spruce among them. Even from a considerable distance this surprising difference is obvious. Normally, and especially just south of the Polar Circle, one would expect to find just the reverse situation. The explanation is not easily guessed – one must climb the mountain to discover it. At the lower altitudes the ground consists of granite washed bare of soil, so that few trees can find a foothold. Above this is a narrow, clearly delimited zone of rough rounded lumps of granite, and directly above that is the zone of abundant vegetation. More than 9,000 years ago the Yoldia Sea came up as far as the worn-down granite boulders; at one time its level was about 300 m above that of the present-day Baltic Sea. The granite boulders formed the shore wall, and the waves of the Yoldia Sea washed the soil away from the granite below. The sea never exceeded that level, so that the layer of humus already present at higher altitudes could continue to develop and eventually support the present forest. Although 9,000 years have passed since the sea receded, the zone once under water has still not regenerated its topsoil layer. Many similar formations can be found in this region. It is likely that tropical rainforest regions once denuded of trees over wide areas a[illegible] cover; they appear not t[illegible] In their place grows so-c[illegible] brush, which can be obse[illegible] many parts of South Americ[illegible] turn is destroyed, the region ma[illegible] resemble a desert. It is que[illegible]ble whether the old climax rainforest will ever be restored; in any case recolonization is impossible within a time relevant to human civilization. In view of this it is hardly likely that such regions can be utilized for agricultural purposes.

A fundamental dogma comparable to the climax concept has been developed in the area of soil science. This is that the soil develops independently of the chemical characteristics of the substratum – that is, independently of the geological history of the region. The gradual formation of the soil depends entirely on the prevailing climate and the associated vegetation. Once sufficient time has passed, the geological substratum no longer affects the distribution of terrestrial plants and animals; the current climate alone is influential.

This dogma is subject to the same criticism as the botanical climax concept; on the other hand, it has repeatedly been demonstrated that locations quite different in their original geological features are colonized in a remarkably uniform manner. In this context we should emphasize that humans have often arrested the process of soil formation, or even brought about a regression, by overexploitation of the natural vegetation and consequent destruction of the soil.

IV. Statics of Ecosystems

An ecosystem consists of abiotic and biotic components. The biotic elements – the organisms – consist of a more or less large number of species represented by varying numbers of individuals (Figs. 130–132). These two measures, number of species and relative frequency of each, have been used to characterize ecosystems since the

Fig. 129. The Aavasaksa Mountains in northern Finland. *Top:* primeval forest at 300 m altitude; *middle:* the shore wall of the Yoldia Sea at just 300 m; *bottom:* sparse pine woods on poor soil – the ancient floor of the Yoldia Sea

Fig. 130. A pictorial representation of a community of organisms: vegetation profile of the Ulmetum davidianae (Japan). (Miyawaki and Fujiwara, 1970)

science of ecology began. As early as 50 years ago Thienemann formulated the fundamental laws of biocenoses, as follows:
1. The more variable the environmental conditions, the larger the number of species present; there are few individuals of each species.
2. The more uniform the environmental conditions, the greater the tendency for a few species to dominate the picture; each consists of many individuals.
On pp. 5 and 6 we discussed the background of these findings. They are physiologically explicable and therefore can serve as a basis for predictions; we can expect them to apply as well to ecosystems not previously studied. But today we would formulate them somewhat differently.
1. The more diverse the environmental conditions, and the nearer to the basic biological optimum, the larger the number of species.
2. The more one-sided the environmental conditions, and the further from the basic biological optimum (though perhaps only temporarily), the smaller the number of species, with certain species more strongly dominant numerically.
But it is relatively hard to work with species number as a descriptor. To be sure of the results, must one really take account of all the species in an ecosystem – from microorganisms to mammals? How great is the risk if one selects only certain groups? Which species should be selected? If one expends enough time and energy one can count almost as many species as one wishes in a system, by including "accidental transients." Some plants grow from seed that happens to fall in the area even though they cannot survive in the long run. Winged insects and birds enter the system but do not stay there. It is extraordinarily difficult to determine whether a species regularly reproduces in a locality. In modern ecology attempts are made to overcome these difficulties by applying techniques of information theory to calculate the diversity of an ecosystem, on the basis of number of species and their relative frequency of occurrence. Diversity calculated in this way is low even with a system

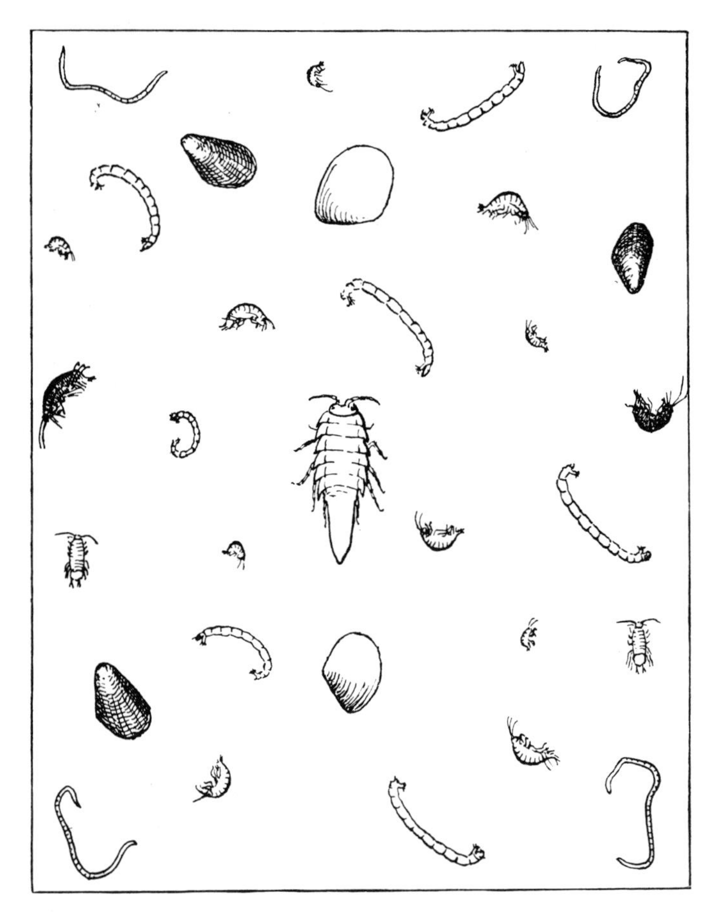

Fig. 131. Inhabitants of the soft floor of the northern Baltic Sea, at a depth of 9–10 m. In 1 m^2 are found Mesidothea, Pontoporeia, Gammarus, Asellus, Macoma, Mytilus, chironomids, and Tubifex. (Remane, 1940)

comprising very many species, if 99% of the individuals belong to one of these species (cf. Block 4). An ecosystem comprising relatively few species with approximately the same frequency of occurrence, on the other hand, has a relatively high index of diversity. That is, a mathematical trick allows one, under certain circumstances, to compensate for differences in the thoroughness of sampling, and the a priori vague concept "abundance of species" can be made more precise.

The actual effect of man's intervention in a natural ecosystem is to enhance certain preexisting factors. As a result, a diverse system becomes more uniform. When the structure of a system is monitored over a period of time, beginning with the onset of human activity, one finds that the consequence of such activity is quite generally a diminished diversity. This is an important finding, particularly where conservation measures are concerned. The number of individuals representing the various bird species is often greater, per unit area, in many urban (and particularly suburban) locations than in most of the more natural systems; and in many cases even the number of species is no less in the suburbs than in habitats still close to nature. However, in such suburbs blackbirds, house sparrows, and great tits account for more than 90% of the population, so that the diversity of these areas is low. Lakes overloaded with nutrients from sewage have a waterbird fauna that in number of individuals

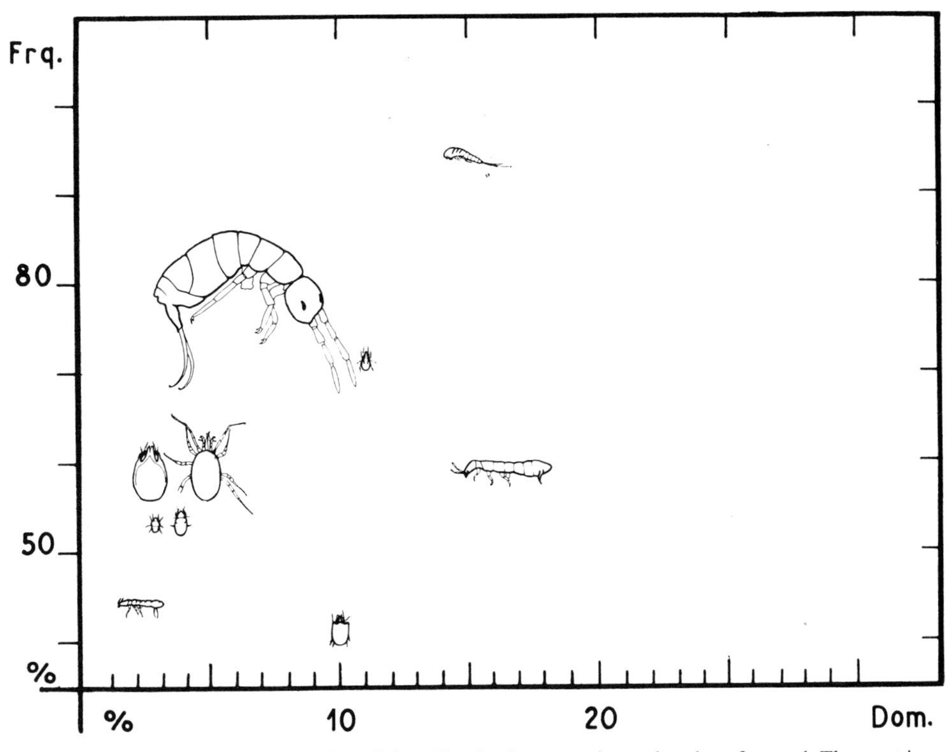

Fig. 132. A quantitative representation of the soil-animal community at the edge of a pond. The most important animals are arranged according to frequency of occurrence (the percentage of samples containing the animal) and dominance (relative number of individuals as compared with the other species). (Haarlöv, 1960)

far exceeds that of neighboring natural lakes (Fig. 133), and in many cases more species are represented. But again the great majority of individuals belong to only a few species – in this case, swans and coots account for more than 95% of the population. Diversity is thus lower here than in unaffected lakes. The story is repeated in fens and in the heavily fertilized green belts derived from them. The same situation arises when the sessile animals on hard substrates in clean and polluted rivers are compared. Here diversity is a good measure of the level of pollution, and a powerful argument in discussions of the need for conservation. It provides a criterion by which one can distinguish systems affected by man from those closer to nature (Bezzel and Reichholf, 1974). But low diversity does not always accompany human intervention. Where the central European beech forests were logged over, the resulting communities (for example, the familiar dry grassland) are rich in both plant and animal species; here diversity has increased. In the succession of sessile organisms on newly available substrates diversity is at first low, then rises to a maximum, and finally falls off to the low terminal level.

All of which is to say that the principle, though valid, is not a universal rule. The number obtained in calculations of the index of diversity is not an absolute value, but a relative measure that conveys information only in comparisons; and the comparisons must be reasonable. The tropical jungle of South America contains a much greater variety of plants and animals than the tropical jungle in Africa. The diversity indices of the two reflect this difference. But it would be nonsense to ascribe the disparity to differences in human influence. Moreover, any comparisons must be based

Block 4. Computation of the diversity index by the method of Shannon and Weaver

Many formulae have been proposed for such calculations. That of Shannon and Weaver is most often used:

$$H_S = -\sum_{i=1}^{S} p_i \ln p_i$$

where H_S = diversity
S = number of species in the group
p_i = relative frequency per unit area (abundance) of the ith species, on a scale from 0.0 to 1.0 (e.g., if the species considered is the second most common it is characterized by $i=2$, and if 10% of all individuals are of this species, $p_i=0.10$).
$\ln p_i$ = the natural logarithm of p_i (this is the most widely used, though other logarithms can also be taken).

The minus sign has been added so that H will be positive.

Example:

First bird group				Second bird group			
Species	p_i	$\ln p_i$	$p_i \ln p_i$	Species	p_i	$\ln p_i$	$p_i \ln p_i$
$i=1$	0.2500	−1.3863	−0.346575	$i=1$	0.500	−0.6932	−0.3466
$i=2$	0.2500	−1.3863	−0.346575	$i=2$	0.1250	−2.0794	−0.2599
$i=3$	0.2500	−1.3863	−0.346575	$i=3$	0.1250	−2.0794	−0.2599
$i=4$	0.2500	−1.3863	−0.346575	$i=4$	0.1250	−2.0794	−0.2599
			−1.386300	$i=5$	0.1250	−2.0794	−0.2599
			$H_S=1.3863$				−1.3862
							$H_S=1.3862$

A warning: In practice, no attention should be paid to differences only in the second or higher decimal place. For example, if the above example is worked out with an ordinary pocket calculator (10 places), exactly the same diversity is obtained for the two groups. It is not reasonable to make the computation more precise than the original data.

The problem lies in the way the data are originally acquired. At the very least, data must be collected with exactly the same methods (cf. Table 11), for it is essentially impossible to give a truly correct numerical description of abundance. For this reason, data from different publications on the diversity of habitats ought not to be compared. (Wilson and Bossert, 1973)

on data obtained with entirely consistent methods of collection, and the effectiveness of these methods must be tested in detail. When different collection procedures are used in a given region the resulting indices of diversity may have practically nothing in common (Table 11). So far the researchers on terrestrial systems have not been able to work out a single method that provides an objective picture of the frequency of all the animal species present. Moreover, comparisons should not extend over too great distances, for several reasons. One is that the number of species present may differ; another is that a given collection method may vary in its effectiveness under different conditions; the nights are cooler in more continental regions, and shorter in northern regions during the summer, so that methods involving attraction by lights do not work well during summer nights in the north. Some methods, if applied year after year, can bring about a genetic change in the population. When

Table 11. The numbers of animals living on the surface of the ground on Spitsbergen, measured with different types of ground traps. The calculated diversity is 2.6 or 1.5, depending on the method used

	Species	Trap type 1 Barber	Trap type 2 Ecclector
Acari	1	1	0
	2	8	1
	3	4	63
	4	66	51
Araneae	1	47	1
	2	10	4
	3	1	4
	4	1	0
Collembola	1	16	0
	2	7	0
	3	12	0
	4	5	0
Brachycera	1	5	0
	2	1	0
	3	0	1
Mycetophilidae	1	12	0
	2	1	0
	3	1	0
Chironomidae	1	5	0
	2	1	3
	3	0	2
Aphidae	–	8	9
Other groups	1	1	0
	2	1	0
Hymenoptera Tenthredinidae	–	1	0
Hymenoptera, parasitic	1	1	0
	2	15	3
	3	4	1
	4	3	1
	5	6	0
	6	3	0
	7	5	0
Total		252	144
H_S		2.6	1.5

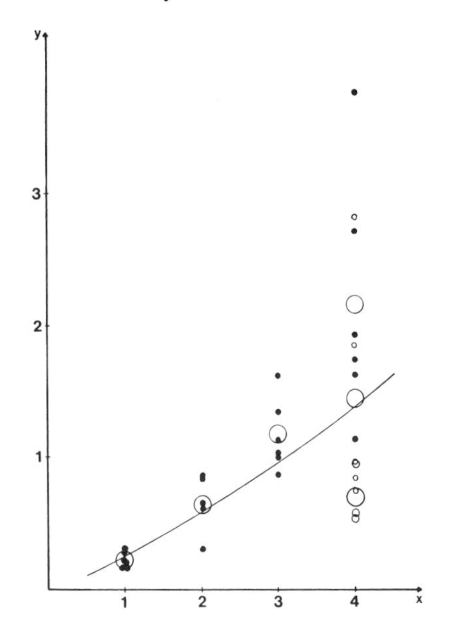

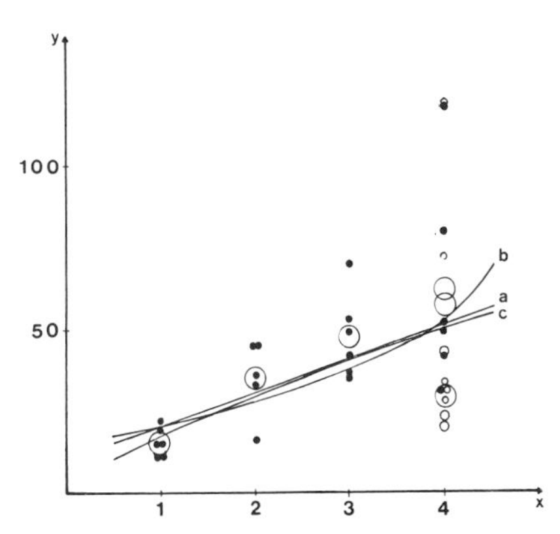

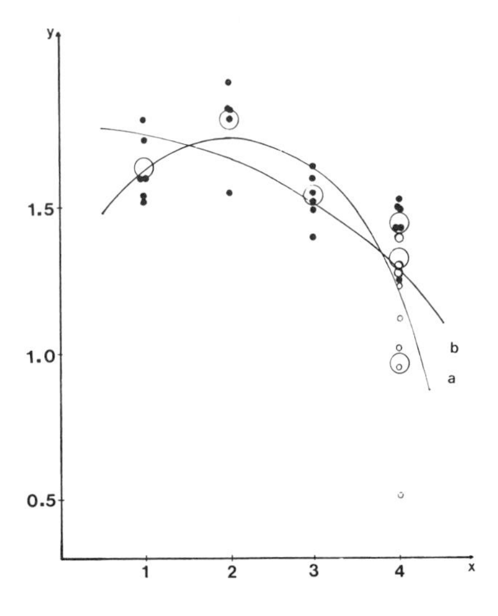

Fig. 133. Mathematical description of an animal community. *Top:* correlation between water quality (abscissa) and density of water birds (ordinate); *middle:* correlation between water quality and water-bird biomass per ha; *bottom:* correlation between water quality and the diversity of water-bird communities. All data from Upper Bavaria. *1* oligotrophic lakes; *2* mesotrophic lakes; *3* eutrophic lakes; *4* polytrophic lakes. Ordinate in top graph: water birds per km distance along shore (Utschik, 1977). The letters *a*, *b*, *c* indicate different methods for calculating the curves

light traps have been used regularly in the same region for a long time the capture rate becomes progressively less, but by using other methods one can show that the species no longer caught are still there. The genotypes that had been subject to the stimulus of the light trap have been largely "selected out" of the population. It is therefore not necessarily justified to postulate a reduction in diversity on the basis of light-trap captures over many years. Finally, the presence of both small and large species can obscure the picture. A small species is almost always more numerous than a larger species, so that cicadas and aphids frequently make it impossible to get a true measure of diversity. The young stages of plants and animals can also mask the real state of affairs. The objective mathematical method is no substitute for a reasoned approach in which counting and identification procedures are flexible. There is no patent solution to these fundamental problems.

However, if the precautionary measures described here are followed, the diversity index can be an extremely good indicator of the progress of a microcosmic or seral succession. It can be an excellent measure of the influence of human activities, for their effect quite generally takes the form of radical amplification of a few factors, which cause marked shifts in the frequency of occurrence of a few species. Diversity data have also proved very useful in the analysis of transects through nonuniform habitats. For example, the diversity of vascular plants in Israel falls off with increasing salinity; that is, from a value of 1.8 in the desert the index decreases as the Red Sea is approached, reaching 0.2 on very salty soil (Danin, 1976).

Many attempts have been made to find correlations among the various parameters of a system that can be so measured. Production has been correlated with dominance and with diversity, and both of these factors correlated with constancy (Fig. 134). When first presented these results were hailed as fundamental, generally

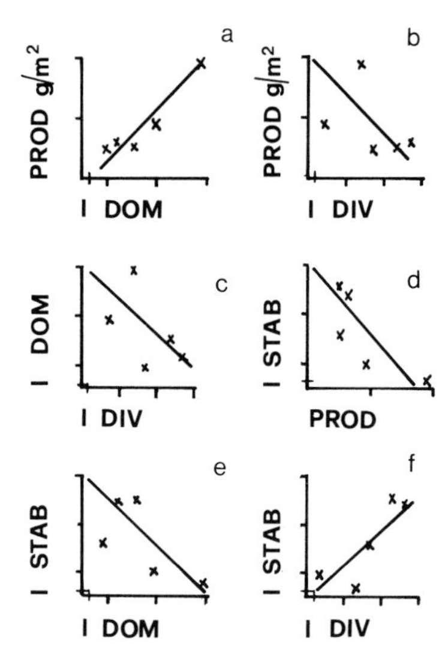

Fig. 134 a–f. An attempt to correlate various parameters of an ecosystem (grassland in India) with one another. The definitions and indices are no longer used in this way. Above-ground primary production, the only measure used here, is not adequate; it has not proved possible to generalize the relationships shown here. The graphs are reproduced here in order to illustrate the remarkable efforts that have been made to understand ecosystems as overall complexes. (Krishnamurthy, 1978; cf. McNaughton; redrawn)

valid laws governing biocenoses; since then, however, the euphoria has given way to a pronounced skepticism (McNaughton, 1967; Krishnamurthy, 1978).

V. Ecosystem Dynamics

1. The Cycling of Matter in Ecosystems

The circulation of biologically important substances depends on their absolute quantity and on the rate of turnover. When turnover is slow the soil may become depleted, if the materials (as happens in tropical rainforests) are stored primarily in living tissue. Slow turnover may give the impression of a very rich soil if the materials are stored to a large extent in the soil or in dead organic matter, as occurs in the humus stratum of European forests. A very rapid turnover with no loss of nutrients has the same effect as fertilization. The layer of dead plants is rapidly broken down to carbon dioxide and water, and the percentage of important plant nutrients such as nitrogen and phosphorus rises. These elements thus become immediately available to the plants. If these plants are regularly grazed, the dung and urine of the animals directly returns to the soil the nutrients that have just been withdrawn. Amount and rate of turnover, then, are problems of the first rank.

However, the word "turnover" does not reflect the full complexity of the situation. The overall turnover in a tropical rainforest is very slow, because most of the material is stored in the large trees; but at the moment when anything at all becomes available turnover occurs with extreme rapidity. As a result, there are essentially no decaying plants or animal structures on the forest floor, and there are no utilizable minerals in the soil. Everything is immediately incorporated into living tissue.

a) The Circulation of Water. Figure 135 illustrates the circulation of water in a region corresponding to the German Federal Republic. About half of the precipitation here comes from water that had evaporated over land, directly or by way of transpiration by plants. The other half comes from the ocean. The illustration does not specify how much fresh water there is on land. This amount can be less than the precipitation of terrestrial origin that falls onto the land in the course of a year. If the turnover of water were very rapid, each drop would evaporate several times and several times return to the earth as precipitation. But it is also possible for the amount of precipitation to be considerably less than the amount of inland water. In this case, turnover would be relatively low. We can learn something else from this illustration, however. When we reduce the absolute quantity of water available on land, then correspondingly less evaporates and returns as precipitation. If turnover is high with respect to the total quantities involved, a reduction in the water has a multiplicative effect in reducing the amount of rain. In Europe, a relatively small area of land surrounded on all sides by ocean, this may play a relatively small role. But in very large land masses far from the ocean, reduction in the absolute amount of water can have dramatic consequences – especially where it is always hot, with a correspondingly high turnover of the available water. If the absolute amount of water in large regions of land is reduced, the amount of rain per unit time falls off exponentially. Because hardly any of the rain falling in large land areas is derived from evaporation of sea water, the rainfall per unit time is directly dependent on the ter-

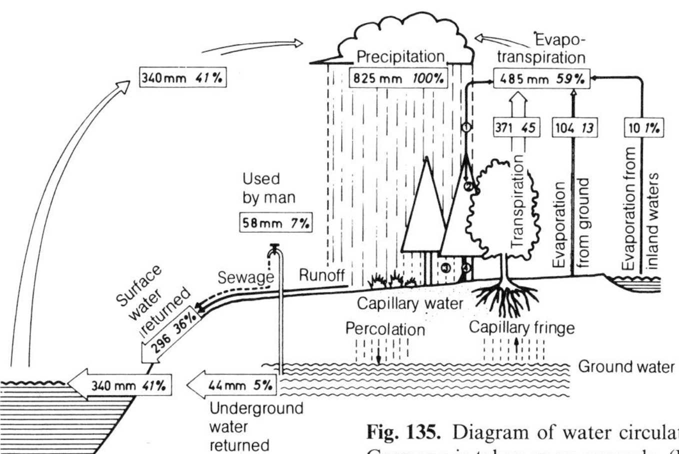

Fig. 135. Diagram of water circulation; the water balance in West Germany is taken as an example. (Larcher, 1973)

restrial water supply. When this water is channeled (to prevent floods and in the construction of urban sewage and drainage systems) so that it rapidly enters the rivers, and when the rivers are contained by dikes so that they are really more like canals, the absolute amount of water available for evaporation decreases and so does the rate of precipitation. Arid regions (that is, steppe regions) that form part of large continents are especially vulnerable to such a reduction in the absolute amount of fresh water. They turn into deserts. But the same effects are felt in the primeval forests of large continents – in the Congo or Amazon basins, for example. Here if the available water is restricted by channeling, as described above, the amount of rain per unit time declines exponentially. And there is another complication. The air temperature over barren ground rises to much higher levels than that over soil covered by plants. Rain falling over a hot desert often evaporates before it reaches the ground, and water circulating within the atmosphere in this way is permanently lost to the system. This is one of the great concerns of ecologists in all discussions of "developing" the last great tropical jungle regions on earth. The problems of water conservation in such places can be inferred from the diagram of Fig. 135, for the circulation of water in Germany.

Forested regions can affect water balance quite differently when the rain derives mainly from the ocean, as happens in western Scotland and in Ireland. The heavy rainfall here, where physical evaporation is relatively slight, can be absorbed into the system and processed only where there are dense stands of trees. Because of the long growing season the trees consume immense quantities of water. If the forest is cleared the rainfall is not used; it accumulates on the ground and forms extensive bogs which prevent regrowth of the trees. This has happened in many parts of Ireland and Scotland. Depending on the origin of the rain and on other climatic features (length of growing season, mean temperature, and so on), forests can be responsible for maintaining or reducing the water supply. It is impossible to find a simple solution at first glance.

In most inland regions the level of primary production is limited by water availability. Under the climatic conditions in Stockholm, according to the theoretical calculations of de Witt, about 2.5 kg/m^2 organic dry matter can be formed per year; under the conditions prevailing in Berlin, the figure is about 3 kg. Taking as a rule of thumb that 500 l water are required per kilogram dry matter, the plants in Stockholm and Berlin would consume 1,250 and 1,500 l, respectively, per square meter annually. Moreover, this water must fall at the right time (the precipitation in winter probably does not count). And there can be no evaporation from the soil, and none can be lost by drainage. In fact, however, only about 700 l of rain per square meter are available in Berlin for evaporation and drainage, over the whole year. It is quite evident that the water factor is strongly limiting; on the earth as a whole, it is a limitation that makes impossible any great increase in yield. Amounts of water sufficient to allow a marked improvement in the harvest can be provided, if at all, only with an extremely high energy expenditure (Figs. 142 and 143).

b) Cycling of Other Materials. The biologically important elements circulate through ecosystems. For each there is a very large reservoir; in the case of carbon, oxygen, and nitrogen this is the atmosphere, whereas phosphorus and sulfur are stored in the earth's crust. Some individual cycles are shown in Figs. 136–138. In part these cycles require a specific sequence of different organisms with different synthetic abilities, such as the nitrite and nitrate bacteria that fix nitrogen in the soil. The chain also involves organisms that fix aerial nitrogen – bacteria, actinomycetes, and probably certain blue-green algae. Where these are abundant, prolonged overuse of resources is less likely to cause severe damage than in regions where they

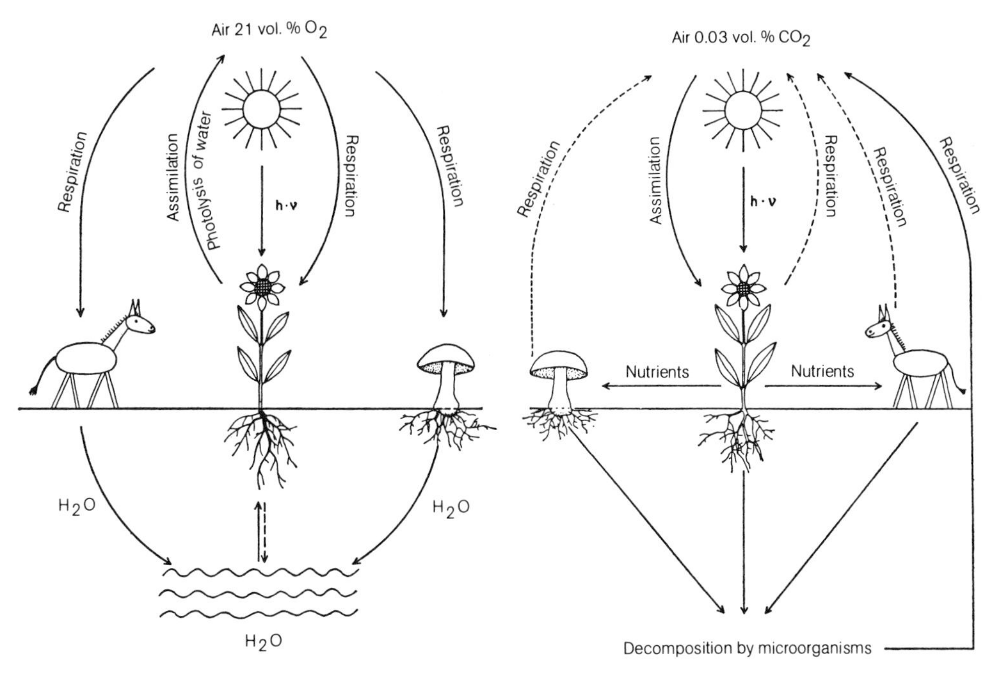

Fig. 136. The oxygen cycle. (Mohr, 1969)

Fig. 137. The carbon cycle. (Mohr, 1969)

play a lesser role. Stands of alder withstood the despoilation of the central European forests fairly well; nitrogen-fixing actinomycetes associate with alder roots. The beech forests also maintained themselves better than others, evidently because their epiphytic blue-green algae fix nitrogen in compounds that then are washed down the trunk by the rain (Denison, 1973). Not a great deal is known about the extent to which such microorganisms can provide other materials necessary for life. In this regard, again, the ecologist must worry about the uncontrolled use of chemicals, for they could exterminate a group of microorganisms crucial to our existence.

Another cycle – that of sulfur – has recently acquired considerable practical importance. Previously a scarce substance, sulfur has become a factor in excess because of the increased release of SO_2, by burning of coal and oil, in the population and industry centers of the northern hemisphere. Acid precipitation (containing sulfuric acid) has become the rule. In southern Norway an average of 4 g SO_4/m^2 per year falls onto the soil with the rain. The pH of the lakes has progressively been reduced; a region the size of Switzerland, over unbuffered granitic ground, has lost

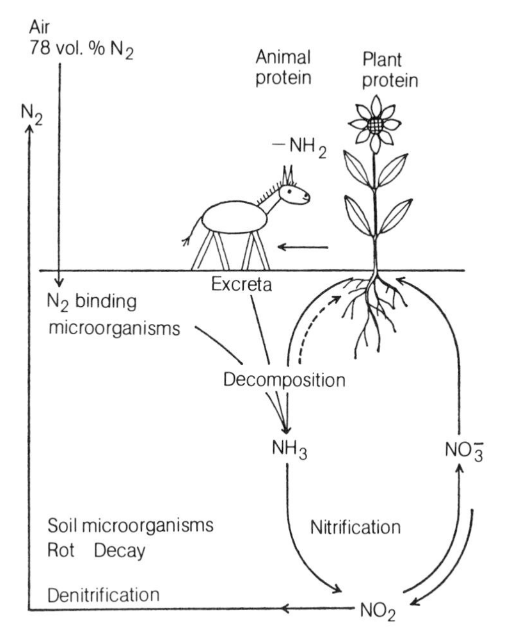

Fig. 138. The nitrogen cycle. (Mohr, 1969)

its entire stocks of fish. Because of leaching by the acid, the already poor soil is still further depleted of Ca and Mg ions, and as a result growth of the forest is impaired (Braekke, 1976).

Human intervention in the carbon cycle can cause a change in the climate of the whole earth. Most of the earth's carbon is stored in organic material. Forests especially, and the tropical rainforests in particular, are vast depots of carbon. The ocean is also a huge reservoir of carbon, most of which is bound in dissolved organic matter. Finally, a great deal of carbon is contained in fossil fuels. When these are burned, and when forests are destroyed on a worldwide scale, these stores are depleted. The chemical energy within them is used by man, and the carbon dioxide thus produced is released into the atmosphere. Since precise measurements of atmospheric carbon dioxide have been made in Hawaii it has been found steadily to increase – by about 5‰ between 1958 and 1976. Every summer there is a reduction, due to the photosynthetic activity of the plants in the northern hemisphere. But each autumn the carbon dioxide content rises sharply again. All climatologists are certain that this increase will profoundly modify the earth's climate. If the process is to be stopped, further destruction of tropical forests must be prevented – but how are hungry people to be convinced that this is necessary? Furthermore, there must be a moratorium on the use of fossil fuels; but at the moment the only alternative is nuclear power, and the consequences of developing this source of energy are not so well predictable as those of the traditional utilization of fossil fuels.

So far no one has succeeded in following these qualitatively presented cycles in complete quantitative detail, through all the microorganisms, animals, and plants of any ecosystem. But the fluxes of materials in a number of ecosystems can be roughly estimated. Figures 139 and 140 illustrate the annual flows of minerals in an old, mature oak-ash forest and in a young mixed

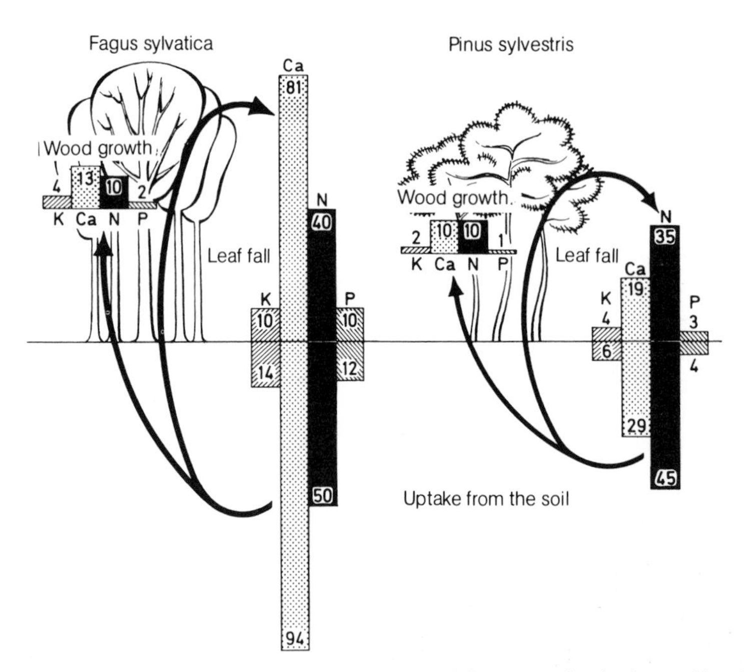

Fig. 139. Annual uptake (data below the ground line), retention in tissues (data in the crowns of the trees), and return (data above the ground line) of the most important minerals (in kg per hectare) by pine and beech trees. Note the large amount of nutrients returned to the soil in the fallen leaves and needles; this is the rotating capital of soil fertility in the forest. (Reichle, 1973)

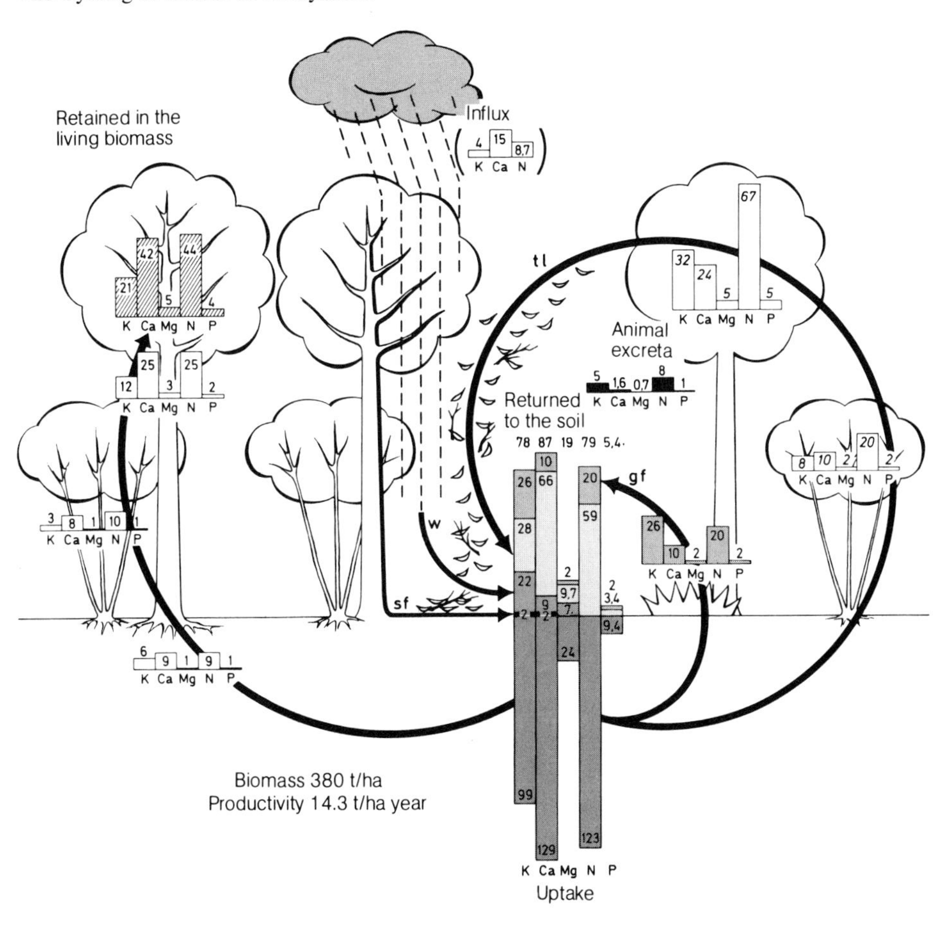

Fig. 140. Annual cycle (in kg per ha) of K, Ca, Mg, N, and P in an oak-ash forest with undergrowth of hornbeam and hazel. (Reichle, 1973)

stand of oaks in Belgium. Of course, numbers of this sort can give only a primitive idea of the situation. Moreover, in these cases the losses by leaching from the system, which are no doubt considerable, have not been allowed for. The fact that one species, a relatively small component of the system, can be of central importance to the system, is shown in Fig. 141. The mussels filter out of the open water large quantities of the dissolved phosphorus and that incorporated in particles, and deposit it on the surface of the sediment. In this form the phosphorus is more readily available to other organisms than when it is in the water. When the cycling of a substance is analyzed, then, the quality of single compartments in the system must be taken into account.

The above proposition, that a single species – not necessarily a common one – can crucially affect the balance of substances in a habitat, can also be derived from a hypothesis of Fittkau (1973). He observed that when the caymans in the estuarine lakes of the Amazon were shot the stocks of fish also declined markedly. He hypothesized that the fish had previously come from the Amazon, where food was relatively abundant, into the comparatively poor Rias lakes to spawn. Here some of them were eaten by caymans, and these animals provided a long-term uniform supply of nutrients to the water of the

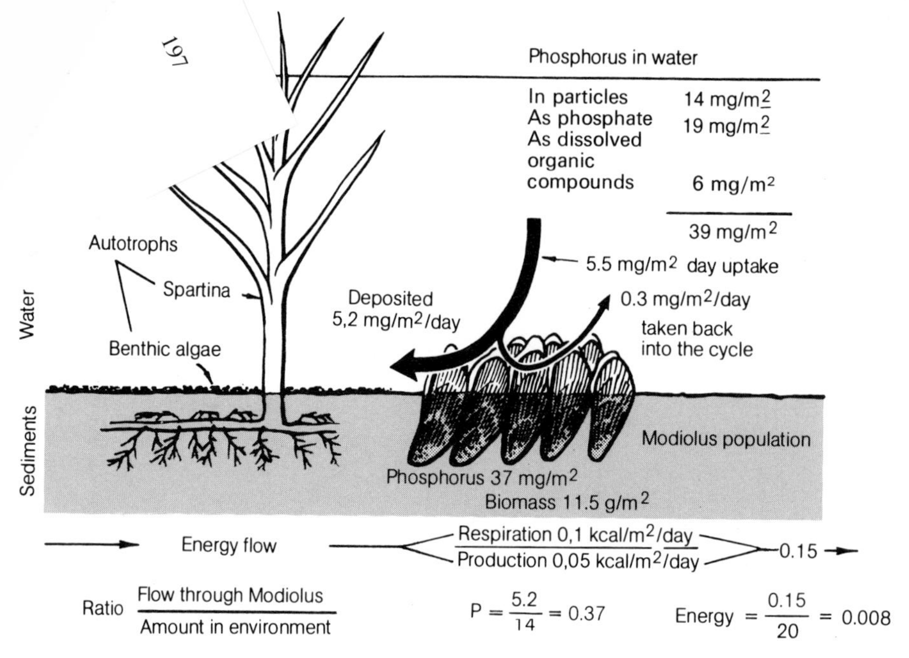

Fig. 141. The flow of phosphorus in a mudflat on the North American coast. The mussel population has quite a pronounced effect on the distribution of phosphorus, even though it is but a small component of the community in terms of biomass and energy flow. (Odum, 1967)

lakes in their excreta. These nutrients allowed a rich planktonic population to thrive, and this in turn served as the first food of the young fish. When the caymans were shot no more fish were eaten; all the fish returned to the main stream after spawning. The food supply in the lakes remained very poor, no rich plankton could develop, and thus only part of the fish brood grew to maturity.

In future we shall have to focus less on global nutrient cycles than on the special compartments in a system. And more attention must be paid to the eventual disposition of the nutrients brought to the ground in rainfall. As a result of industrialization, more and more nutrients enter the air and are then carried down by rain; overfertilization can occur, and in some places the consequences are undesirable. At present the high-moor vegetation in central Europe , under the influence of mineral-rich precipitation, is being progressively changed toward a meadow or forest vegetation. Sulfur-containing precipitation can be very hazardous, especially on acid soils and acid bodies of water. When there have been heavy falls of sulfurous snow during the winter, the spring thaw can suddenly lower the pH of large lakes to about 3. This naturally kills off all the animal and plant life. It is in this way that the lakes of a region of southern Scandinavia as large as Switzerland have lost all their fish; the lake bottom, of primitive rock, provides no buffering.

2. The Energy in Ecosystems

a) Productivity. Productivity is the amount of organic substance acquired by an individual, a population, or a system per unit time. The ecological concept, then, differs from that of agriculture, in which "production" is considered to be only that material utilizable by humans in some form. In the agricultural sense, any roots or above-ground parts that cannot be used as food or fodder are ignored. Many misunderstandings in discussions between ecologists and agronomists are based on this discrepancy.

A fundamental distinction must be made between the numbers of organisms in a habitat and the amount of organic matter they produce per unit time. A large population by no means implies high production. Indeed, the suggestion has been widely defended that a large population, a large biomass per unit area, is a typical strategy by which organisms adapt to low nutrient availability; a large population is better able to bridge periods when food is scarce than a small one. The turnover – that is, the production – of such a population is comparatively small. It follows from this hypothesis that small biomass is characteristic of habitats with fairly large amounts of available nutrients, and this small biomass should show a high turnover. Habitats with large biomass and limited available nutrients are thought to contain very many species, whereas those with little biomass and high turnover can house but a few species.

We can be quite certain that this hypothesis, in this simple form, is not generally applicable. It is presented here in order to underline the difference between biomass and production in a habitat, and because it is definitely worth considering, at least, for a number of habitats. In testing its validity, one will have to ask which kind of organism (plant, animal, microorganisms) the large population comprises, and here, again, divide the system into compartments.

The amount produced, as a rule, is smaller than the mass of the producing organisms. Consider a forest, for example. The amount of organic matter contained in the stand of plants has, under some circumstances, been developed over centuries; the production in a single year is distinctly lower. But where small organisms are concerned the relationships can be different. This is the case with bacteria, for example, and is an especially familiar feature of planktonic algae. The production of new algae per unit time is considerably higher than the stand of algae present in the body of water at any time.

Production falls into two major categories. One is primary production, by green plants and photosynthetically active bacteria, which produce organic matter from solar energy and inorganic matter. The other is secondary production, by organisms not capable of photosynthesis – bacteria, fungi, and animals, which convert the organic matter of the plants into the substance of their own bodies. Each of these categories can be subdivided into gross and net production. Using light, a plant produces organic from inorganic matter. This is gross production. But part of this matter is respired, used up by the plant's metabolism. Gross production minus respiratory losses is called net production. It is only the net primary production that we can realistically measure; gross primary production is usually simply a logical construct.

Gross production and net production by animals are easier to distinguish. The respiratory losses of animals are relatively accessible to measurement, and the factors determining them are relatively well known. Net production is of course the most important to ecosystem research, for it is this that can be passed on to the next trophic level. To humans, the animals of especial interest are those with net production not much lower than gross production. The ecological efficiency of animals has therefore been widely studied (cf. Table 4 p. 50).

For the reasons presented on page 54, net primary production (called simply primary production in the following discussion) is largely independent of the species of plant. Primary production of a system is affected chiefly by the duration of the growing season and by the water supply. In natural ecosystems it is usually a valid starting assumption that nutrients are present in adequate quantities. The nutrient poverty of many present-day cultivated areas has

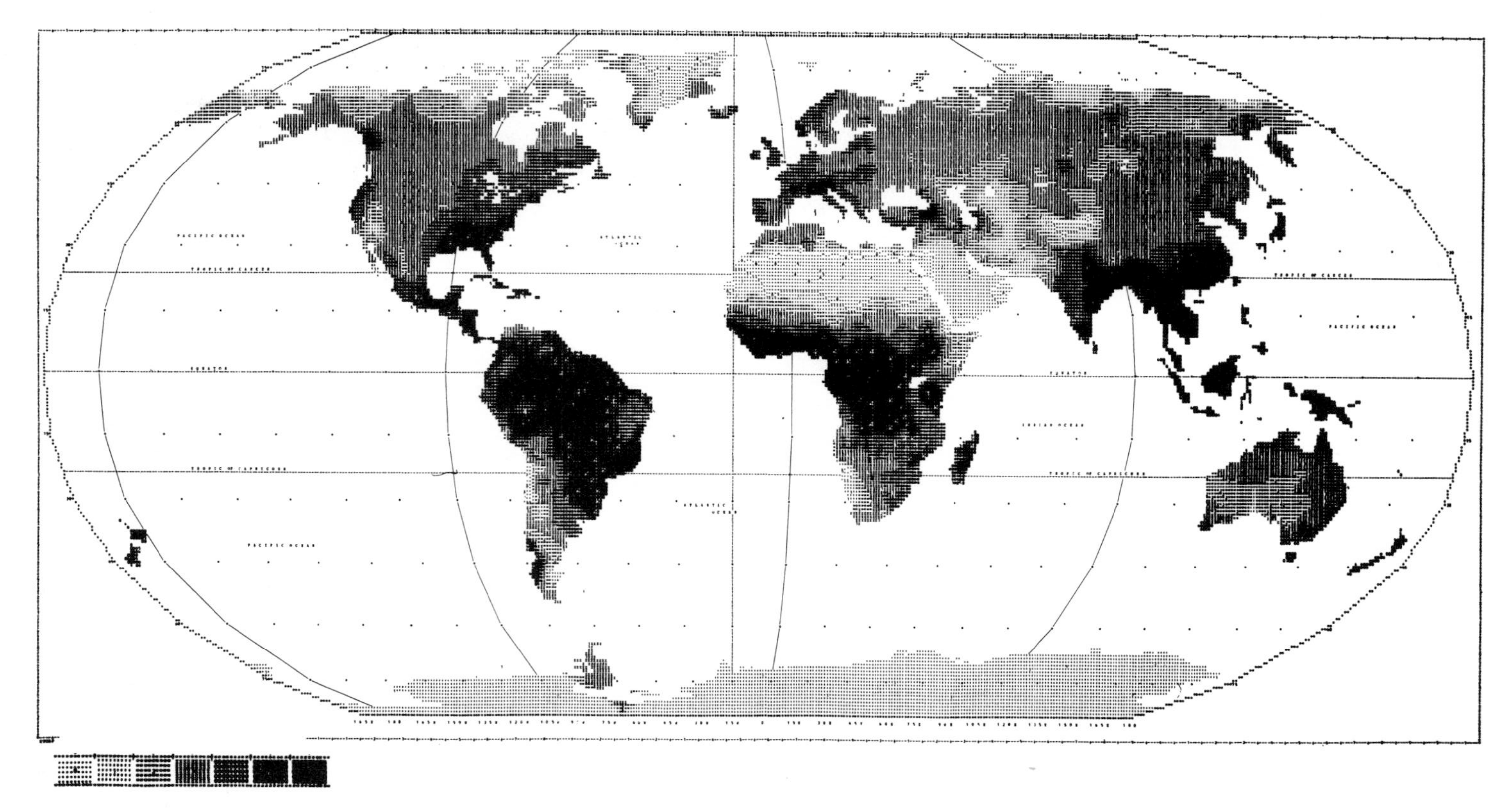

Fig. 142. Calculated primary production by the world's land ecosystems, based entirely on precipitation and length of growing season (Lieth and Whittaker, 1975). In moist temperate and cool regions the representation can be made more precise by taking temperature into account as well

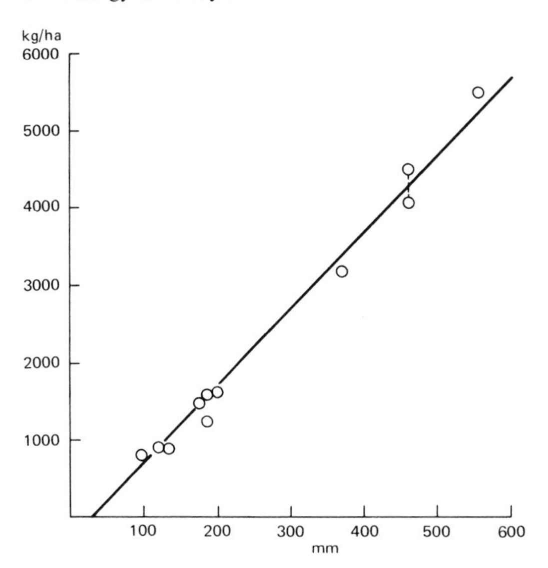

Fig. 143. The dependence of primary production in steppe and desert regions of southwestern Africa on amount of precipitation. (Walter, 1973b)

come about through decades or centuries of overuse. Fertilization in many cases does no more than restore these areas to their original fertility. Water supply and length of the growing season, then, are a sufficient basis for a first-approximation estimate of primary production in land ecosystems (Figs. 142 and 143).

It is impossible to make direct comparisons of this primary production of a system with "the" secondary production. Secondary production by a primary consumer – a herbivore – is necessarily higher than that by a secondary consumer that feeds on this herbivore. And production by a tertiary consumer is necessarily still lower. "The" secondary production of an ecosystem therefore in every case approaches zero – a self-evident fact that need not be discussed further (cf. p. 51 and p. 207). The significant point with regard to the human nutritional situation is that as each successive trophic level is passed secondary production decreases drastically. From an energetic point of view it would be more sensible to use fish and flesh refuse directly as human food than to return them to the cycle as food for animals.

A number of different methods have been used to estimate the level of net primary production. The simplest is the harvest method, in which the whole stand of plants, together with the roots, is collected. When the plants are annuals, the year's production can thus be measured directly. With plants that live more than one year, one faces the problem of distinguishing the organic matter originally present from that added during the year under study. This distinction is extraordinarily difficult in the case of the root system. For this reason the net primary production given is often that of the parts above-ground alone. Furthermore, the above-ground net primary production is the most important in general for zoological and ecological purposes. One way of estimating the annual growth is to harvest the stand both at the beginning and at the end of the growing season. More precise data are provided by gas-exchange measurements. Part of a plant, or the whole plant, is enclosed in a cuvette within which the air is kept at the same humidity and temperature as that of the surroundings. Air is passed through the chamber and the concentration of carbon dioxide in the entering air is compared with that at the exit. The difference is the carbon dioxide that has been incorporated into the plant tissues. The method is very exact, but because of the accuracy with which the microclimate within the cuvette must be controlled it is technically complicated and laborious.

Production in water is measured chiefly by the dark-bottle method. Two bottles are filled with the water to be studied, and they are hung at the desired depth in the water. One of the bottles is made light-tight by covering it with aluminium foil. In the lighted bottle the planktonic algae continue to take up carbon dioxide, whereas in the dark bottle they only respire. The oxygen content of the two bottles is then determined by the Winkler method, and the difference between the two indicates the rate of photosynthesis and thus the amount of carbon dioxide incorporated. In bodies of water with abnormal chemistry the method is not very accurate; the milieu within the bottle changes. Moreover, the endogenous diurnal rhythm of the algae can affect the results. In many cases, therefore, a more elaborate method using radioactive labeling has been adopted. Here the water is injected with $NaH^{14}CO_3$. Under the assumption that the planktonic algae assimilate $^{14}CO_2$ to the same extent as $^{12}CO_2$ and do not deposit ^{14}C in a special way, measurement of the amount of ^{14}C taken up by the end of the experiment can be used to determine the total amount of carbon assimilated. An objection to this method is, for example, the fact that C_4 plants, because of the higher affinity of PEP for carbon dioxide, incorporate ^{14}C to a greater extent than do C_3 plants, so that errors can be introduced. Furthermore, N, P, or Si deficiency can rapidly develop in closed vessels. When this happens protein synthesis is largely blocked, while photosynthesis continues. These cells excrete a considerable percentage of their primary production in the form of soluble compounds; this does not appear in the incorporated ^{14}C activity.

The diversity in methods of production measurement is necessarily accompanied by a variety of units for production. Sometimes it is expressed in g dry weight, and sometimes in the amount of carbon dioxide taken up. Because inorganic material is often deposited in plants, it is customary to indicate the amount of energy (in calories or joules) bound in the organic matter (Block 5).

Very extensive summaries of the energy content of organic matter are now available (Cummins and Wuycheck, 1971). From these data, knowing the general possibilities of error in the methods used, one can derive rules (d'Oleire-Oltmanns, 1977) that allow one to omit certain measurements. On the average we can take it that 1 g (ash-free) animal substance corresponds to about 5.6 kcal, and 1 g of plant substance has about 4.6 kcal. The values for seeds are considerably higher.

There is thus no single best method for determining primary production; under different conditions, different methods will be required. All of these are subject to error, and they are not directly comparable. However, the errors have now been reduced to an order of magnitude low enough that estimation of world-wide production is possible.

Although the determination of primary production, risky though it remains, is in principle so well founded methodologically that "recipes" can be given, the same cannot be said of secondary production. Here the difficulties begin as soon as one wishes to transfer the growth rate of a single animal, measured in the laboratory, to field conditions (see p. 45f.). In the field mortality (see p. 117 f.) in the population complicates the issue. Two opposed processes are operating, the growth – the production – of the individuals and the continual decrease in number of individuals by mortality. As has been discussed, the two processes do not occur uniformly in time. In insects, for example, production is highest during larval development; the highest mortality occurs in the youngest stages. With species that have sharply separated generations or clearly distinguishable stages in the life cycle, so that they can be subdivided into cohorts, regular monitoring of the population (cf. p. 206 ff. concerning the difficulties of measuring a pop-

Block 5. Data for the conversion of various units in production ecology

Production is commonly indicated by a variety of units. So that approximate orders of magnitude can be estimated for purposes of comparison, a number of conversions are listed here. The comparisons themselves, of course, can also be only approximate.

1 g C corresponds to	2.2 g	organic plant dry matter
	3.3 g	dry matter of phytoplankton (i.e., including inorganic components)
	42 g	fresh weight of phytoplankton
	1.7 g	organic animal dry matter
	8.3 g	fresh weight of zooplankton

1 g plant dry matter, ash-free, corresponds to	4,000–5,000 cal
1 g animal dry matter, ash-free	5,000–6,000 cal
1 g carbohydrate	3,700–4,200 cal
1 g fat	9,500 cal
1 g protein	3,900–4,150 cal

1 g oxygen uptake corresponds to 3,280 cal = 0.345 g fat
= 0.728 g plant matter
= 0.596 g animal matter

1 g carbon-dioxide release corresponds to 1,800 cal
1 ml O_2 uptake corresponds to 4.687 cal
1 ml CO_2 release corresponds to 3.558 cal

1 cal = 4.186 joule   1 joule = 0.23889 cal

ulation) allows production to be estimated (Fig. 33). But with species having many intermingled generations with ill-defined stages, and within which by no means all individuals proceed to reproduce – as appears to be the case among most mammals and birds, at least – things are considerably more difficult. Special procedures must be adapted to meet each case (cf. Petrusewicz and MacFadyen, 1970).

An especially problematic task is that of establishing the numbers of microorganisms. Counts made with different staining methods in many cases give discrepant results. Furthermore, such counts give no information about the species (and thus the special capabilities) of the organism, nor can one discern whether the cells counted were still alive when the sample was taken. Methods employing vital stains such as acridine orange have also proved not to be reliable enough. Recently researchers have turned to a method in which the content of ATP (a compound that breaks down almost immediately when an organism dies) is used to infer at least the biomass of the living microorganisms. This procedure assumes a constant relation between biomass and amount of ATP, which seems to hold only very approximately. In many cases no attempt has been made to estimate biomass; rather, the activity of certain key enzymes is used as the sole indicator of the activity of the microorganisms. Wieser and Zech (1976), for example, showed that there was particularly high dehydrogenase activity at the surface of sand grains on the ocean shore; in the interstitial water the activity of this enzyme, an important part of the electron-transport system, was much lower (only about 10%–20% of the total activity). And yet another problem is generally encountered. This is evident from the fact that, with even a minimal rate of division of the microorganisms under the prevailing conditions, the consumption of matter ought to be much greater than it actually is (assuming that the numbers of bacteria derived by various methods are minimal values). One can only conclude that the majority of microorganisms in the soil are inactive, whereas others seem to be excessively active. Animals of the ocean floor have been shown to engage in "gardening." They accumulate bacteria in front of themselves by special secretions and stimulate them to rapid growth. Perhaps this observation is comparable to the fact that in most bird populations a large fraction of the individuals do not reproduce – they are not involved in production (cf. p. 112). But serious research into these problems has only just begun, and the field is still in a state of flux.

Normally one would expect a close relationship between the level of primary production and the density of animal populations. Such a relationship, on a general scale, can be demonstrated for marine habitats. The regions of the ocean with highest primary production to a large extent coincide with those where the fish catch is largest and where the bird populations are most dense (Fig. 144). Similarly, Hinz showed that on Spitsbergen the regions of highest plant productivity gave the largest numbers of animal captures. But here we encounter a number of problems. No animal-capture numbers are absolute indicators, for they refer selectively to particular groups. If the number of mammals is compared with the primary production of forested and treeless regions, no sensible correlation emerges; the number and productivity of mammals in forests are considerably lower than in any region lacking trees, from the tundra to the steppes at tropical latitudes. In part this simply results from the facts that large mammals cannot move freely in the dense jungle and that most of the production of a primeval forest occurs in the crown stratum, out of reach of the large mammals (Table 12). This interpretation accounts for the fact that a poor arctic tundra sometimes supports a hundred times as many mammals as a primeval forest in central Europe. At the moment it is impossible to tell whether taking *all* animal groups into account would give appreciably different results.

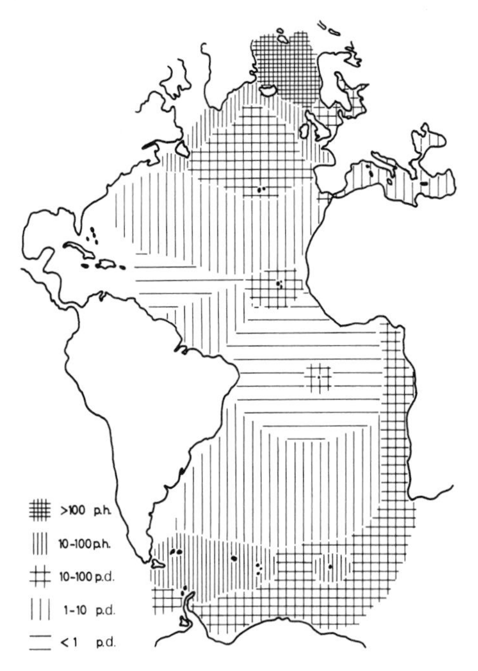

Fig. 144. Density of marine birds in the Atlantic Ocean, as counted from ships. The density ranges from more than 100 birds per hour (p.h.) to less than 1 bird per day (p.d.)

There is another reason why the level of primary production cannot be closely correlated with animal productivity; from one year to the next there can be extreme differences in the amount of accumulated plant substance, depending on the differences in the weather conditions from year to year (cf. Figs. 18, 106 a–c, and 167). The number of animals present is not necessarily controlled by the same factors and in the same way as the productivity of plants. Thus there are differences in the rate of utilization of the primary production in different years. In central Europe a year very favorable to grasshoppers is not noticeable as such until the following year, when the new generation emerges from the abundantly produced eggs and begins to grow. Plants, by contrast, tend to respond immediately to the prevailing conditions. The situation becomes quite difficult when comparisons over large geographical re-

Table 12. Degree to which primary production is utilized by warm-blooded herbivores. (Remmert, 1973)

Type	Harvestable primary production per ha per yr, kg dry matter	No. sheep per ha per year this energy can support	No. of animals per ha	Degree of utilization in % (rough estimate)
Tundra, Spitsbergen	(270–500 kg/ha)	1	Reindeer 0.006 Wild geese (90 days) Ptarmigans Musk oxen	1–2
Steppe, Africa (Serengeti)	7,500 kg/ha (30×10^6 kcal)	20	0.5 animals of many species (2.5×10^6 kcal/ha/yr)	8
European mountains	9,000 kg/ha	24	Red deer 0.006 Roe deer 0.02 Hares	0.02
European mountains, herb and shrub strata only	500 kg/ha	1.4	Red deer 0.006 Roe deer 0.02 Hares	1.0
Hirta St. Kilda, maritime mountain meadow	2,000 kg/ha	5	Soay sheep 1.72	31
Carpathian mountain meadow	6×10^6 kcal/ha/yr	4	15 Muridae	1

gions are attempted. Because there are no herbivorous ectothermic animals in the far north, these regions cannot be directly compared with steppes with respect to the ectothermic fauna. Nor can marine and terrestrial regions be compared. On land the primary production is used by animals of direct interest to man (plant-eating birds and mammals). In the ocean, on the other hand, the fish interesting to humans are all secondary, tertiary, or quaternary consumers. They feed on large clams or fish; only very few species are able to live on planktonic crustaceans – that is, primary consumers. For this reason the production of animals important to man in the ocean must be well below that on land, even for a given level of primary production.

Finally, special features may be overlooked when large-scale comparisons are made. Though primary production is about the same in the two bodies of water, the western Baltic Sea provides far higher yields of food fish than does the Skagerrak. The reason is that the two differ in salinity. The Skagerrak has normal marine salinity, with the normal marine fauna. In the western part of the Baltic the salt concentration is reduced, and a great many animal forms have disappeared. These – irregular sea urchins, sea stars, and large gastropods – are competitors of other marine animals, especially clams. Clams (Cyprina islandica above all) are excellent food for the cod, whereas the animals that have vanished from the Baltic would not have been eaten by the fish in any case (Arntz and Hempel, 1971).

Furthermore, there is the following consideration. Two bodies of water may exhibit identical levels of primary production, although in one the producers are very small planktonic algae and in the other very large species. The very small algae in the first habitat can be utilized only by very small planktonic animals, which in their turn are the food of the larger planktonic animals on which fish feed. By contrast, the larger planktonic algae in the second habitat can be eaten directly by large planktonic animals; in this case the food chain is shorter by one link. At a rough estimate, this shortening increases fish production by a factor of 10. Where vascular plants on land are concerned, it is self-evident that only part of the production is normally used by animals, or that different parts of the plant tissue produced are used by different animals – some eat roots, others leaves, others seeds, and still others the nectar of flowers. Here the central issue is the part of the plant by which production is represented; the overall level of production is not the first consideration (cf. p. 145).

For all these reasons, it is inappropriate to try to find a direct relationship between the level of primary production and the number of animals – especially if one wants anything but a very rough indication, and if one is interested primarily in the animals important as food for humans.

The differences in production within a system can be well illustrated by a food pyramid, in which the individual trophic levels are set one above another. But this form of presentation conceals awkward problems. Should the numbers of individual organisms be indicated, or their biomass? A great many Orchestes fagi can live on a single beech tree, but if the biomasses of the green leaves and of the beetles are compared, the picture looks quite different. And if production is judged entirely on the basis of physiological experiments, still another picture emerges. For example, a certain amount of hay can feed a certain number of rabbits for a certain period of time. In theory, a certain number of martens can feed on this number of rabbits for the same length of time, and a certain number of wolves can live on these martens for the same period. Finally, it has unfortunately become accepted practice to limit such a pyramid to systematic groups; a certain number of plant-eating birds is associated with a certain number of carnivorous birds, and these with a certain number of birds that prey on the carnivores. This approach is unreasonable, as the example of the birds – hardly any of which are plant-eaters – emphasizes.

The only kind of pyramid that should be published is one in which the net production at each trophic level is indicated. Such a "production pyramid" is considerably more difficult to construct than one based on number of individuals or biomass, but has the advantage of being really unambiguous. The "upside-down" food pyramid for open water should be dispensed with entirely. This structure has been derived from the fact that the biomass of the consumers in open water is greater than that of the primary producers, the planktonic algae. But the productivity of the planktonic algae – their rate of cell division – is so enormous that the situation, as expressed by a production pyramid, appears entirely normal (Fig. 145).

b) Measurement of a Stand. When a stand of plants or population of animals is to be quantitatively evaluated, a distinction is made between number of species and number of individuals. In general species number is relatively easy (though by no means simple) to determine. Models derived from the island theory of MacArthur (see MacArthur and Connell, 1970) actually allow one to estimate the number of species present fairly reliably on the basis of the area covered by the region of interest, if the number of species in comparable regions is known (Fig. 146); as area increases the number of species is greater, even if the larger area contains no new habitats.

The determination of number of individuals – an indispensable item of information in ecosystem research – is far more difficult, especially in the case of land plants. Estimates of root mass, in particular, are frequently challenged. Only very recently have reliable data been published, chiefly by Kummerow. In general rough approximations are offered, and these are not dependable; under different conditions a given plant can have quite different ratios of root-system size to phytomass above ground.

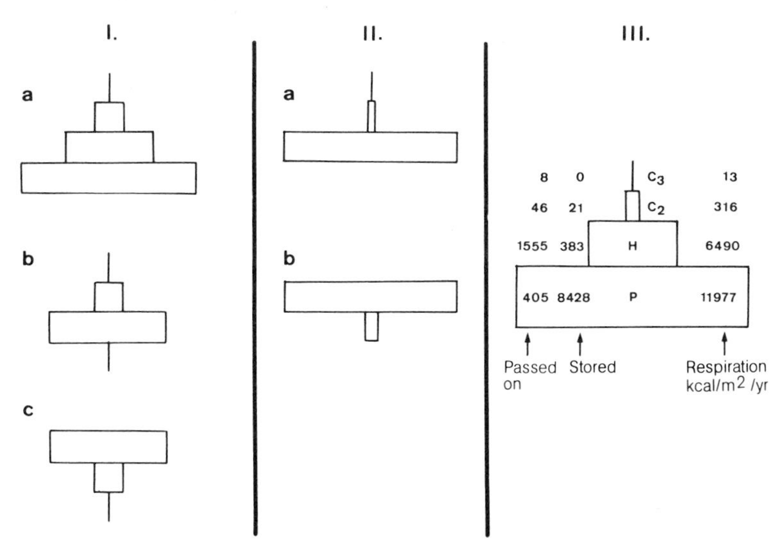

Fig. 145. Various types of food pyramids, summarized from Phillipson (1966). At the base are the producers, and above them are the consumers. *I.* Numerical pyramids: *(a)* primary producers are small; *(b)* primary producers are large; *(c)* pyramid for plants and plant parasites, which in turn are infested by hyperparasites. *II.* Biomass pyramids: *(a)* fallow field in Georgia, USA; *(b)* English Channel, where the amount of planktonic algae is smaller than that of planktonic animals. *III.* Energy pyramids (Silver Springs, Florida, USA). All data in kcal/m²/year. *P* producers; *H* herbivores (=consumers I); C_2 consumers II; C_3 consumers III

There is therefore also a high probability of error in constructing a simple energy pyramid based on numbers of individuals per unit area (or their biomass). So far, as discussed on p. 190 ff., there is no method for the quantitative determination of animals on land; enormous errors can occur. But we must know precisely the number of animals in an area if we want to say anything quantitative about the ecosystem. We also need to know the number of offspring that the animals in the system produce per unit time, and the mortality of the progeny (when birds' nests are deserted or robbed, for example, or caterpillars are infested by parasites), but this information is almost never available. Indeed, it is practically impossible to obtain a simple count of the adult population of large, relatively conspicuous animals. A discouraging example is the recent attempt to quantify the roe deer population in the cultivated areas of central Europe. Danish studies, in which the area investigated was fenced off, showed that the real population was at least three times as large as all the other counts by experts had indicated. This fundamental discovery – that the true number is three to five times greater than the best estimates – may well apply to all the roe deer populations in central Europe. Even enclosed populations in which every individual is marked can never be completely counted by the professionals in charge of them. There is an annual cycle in "observability" (Ellenberg, 1978).

If it is so difficult to obtain a reliable count of the roe deer in central Europe, a region relatively easy to inspect, one can imagine how far wrong the numbers given for other animals and birds may be – apart from animals so extremely rare that they deserve protection, such as the peregrine falcon, golden eagle, and eagle owl. The counts of the various mouse species per unit area so far available are exceedingly unsatisfactory. Almost all data on mass reproduction of mice are based on relative studies; traps are set up at the same spots year after year, according to a maintained pattern. The differences in number of animals caught are used as indicators of fluc-

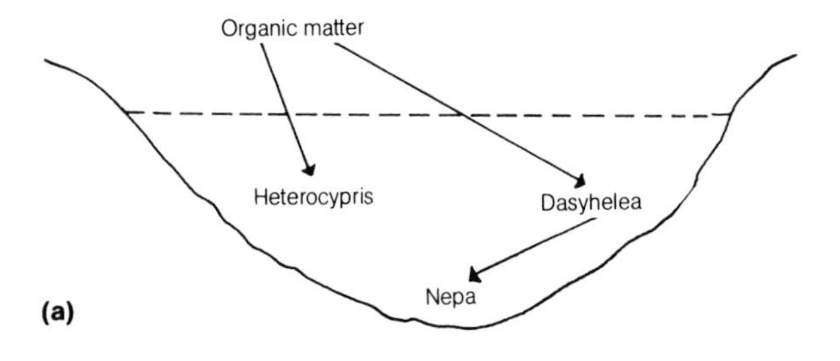

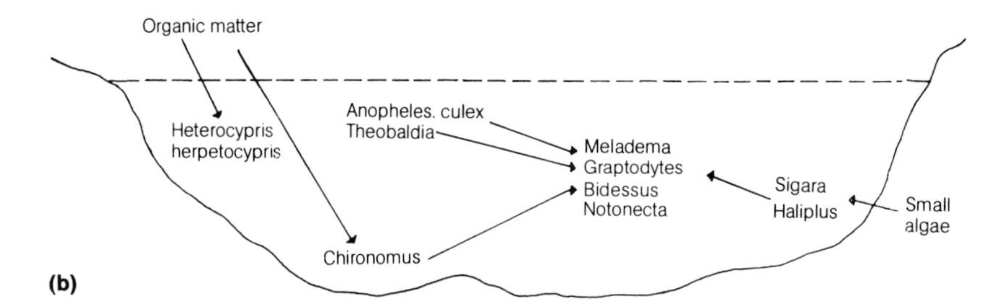

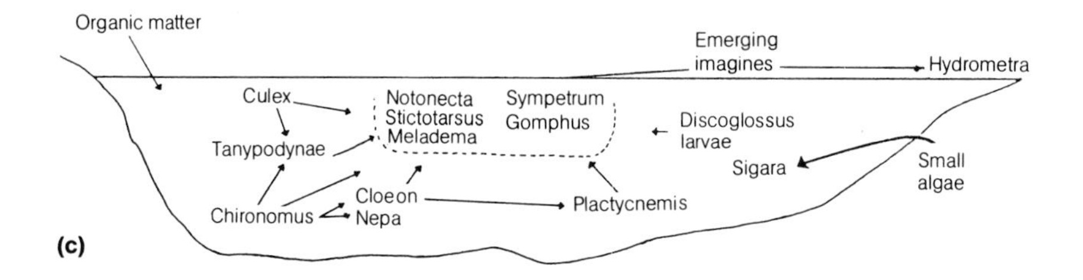

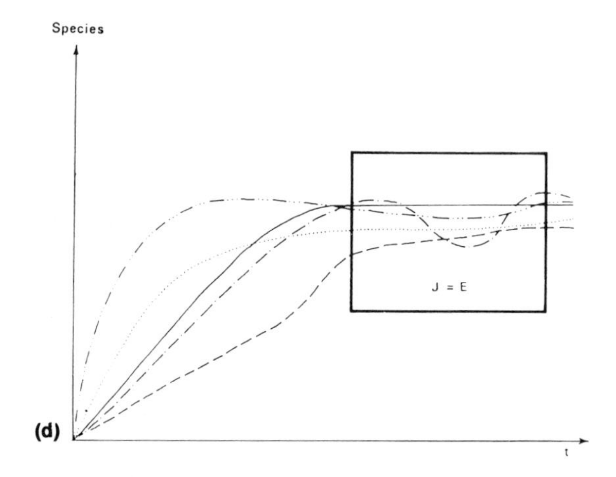

Fig. 146a–d. As rock pools increase in size, the food relationships become progressively more complex, for the number of species becomes much greater (rock pools in a dry river bed in southern France) (Remmert and Ohm, 1977). **a** Water surface area ca. 0.05 m^2, depth 8 cm. **b** Area ca. 3 m^2, depth 30 cm. **c** Area 3.5 m^2, depth 40 cm. **d** The relationship between number of species and area of a habitat in the course of time. After some time an equilibrium is reached between area and species number; the immigration of a new species causes one of those already present to die out. (Müller, 1974)

tuations in population density. A method that ought in theory to work is to catch as many members of a species as possible, mark them, and then let them go. Then there is a second trapping session, and one can compute the population size from the ratio of marked and unmarked animals. But not all animals are caught so easily a second time, not all can be permanently marked, and marking can put the animals at additional risk. And special difficulties arise where the animals are clustered, territorial, or randomly distributed – that is, everywhere. None of the techniques for counting insects gives anything like a satisfactory result. The most reliable counts seem to be obtained for terrestrial insects, those of the soil as well as in the herbaceous, shrub, or canopy strata, by collection by hand from a test area. It is necessary to distinguish precisely among the different species, and in many cases to distinguish the young and adult stages of a single species. Remember that animals of different size can vary greatly in metabolic rate, and thus in the amount of food eaten per unit time and in productivity (p. 43f., Fig. 31). It is therefore pointless from the outset to lump animals at the same trophic level in a single biomass. Bearing all this in mind, we must regard the data in Table 12 as very rough indicators.

There have been an especially large number of studies on the density of bird populations. Some of the results are summarized in Table 9. But here, too, the variations are huge, as Fig. 171 illustrates, and there is a very high probability of error. Nevertheless, at present these bird counts are probably the most reliable data of all on the density of terrestrial animals. Because birds have been counted over a very long time in very many parts of the world, and because there are large numbers of extremely knowledgeable amateurs in the field of ornithology, estimates of bird population size have attained a special significance. Birds seem to be useful as "bioindicators;" routine monitoring of the bird population can provide evidence about the state of the environment. For this reason programs have been instituted in several countries (the "Common Bird Census" in the British Isles, for example) in which quantitative determinations of the bird population are regularly carried out (Svensson, 1970). Despite the likelihood of error this method still gives us the best information about changes occurring on our planet.

The situation in water seems to be simpler. Here the level of primary production evidently depends chiefly on the availability of nutrients and light. When larger amounts of nutrients are provided, productivity can be decisively increased. The number of algae in a body of water rises sharply, but at the same time the transmission of light through the water is reduced. At the deeper levels, where photosynthesis had been possible when minerals were scarcer, autotrophic algae can no longer exist. Now oxygen is used up at these depths, but no more is liberated. As a result, in the deeper parts of nutrient-rich waters anoxia kills off both animals and plants. The eutrophication of lakes and regions of ocean is dangerous because of the associated reduction in available oxygen. On land, where oxygen is always available, overfertilization does not have such consequences.

The process in water can be followed closely by filtering a certain amount of water through a simple plankton net. The number of individual planktonic algae and animals in a whole body of water can be readily derived from these results. Even planktonic microorganisms are now relatively easy to count. Bottom-dwelling organisms are also accessible; grabs can be used to bring to the surface a piece of bottom material of known area, and with enough samples of this sort the number of animals colonizing the bottom can be determined. As always in biology, the identification of species can be difficult, but here again the problems in the aquatic realm are slight compared to those encountered on land. For these reasons there have been many more studies of aquatic populations and ecosystems than of terrestrial ones. In many cases, the results have been applied directly to land communities; but it has turned out that such application is almost always incorrect. The situation on land is not only much more difficult methodologically, but the general functional principles of populations and ecosystems on land are more complicated than in the water.

c) Food Chains and Food Webs. The distributions of plants and animals derived from such measurement procedures can be

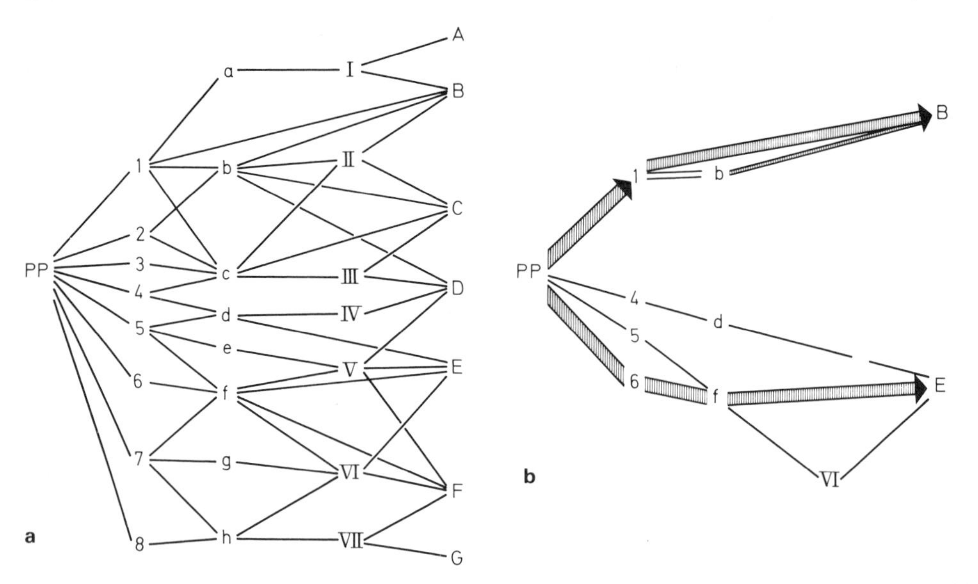

Fig. 147. a Hypothesized relationships within a food web. PP, primary production. **b** The same system, with the food flow quantified and all flow paths accounting for less than 0.001 of the total omitted. The remainder essentially follows two major paths

arranged in terms of function; the ordering most commonly used is based on the flow of food. We can distinguish food chains, food webs, and – associated with these – trophic levels. Food chains are found primarily in habitats with very few species. In the eastern African Lake Nakuru, a soda lake, one species of blue-green alga [Oscillatoria (Spirulina) platensis] is the producer for two consumers, a copepod and a flamingo (cf. p. 250). When the systems are more complicated, the simple food chain tends more and more to become a food web. This tendency can be followed very well in ponds of different size (Fig. 146). In very complex habitats the networks become so complex that they cannot be traced in detail. Whereas the food chains consist of readily distinguishable links – primary producer, consumer I (which feeds on the primary producer), consumer II (which feeds on consumer I), and sometimes still higher "trophic levels" – such distinctions are no longer possible in food webs. One can make at most an approximate subdivision into different trophic levels.

Such food webs have been presented as descriptions of the "community nexus" in a wide variety of aquatic and terrestrial habitats. They have profound implications for the stability of habitats (see p. 240). It is a priori clear that a system comprising a network has greater stability than one composed of independent chains. However, no such web has so far indisputably been shown to exist. Proof of its existence would require that the quantitative relationships be known. Figure 147 shows a hypothetical food web as it is classically presented, and with it – hypothetically! – the precisely quantified food relationships. The quantification reveals that in actuality we are dealing not with a web, but with chains. In reality, the netlike arrangement plays no role at all. The frequently drawn conclusion that web structure implies stability, then, is not reliable in the absence of quantification. Because no ecosystem on earth has been described in really quantitative terms, there is as yet no confirmation of the web structure. Before the question of stability can be argued further, we must apply ourselves to quantification.

Such studies as are available do show that stands of cultivated plants are less vulnerable to mass outbreaks of "pests" when they are not monocultures, but comprise two or more species (for example, the maize and sweet-potato plantations in Costa Rica). The leaf beetle Phyllotreta cruciferae, which is a severe pest on cabbages in the USA, did considerably less damage and occurred in considerably less dense populations when the cabbage fields also contained other plants, such as tomatoes, tobacco or ambrosia (Ambrosia artemisiifolia). The reason was not a decrease in the numbers of predators or parasites, but the fact that the insects find their food plants by following the chemical stimuli the plants release; these are not as easily recognizable in the polycultures (Tahvaneinen and Rood, 1972). At the same time, this example shows that associations based on food relationships and thus on energy flow are only one facet of the complex ecosystem; in this case food relationships are drastically altered by parameters of sensory physiology.

d) Energy Flux. On p. 45ff., in discussing the ecological significance of food supply, we considered the flow of energy through a single organism. Now we must turn to the flow of energy through an ecosystem. Solar energy is converted to chemical energy by the green plants; this chemical energy is passed on through some number of stages, and in the process is converted in a stepwise manner to heat energy. Eventually, the chemical energy ought to fall to zero. But in asking ecosystem research to verify this expected result, we are making a demand comparable to that for quantification of the community nexus. The only data available so far are more or less applicable estimates for overall systems (chiefly aquatic) and quite detailed studies of the flow of energy through individual animal populations. Even though this is not a very satisfactory base of departure, a relatively consistent pattern results (see Figs. 148 and 149).

The herbivores, consumers I, use up only a fairly small part of the food theoretically available to them. All the plant-eating in-

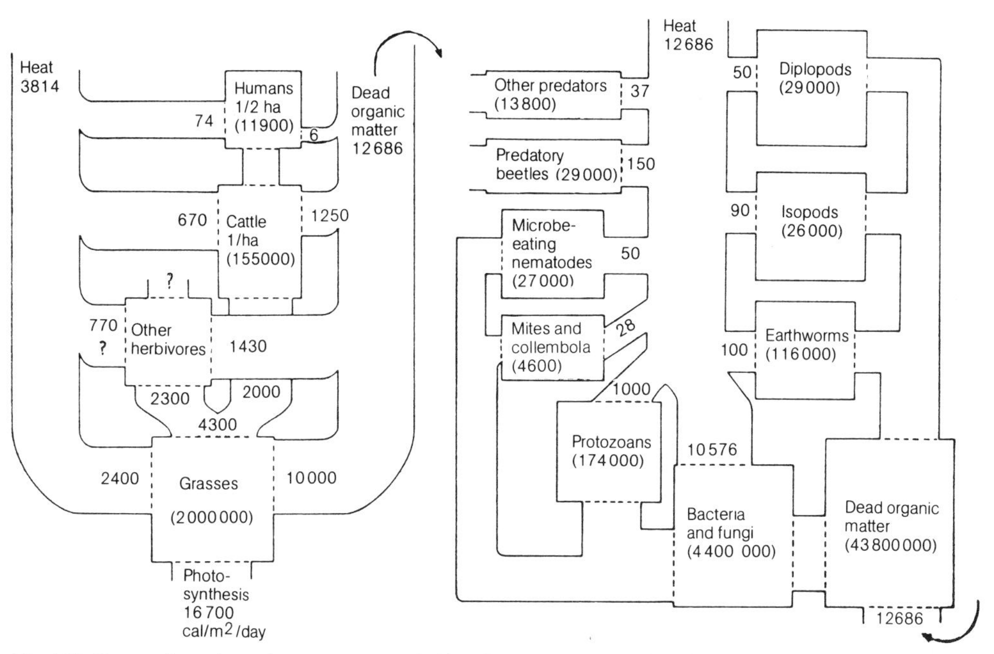

Fig. 148. Energy flow through a cow pasture (with reference to MacFadyen in Tischler, 1965). The dead organic matter in the left part of the figure, which shows the herbivores and the predators that feed on them, enters the right part of the figure, which shows the flow of energy in the soil

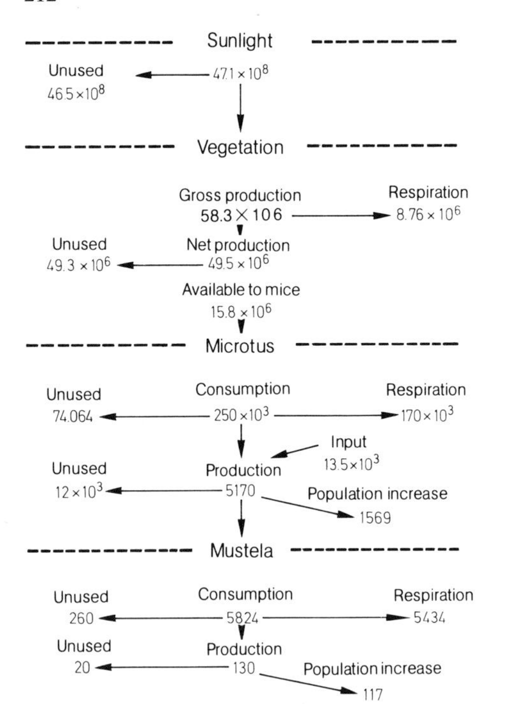

Fig. 149. Energy flow in a fallow field in the USA. (Golley, 1960)

sects in a beech forest together consume at most 10%–15% of the substance produced by the beeches in a year (Funke, 1973); other animal groups can be neglected as herbivores in this habitat. Grasshoppers in the meadows of central and northern Europe consume no more than 10% of the plant tissue formed (Gyllenberg, 1974). The warm-blooded herbivores of the tundra – reindeer and musk ox above all – utilize only 5%–10%, and the large grazing animals on the African steppes barely exceed 10% utilization (Table 12). Still less of the plant matter in the forest is consumed by mammals, for the main productive stratum, the canopy, is out of their reach. Small mammals in general consume barely 1% of the net primary production; only in a very few cases can this figure rise to as much as 6% (Ryszkowski, 1975). However, particular parts of the plants, such as the seeds, can be consumed in higher proportions – 30%, 75%, or even 90% (cf. p. 134). It is generally true that on land a maximum of 10%–20% of the substance formed by plants is directly consumed by herbivores. All the rest eventually (very much later, in the case of trees) enters the food chain responsible for the decomposition of dead matter. By far the greatest energy turnover on land, therefore, as chemical energy is used up, occurs in the soil layer. One must bear in mind that excrement, exuvia, and other things derived from the bodies of the first consumers (hair, feathers, shed antlers) are also passed on directly to the food chain in the soil. Now the question arises, how large a percentage of the net production of the first consumers enters the secondary consumers – predators and parasites? Here, again, the value seems in general to vary between 5% and 15%. The greater part of the substance produced by the first consumers, then, also enters the soil stratum, from the dead bodies.

Three consequences follow from these considerations.

1. We must consider the soil in greater detail, for it is the chief site of turnover in terrestrial ecosystems.
2. We must ask whether the picture we have drawn is really sufficient, or whether a finer subdivision is required.
3. We must inquire into the causes and the ecological effects of the surprisingly low rates of utilization observed – in the utilization of primary production by herbivores as well as that of the net secondary production of any animal population by animals at the next trophic level.

We shall begin with the third question. In regions where there is a marked change of seasons, the annual production is of course not always at hand; there are times of very high productivity, and other times in which little or practically no food is available. The minimum, during the dry season or in winter, determines the density of the large-mammal population during the main production season, and thus the degree of utilization (Fig. 150). The longer and more severe the winter or dry season, the smaller the percentage of production likely to be utilized. This tendency is to some extent compensated by migrations, like those

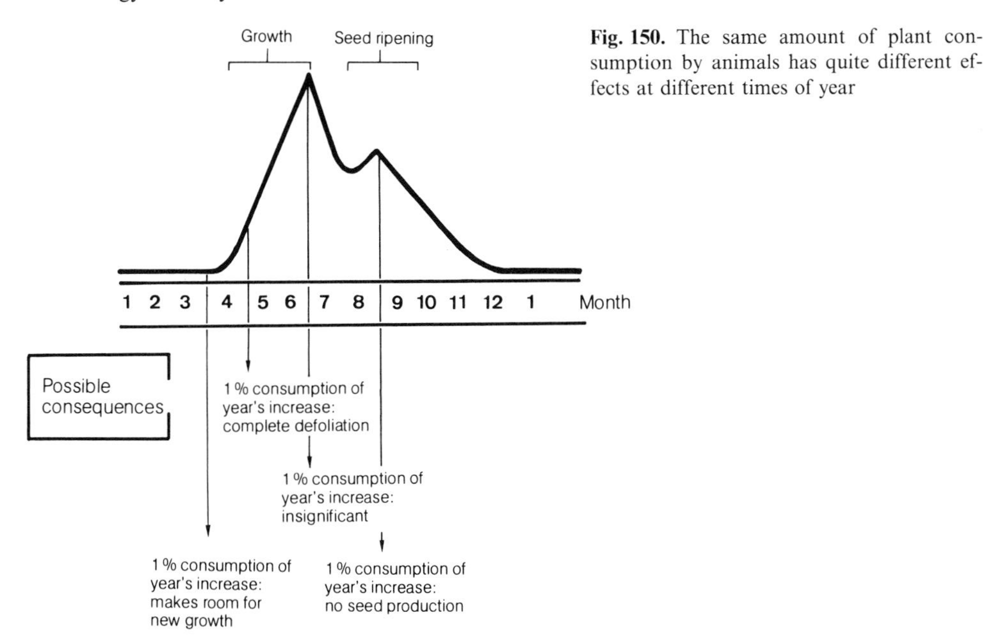

Fig. 150. The same amount of plant consumption by animals has quite different effects at different times of year

which used to bring red and roe deer down from the mountains in winter. In the tropical rainforests there are no seasons. But here the productive zone, the crowns of the trees, is largely inaccessible to the herbivorous mammals. In treeless regions where seasons are absent or not well marked, it may be that the proportion utilized by mammals is much higher. This could happen in temperate-zone heaths near the coast, where because of the prevailing winds trees cannot grow. The situation is similar on high-altitude steppes near the equator; here the very small territories of the vicuna suggest that their rate of utilization is very high. Furthermore, the small mammals and insects with high rates of reproduction appear unable to "keep up" with the rapid new growth of vegetation in the spring. They are directly dependent on plant food, and cannot survive warm periods without food. Their lives would be endangered if they were to emerge before the leaves sprout in the spring. If they appear after leafing-out has begun – even immediately after – as a rule the plants have gained an advantage in development that cannot be overcome. But if leafing-out and the appearance of the insects are simultaneous, the trees can easily be stripped bare; this happens, for example, during mass outbreaks of June beetles. As yet we cannot specify a percentage utilization in such cases. All the trees so damaged in temperate zones can put out new leaves and exhibit full production after the June beetles have died out. However, another example shows that mass outbreaks in natural regions of vegetation can have extremely dangerous results. In the Finnish and Swedish birch forests, mass outbreaks of the moth Oporinia autumnata cause widespread destruction; the forest boundaries in many cases are pushed back by 30 km (Fig. 151). Ultimately we must admit that the causes of the low rate of utilization of the permanently available food reserves in the tropical rainforest by herbivorous insects are not known.

Though low utilization is the rule, we must always be prepared for surprises. In the two examples that follow, plant-eating animals use up nearly 100% of the primary production. The first of these is the only native herbivore of central Iceland, the pink-footed goose. The contemporary populations of this bird here are so dense that they consume nearly all of the vegeta-

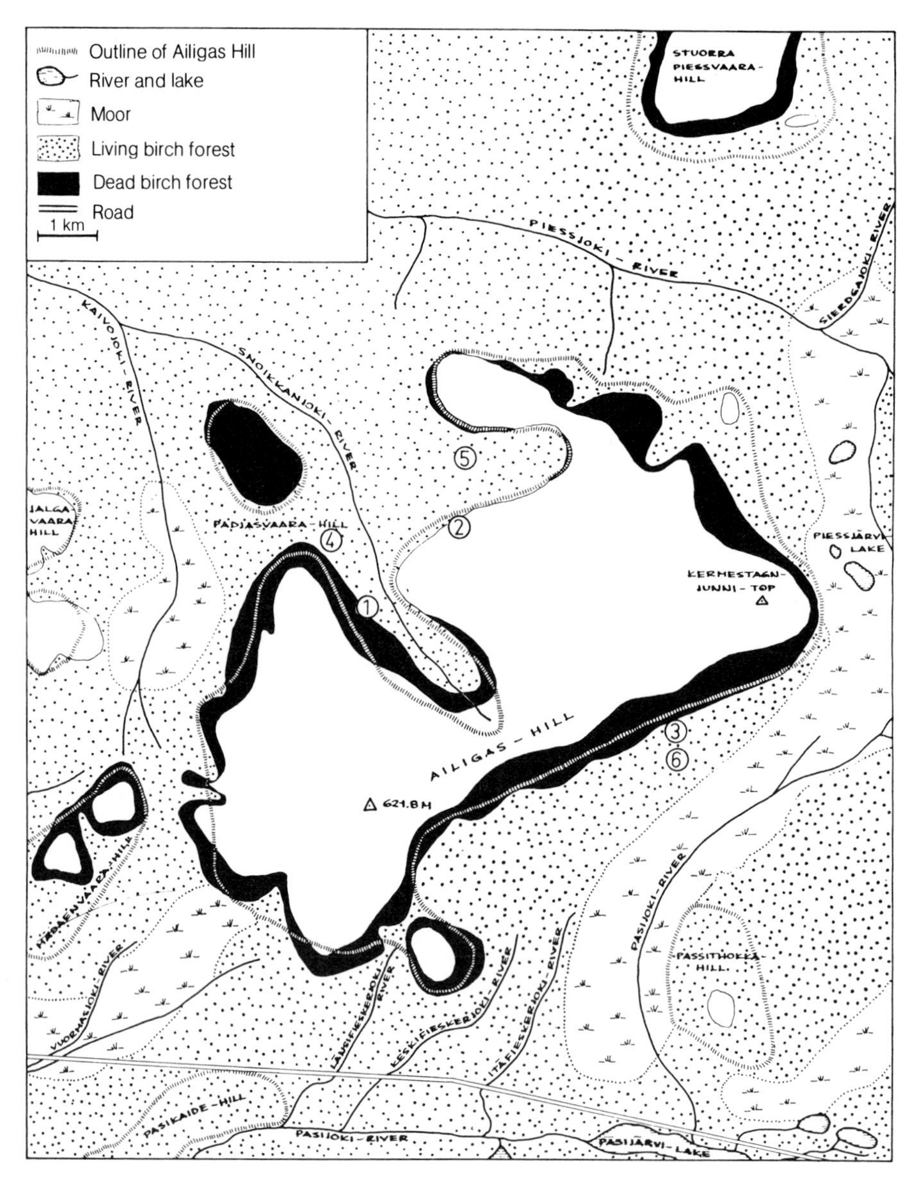

Fig. 151. The tree line in a mountain region of northern Finland has been pushed back by the caterpillars of Oporinia autumnata. (Nuorteva, 1963)

tion of the few oases during the breeding season. This is possible only because the geese can migrate so rapidly; they arrive only after the vegetation has sprouted and leave after they have eaten everything up. Another likely contributing factor is the very favorable conditions under which this population spends the winter. The losses that probably were once routinely suffered in the overwintering region no longer occur (Gardasson and Sigurdson, 1972, 1974). The second example is provided by the studies of Reichholf (1975, 1977) on the lakes produced by damming the lower reaches of the Inn River. He showed that the large flocks of ducks that overwinter there consume almost all the aquatic plants. This happens only when the animals are not disturbed during the winter, so that they can stay in the region all day and night. Here, again, the high rate of utilization is possible only because the birds migrate, arriving after the vegetation has grown tall and leaving the lakes when the water plants have been grazed off. The consequences are important ecologically. The energy contained in the plants is respired above the water by the ducks. If it were not for the ducks, the plants would sink to the bottom in the winter and rot there, withdrawing oxygen from the water in the process. Without the ducks anoxic conditions would develop; with them, they do not. In this sense the ducks are responsible for the large stocks of fish in the lakes.

In evaluating energy flow, one encounters a special problem with regard to the food not used by the animals. Here we do not mean the part that leaves the gut undigested – although this element is frequently underestimated. For example, the dipteran larvae feeding in acid beech forests chew up about 30% of the litter from the stand, whereas they actually utilize only a few percent. What is meant by "unused food" is the amount destroyed, but not consumed, by animals during feeding. The larvae of the alder-leaf beetle eat only the photosynthetically active layer of the leaf; most of the leaf, with all the veins, remains behind and turns brown. The amount of food eaten is thus considerably less than the amount of plant matter destroyed. In this respect there are large differences among species, so that it is impossible to present a generally valid picture. Grasshoppers and hawk-moth caterpillars, for example, all begin to eat a leaf at its tip. These animals consume all the plant matter they destroy.

Let us turn to the second question, whether these results really suffice to quantify the biocenotic interactions. The answer is an unequivocal "no." A much more refined subdivision of the system is necessary for such quantification. We cannot speak of "the primary production" of which a certain percentage is utilized. It is possible, for example, that certain species, or certain parts of a plant, are 100% utilized, whereas others are essentially ignored by the herbivores. Oaks at the edge of a forest are exploited to a considerably greater extent by all sorts of herbivores than beeches similarly located. Willows, poplars, and blackthorns are relatively frequently and obviously attacked by plant-eating insects, whereas pines and spruces are heavily infested by insects only in exceptional cases. But these are qualitative observations, which have not been quantified. Among the large ungulates of Africa, as we have seen, the different species exhibit strict specialization to the different strata of vegetation.

But even if we distinguish among species we have gained relatively little. The individual plants are also differentially vulnerable to herbivores. These differences are in part genetically based, and in part associated with the habitat. Alders or spruces in dry habitats (or in dry years) tend to be attacked by herbivores more extensively than they are under moist conditions. And finally, the separate parts of a plant are used as food by herbivores to widely differing degrees. So far we have been discussing the parts of plants that grow above ground, disregarding the fact that there is about the same phytomass underground. We know nothing about the degree to which this part is utilized. Moreover, different above-ground organs are formed in different years. The irregular production of seeds by forest trees is an example of this effect. In some years the accumulated production of many years is evidently channelled into seed formation (Figs. 152 and 106). As yet we cannot say whether more plant matter is consumed by herbivores in a seed year than in those when no seeds are produced. Presumably consumption is higher in a seed year, but it would make sense to give a percentage figure only if a mean value for many years could be cited; voluminous acorn production is possible only because reserves have accumulated in the oak tree during many previous years.

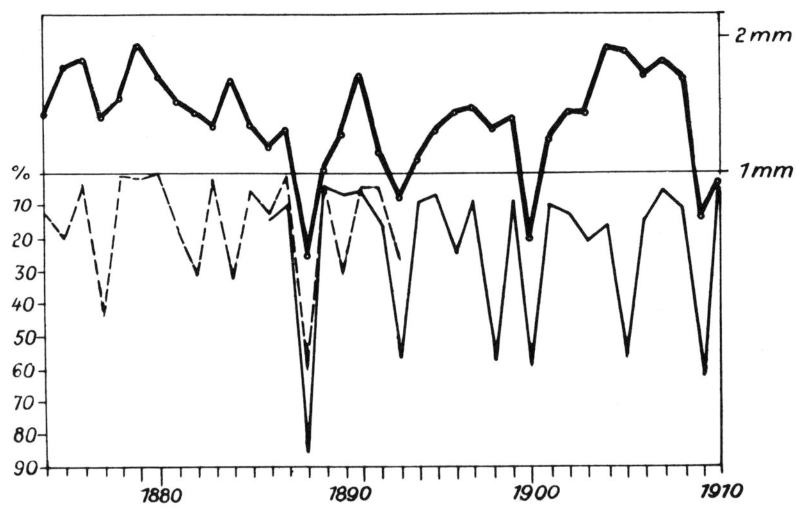

Fig. 152. The relationship between seed yield and thickness of annual rings in beech trees (Maurer, 1964). *Upper curve, thick line:* annual-ring width; *lower curve, thin lines:* seed yield in % of maximal yield

Most of the organic matter formed by the plants is not consumed by herbivores, but eventually reaches the ground. Normally, these substances are continuously decomposed and worked into the soil. In terrestrial habitats, therefore, by far the highest energy turnover occurs in the soil stratum. In spruce and pine forests the amount of litter falling per year has been estimated as 1,500–3,000 kg per hectare, and in good mixed deciduous forests of central Europe as 4,000 kg/ha or more. In a balanced system, then, this amount must be decomposed each year; otherwise organic matter would accumulate on the forest floor. This is not to say that a particular beech leaf is decomposed in the course of a year. In general, a fallen leaf in an acid beech forest requires about 5 years before it is no longer visually identifiable.

This decomposition is not a rapid and complete breakdown into CO_2, H_2O, and minerals. The process is slow and involves complicated resyntheses, in the course of which humus is produced. Humus substances, because of their great stability, are of very special significance in the preservation of soil structure.

As far as the participation of organisms in the breakdown of litter is concerned, there are both qualitative and quantitative considerations. In view of the enormous number of soil-dwelling animals it is a priori likely that they play a central role in decomposing plant matter (Fig. 153). Generally, for practical purposes, macrofauna are distinguished from microfauna; the former is understood to include chiefly the earthworms, enchytraeids, isopods, diplopods, and large insect larvae that in Europe are especially important in the breakdown of litter. Evidently the litter is first eaten by the larger animals, which can attack even the less digestible elements with the aid of microbes. Isopods, snails, and dipteran larvae are equally important. The significance of these primary decomposers is evident in Fig. 153; in their absence, the rate of decomposition is almost halved. During this first passage through a digestive system, little energy is withdrawn from the material. All animals take up very large amounts of fallen leaves and produce very large amounts of excrement; they utilize the litter relatively poorly. Nevertheless, this first passage is of fundamental importance to the system, as follows.

1. The water-binding capacity is increased. The excreta of the primary de-

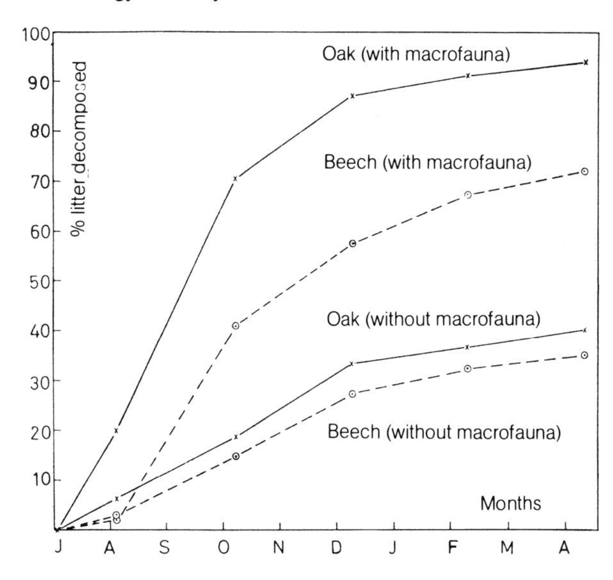

Fig. 153. Decomposition of beech and oak litter in the natural habitat, with and without the action of the larger decomposers. (Thiele, 1968)

composers holds moisture more uniformly than the original litter.

2. Having been comminuted, the material becomes accessible to a large number of so-called secondary decomposers – primarily springtails, mites, and nematodes. The excreta of some primary decomposers are more readily consumed by the secondary decomposers than those of others, so that if the process of decomposition is to proceed evenly there must be an undisturbed, well-balanced community of animals in the soil.

3. The excreta of both primary and secondary decomposers provide particularly good conditions for the development of microorganisms, above all actinomycetes and bacteria. Their presence can be inferred by comparison of the metabolic activity (CO_2 release) in fallen leaves with that in excrement.

This first passage through the gut is followed by repeated passages, through the same or other animals. Wieser (1965) showed that isopods flourish particularly well when they can eat their own excreta over and over again. The microorganisms break down those plant parts that are hard to digest, and the animals are nourished primarily by these microorganisms. Would decomposition proceed more rapidly if only the microorganisms were present? After all, the activity of the animals appears to cause a considerable reduction in the numbers of microorganisms! This is the sort of false conclusion that is so commonly encountered in analysis of predator-prey systems. If the animals were not present, the microorganisms would very rapidly reach an extremely high population density and then almost stop reproducing; their cultures would be in the stationary phase. By virtue of the fact that their population density is kept low, they maintain maximal production – and thus optimal performance – over the long term (cf. p. 153 ff., Figs. 153 and 154).

The importance of the comminution of litter is also shown by certain experiments done in Russia, and others by Herlitzius and Herlitzius (1977). Completely intact beech foliage on acid forest soil is very resistant to attack and hard to break down. The excrement of leaf-eating beetles that falls down from the canopy and the dead bodies of these beetles provide nuclei in which microorganisms can begin to grow (cf. pp. 227, and see Fig. 155).

During the repeated passages through animal bodies, easily decomposable substances are withdrawn from the substrate. The carbohydrates are predominantly bro-

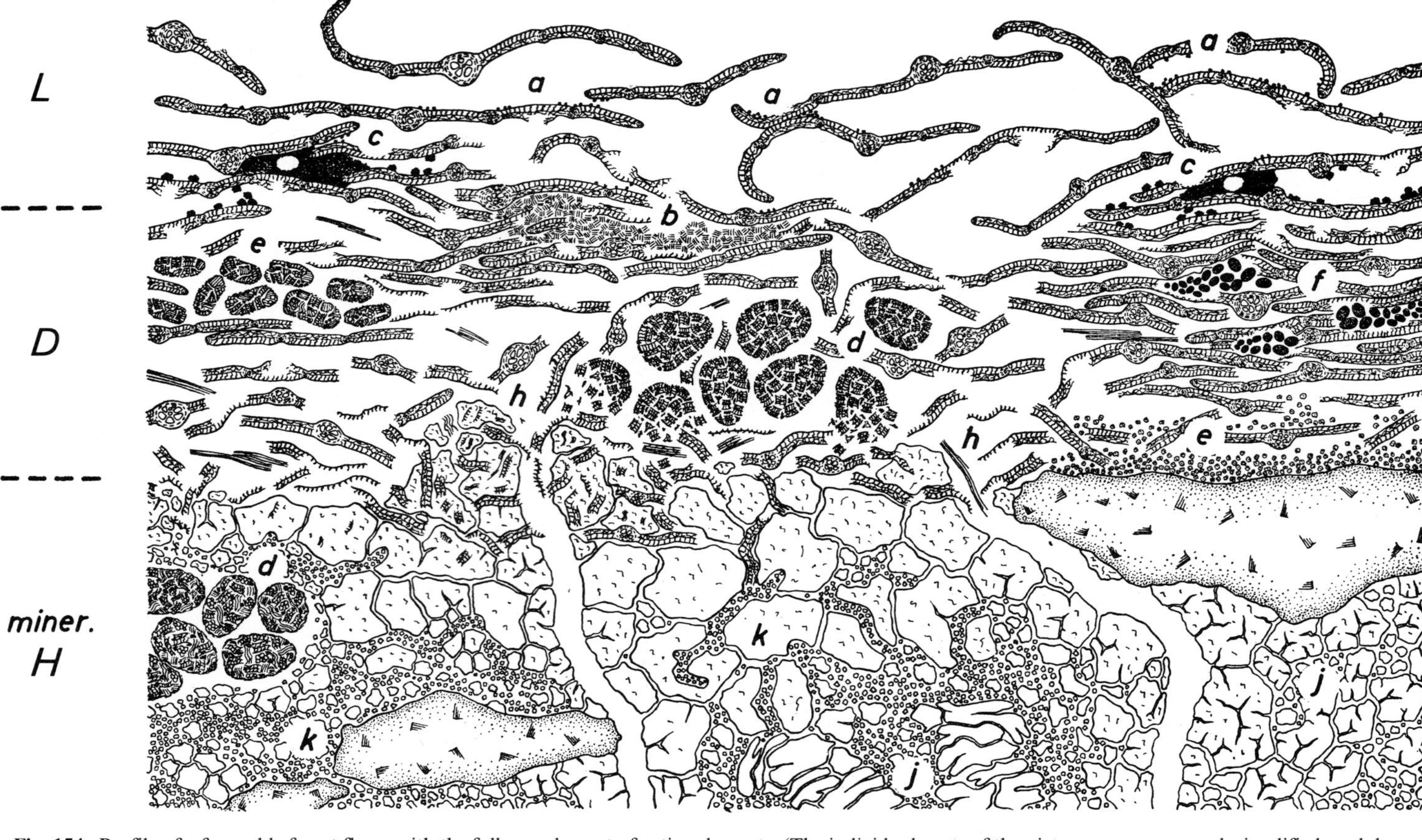

Fig. 154. Profile of a favorable forest floor, with the full complement of active elements. (The individual parts of the picture are compressed, simplified, and drawn to different scales.) Combined from various observations. *L*-horizon, litter; foliage that is slightly decomposed. *D*-horizon, duff; largely decomposed debris. *H*-horizon, humus, mixed with mineral particles. *a* leaves with the excrement of fairly large Collembola; *b* fecal mass of small dipteran larvae; *c* scattered tubes of excrement from Dendrobaena, between leaves in the lower L horizon. *d* *(middle)* fecal balls of large diplopods; *(left)* fecal balls of tipulid larvae. *e* *(left)* Dendrobaena excrement; *(right)* excrement of enchytraeids. *f* packet of slowly decaying leaves. *h* burrow of Lumbricus terrestris; *j* fecal mass of Allolobophora; *k* passages dug by enchytraeids, containing their excrement. (Zachariae, 1965)

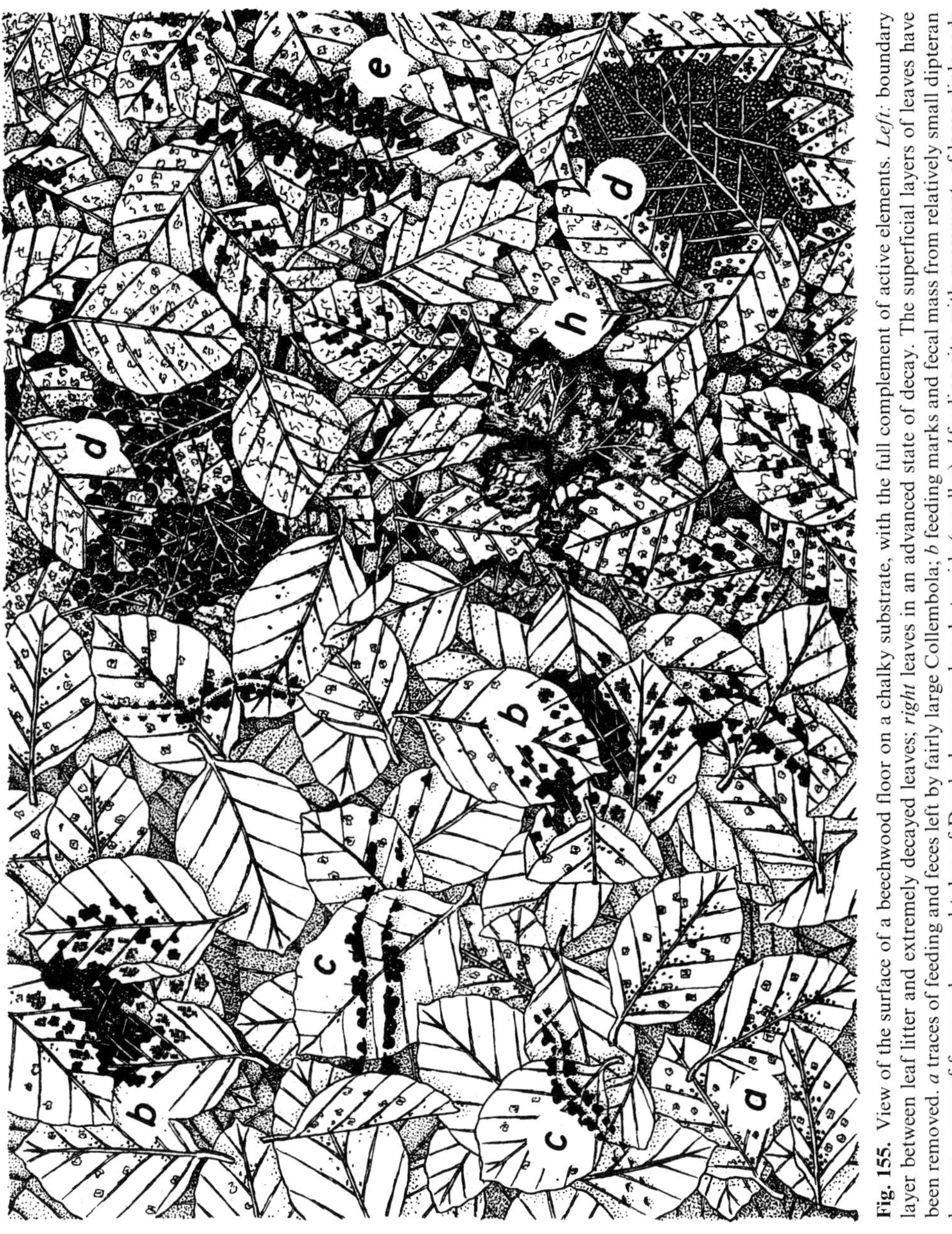

Fig. 155. View of the surface of a beechwood floor on a chalky substrate, with the full complement of active elements. *Left:* boundary layer between leaf litter and extremely decayed leaves; *right* leaves in an advanced state of decay. The superficial layers of leaves have been removed. *a* traces of feeding and feces left by fairly large Collembola; *b* feeding marks and fecal mass from relatively small dipteran larvae; *c* rows of feces and opened tubes of Dendrobaena and enchytraeids (smaller). *d* feeding site, with excrement, of large diplopods *(above)* and bibionid larvae *(below)*. *e* pile of excreta from Dendrobaena; *h* earthy fecal masses of Lumbricus. (Zachariae, 1965)

ken down into carbon dioxide and water. The percentage of the most important plant nutrients (nitrogen, phosphorus, potassium) rises, and lignin and tannins – products of the polymerization of phenols – accumulate in the excreta. In this neutral or weakly alkaline milieu these substances come into close contact with nitrogenous metabolites produced by the animal decomposers. The conditions thus favor the microbial breakdown of the lignins to phenols, at a rate that depends on the amount of nitrogen available to the microorganisms.

The analysis of humus fractions from different soils, together with model studies, shows that in the milieu described, when nitrogen and sufficient oxygen are present, phenolic substances are oxidized to oxyquinones, which polymerize to form larger molecules. The process is accelerated by phenoloxidases of the microorganisms. The end result is the production of humic acids – polymers of complicated structure which are extraordinarily stable (Fig. 156). Breakdown of humic acids in the soil occurs very slowly; it is likely that the chief agents of their decomposition are mycorrhiza fungi. Humic acids, then, are an extremely stable element of the soil, especially because they can form clay-humus complexes by combining with aluminum and iron hydroxides. These complexes ultimately determine the fine structure of the soil, the friability required for good ventilation and water retention. The proportion of clay-humus complexes in the soil is thus responsible for soil fertility – at least in the majority of soils (Figs. 150 and 157). The quantitative division of the labor of remineralization, between herbivores and soil organisms, has probably resulted from a many-faceted coevolution that has led to optimization of the system (cf. p. 179).

When the decomposition of litter proceeds as described, the result is the ideal form of humus, mull. But it does not take the same course in all habitats; there are wide differences in rate of decomposition in different types of forest. The quality of the humus, the rate of breakdown of leaf litter, and the

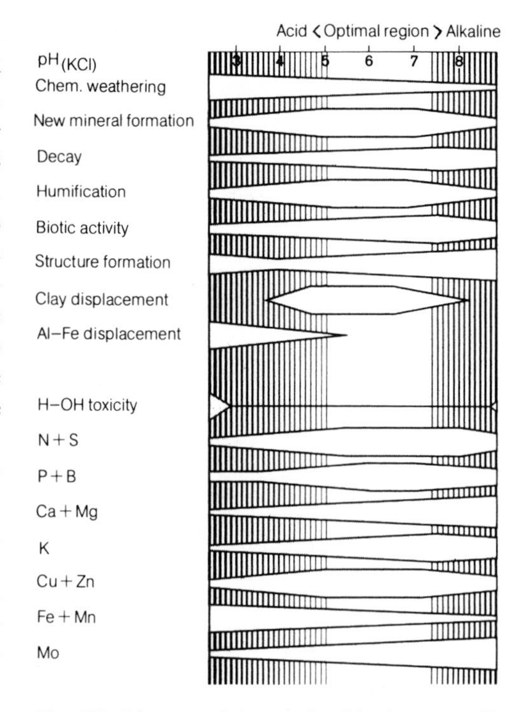

Fig. 157. Diagram of the relationships between pH and other properties and processes in the soil. Among the magmatic rocks, for example, granite is considered acid and gabbro (with basalt and diabase, or the sands formed from them) basic. (Schröder, 1972)

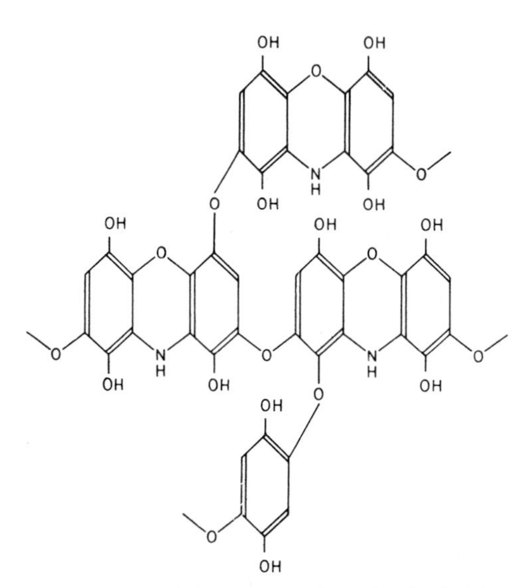

Fig. 156. Part of the structural formula of a humic-acid molecule. (Thiele, 1968)

number of litter-consuming animals in a habitat are closely related. Comparison of three forest habitats with decreasing humus quality (mull on lime soil, mull on acid soil, and mor on acid soil – the last is a very unfavorable form of humus) showed that the rate of decomposition decreased in this sequence, and the number of animals was related about as 3:1–5:1. Soil pH and the organisms supported are frequently correlated, and even though the dominant plant species are in many cases the same (beech on chalk and variegated sandstone; pines on sand and chalk; birch forest on quite different geological substrata in Greenland, Iceland, and Scandinavia) the differences are striking – particularly with respect to the density of the animal species present. A number of other factors are associated with pH (Fig. 157; see also Schröder, 1972), which together make acid soils generally poorer in nutrients.

Humus, as an organic substance, does not continue to exist indefinitely. It in turn is eventually broken down by microorganisms. Naturally, this process occurs especially rapidly under hot, humid conditions (in a tropical rainforest). Although humus is not a plant food, in most soils it is essential to plant nutrition. The humic acids can adsorb molecules that otherwise would soon be leached out by the rain. Moreover, humus substances can retain large amounts of water. Humus therefore guarantees the plants a steady supply of minerals and water. The destruction and disappearance of humus in many soils is responsible for a decline in primary production (the relationships described have been analyzed by Thiele, 1968; see also Schröder, 1972).

In principle, the same relationships are found in water; only a relatively small fraction of the producing planktonic algae is consumed as living cells by herbivores (Fig. 158). Most of the algae die, sink toward the bottom, and are eaten in deeper water by consumers in the detritus food chain. On land as in water, then, we have what has been called a Y-shaped model of energy flow (Fig. 148). In Fig. 147 we took a diagram of biocenotic interrelationships which, while unquantified, appeared to be a very complex food web, and found that quantification produced two nearly independent food chains; although the latter was a purely hypothetical model, the above considerations show it to be quite realistic.

Most of the production in water is by phytoplankton rather than by sessile plants. It is therefore restricted to water levels near the surface. The more transparent the water, the deeper the producing layer can be. The higher the nutrient content, the more dense the phytoplankton, so that light penetrates less deeply and the producing layer is shallower. In this sense a large body of water can be compared with a forest, where production is also restricted to the upper surface; in the shaded region near the ground the chemical energy formed is used up. In deep bodies of water where oxygen is abundant the rain of detritus is essentially all consumed before reaching the bottom. This is generally true in the oceans; in the Baltic Sea, for example, it has been documented by analyses of protein content (Figs. 159 and 160). In such waters the bottom contains very little organic matter, and as a result the fauna is relatively diverse and in many cases consists of fairly large animals which presumably develop very slowly. In waters with little oxygen at depth consumption occurs anaerobically or not at all; the bottom becomes covered with decaying mud in which H_2S is formed (Fig. 161). Oxygen deficiency occurs in the depths when the water becomes opaque because the nutrient content is too great, and/or when the deeper water of an ocean or lake is never exchanged with the more superficial water because autumn and spring storms fail to churn the water sufficiently. Under such conditions the water becomes increasingly uninhabitable down to a certain depth. At great depths more and more organic material accumulates. These large-scale relationships probably can serve as models of the origin of petroleum. Conditions are like this in the Black Sea, and it is feared that they will become so in the Baltic. But the situation in the Baltic is complicated by irregular inputs of oxygen-rich, high-salinity water from the North Sea; these masses of water, being more saline and thus heavier than the Baltic water, move into the Baltic along the bottom, displacing the water that is low in oxygen and salts. So far this process has repeatedly brought about complete decomposition of the deposited organic matter. In shallow bodies of water the organic matter produced reaches the bottom be-

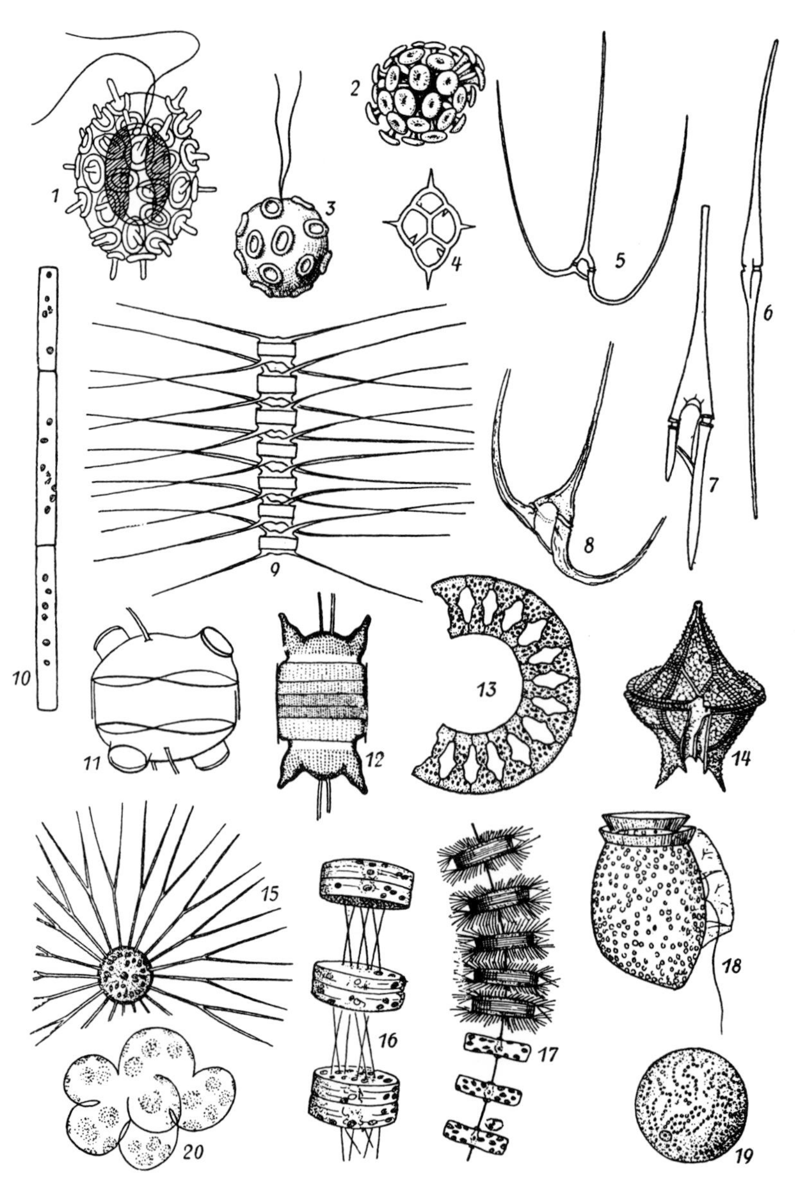

Fig. 158. Plants of the North Sea plankton. *1* Syracosphaera subsala; *2* Discosphaera tubifex; *3* Pontosphaera huxlei; *4* Dictyocha fibula; *5* Ceratium macroceros; *6* C. fuscus; *7* C. furca; *8* C. longipes; *9* Chaetoceras boreale; *10* Leptocylindrus danicus; *11* Cerataulus turgidus; *12* Biddulphia aurita; *13* Eucampia zoodiacus; *14* Peridinium divergens; *15* Bactiastrum varians; *16* Coscinosira polychorda; *17* Thalassosira gravida; *18* Dinophysis acuta; *19* Halosphaera viridis; *20* Phaeocystis puchettii. (Gessner, 1957)

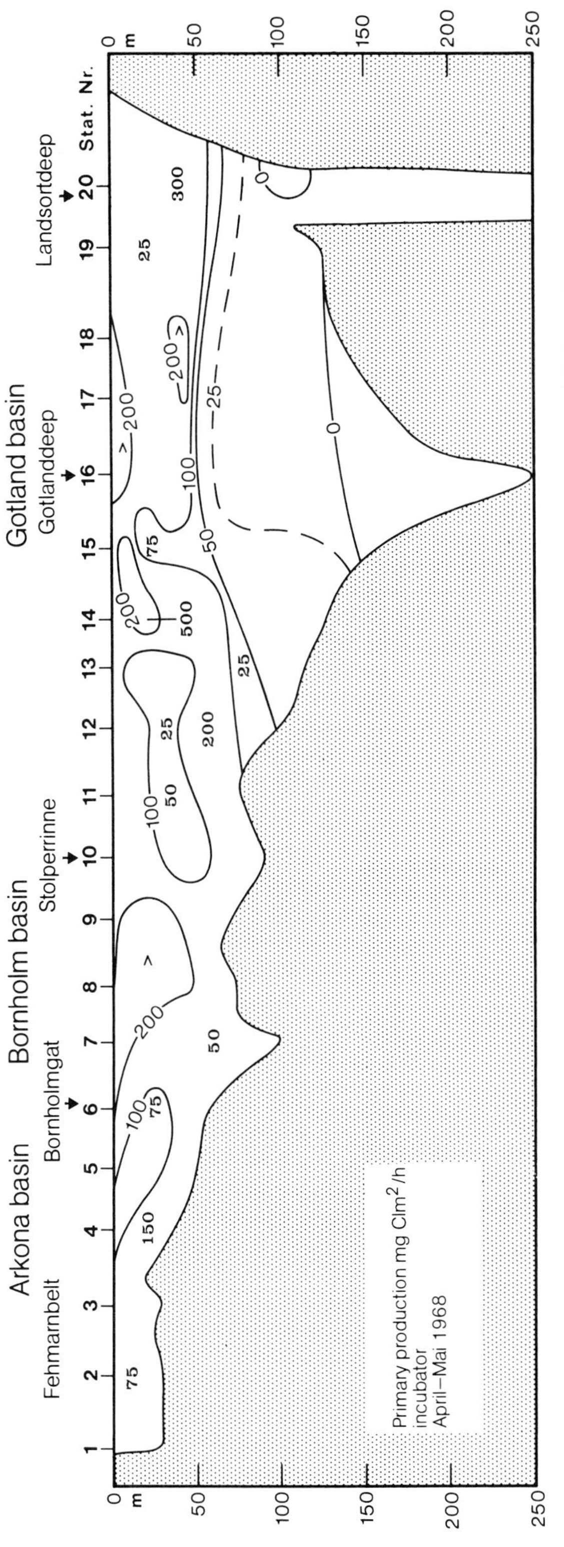

Fig. 159. The level of primary production in the western Baltic Sea, measured by the incubator method. (Magaard and Rheinheimer, 1974)

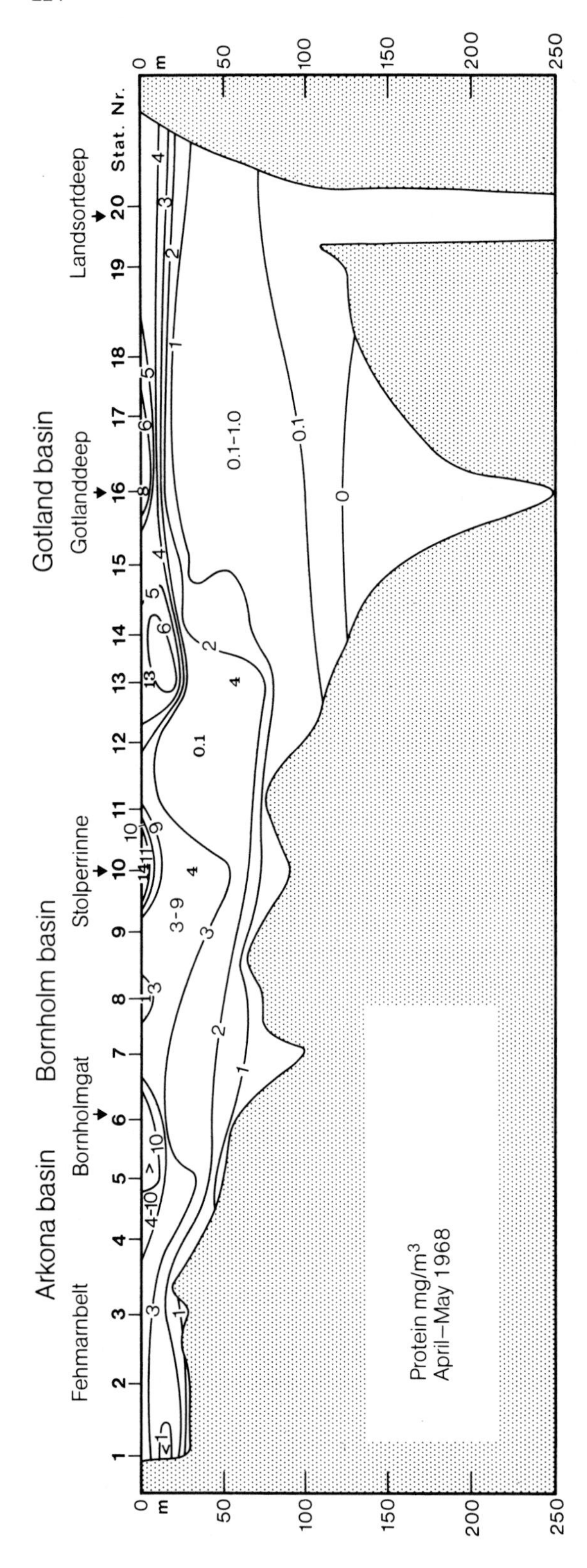

Fig. 160. The protein content of the Baltic Sea, as a measure of the living biomass. (Magaard and Rheinheimer, 1974)

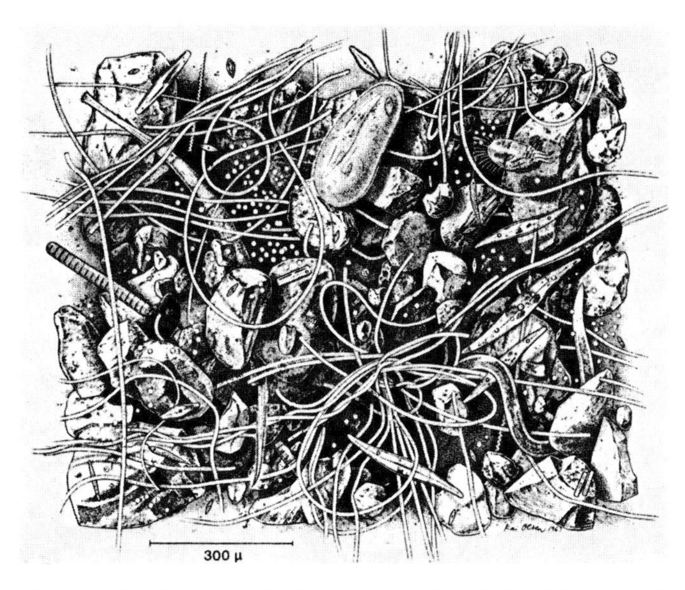

Fig. 161. Microflora and microfauna on strongly reducing Baltic-Sea sediment. Spaghetti-shaped: Beggiatoa (Cyanophyceae); small round dots: Thiovolum (sulfur bacteria). Ciliates: long and slender, in the left half of the figure, Trachelorhaphis; oval, upper middle, Frontonia, and upper right, Diophrys. Boat-shaped diatoms are also shown, and a nematode is at the lower right. (Fenchel in Jansson, 1972)

fore it is decomposed. The abundance of this food source is evident in the large numbers of filter-feeding animals that inhabit fresh water and the shallow parts of the oceans – barnacles (Balanus), mussels, bryozoans, a variety of polychaetes, and many others.

Let us consider these relationships in detail. Primary production is entirely performed by unicellular algae of several sorts – peridinaeans, diatoms, and the very small flagellates of the nannoplankton. In cooler regions or seasons the diatoms predominate, whereas under warmer conditions they are less in evidence. In warm oceans primary production depends heavily on the nannoplanktonic flagellates. The significance of this fact is enormous, even though the level of primary production may be identical in the two cases. The organisms of the nannoplankton are so small that they escape not only ordinary plankton nets but also most of the animals that eat plankton; only highly specialized animals, usually very small, are capable of utilizing the nannoplankton. The best known example is the mantle of the planktonic tunicate Oikopleura, an extremely complicated structure within which the animal encloses itself. The mantle presents a very fine-meshed lattice to the flow of water generated by the tunicate, and the nannoplankton that accumulate on it are used by the animal as food. In fact, it was this observation that first led to the discovery of the nannoplankton, which until that time had been overlooked. The nannoplankton, then, serve as food for small to very small planktonic animals, and these in turn are eaten by larger zooplankton. But if the primary production is by relatively large phytoplankton, the large planktonic animals can utilize it directly. By the rule of thumb that only about $^1/_{10}$ of the energy is retained at each transition from one trophic level to another, it is clear that this extension of the food chain when the nannoplankton are the primary producer must have a drastic effect on secondary production by larger inhabitants of the open water; the necessary consequence is a paucity of fish. Recent studies indicate that the poverty of tropical oceans is not necessarily all a matter of mineral deficiency; to some extent it results entirely from the fact that the nannoplankton are responsible for primary production. (Evidently this nannoplankton population is extraordinarily resistant to a great variety of influences, so that if it were to replace the primary producers in other parts of the ocean fish production, and thus the world's fisheries, would be endangered.) For this reason the precise analyses of the requirements and reproductive rates of phytoplankton currently underway should have high priority (see, e.g., Schöne, 1977; Werner and Roth, 1977).

The ocean's planktonic animals comprise two fundamentally different groups. The animals in one group spend their entire lives in the open water, whereas those in the other are the larval forms of bottom-dwellers. The latter are naturally seldom encountered in the large oceans. Planktonic larvae are characteristic of the smaller seas – North Sea, Baltic Sea, and Mediterranean, for example – but are almost nonexistent in fresh water. The planktonic animals in lakes spend their whole lives in the plankton, with at most certain temporary stages (as happens among the water fleas and their relatives) resting on the bottom. There are probably great differences in the food requirements of the zooplankton, as has been suggested by the studies of Lampert (1976; cf. p. 49). But we know almost nothing about them. The next trophic level is represented by small or somewhat larger plankton-eating fishes, such as herring (Clupea) and whitefish (Coregonus), and these in natural systems are eaten by large predatory fishes, mammals and birds.

Here, again, we may take it that in general only about 10% of the algae are directly ingested and digested by animals (ingested algae, if undamaged, are often passed out again in the excreta and then simply continue to grow).

Yet another factor is involved. Planktonic algae "lose" a large part of their photosynthates (though there is no reliable, quantitative estimate of how large it is) into the water, where they drift about as relatively short-chain carbohydrates, as fatty acids, and as amino acids. These losses are greater if, while photosynthesis is proceeding rapidly under strong light, certain substrate materials fall below a critical concentration; then subdivision can no longer occur. If the nitrogen concentration is below the critical level no more amino acids and protein molecules can be synthesized, but carbohydrates continue to be produced, and these must be eliminated. As a result, the water accumulates large amounts of free organic molecules. The same thing happens when planktonic organisms die or are injured. For this reason the amount of dissolved organic compounds in the ocean is considerably higher than that of all organic compounds bound in plants and animals. The compounds dissolved in fresh water are all taken up, incorporated and broken down by bacteria; in the ocean there are many organisms that are either obligatory or facultative feeders on such dissolved organic compounds (cf. 20, 42).

In water, then, the formation of a soil such as occurs on land is neither necessary nor possible. The principle of energy flow in water is based on complete breakdown of organic matter. There is no question of humus formation. If organic matter is deposited on the bottom and not decomposed, the first result is a decaying sludge; then firm deposits of organic matter are formed which eventually can become fossil fuels (peat, coal, petroleum). These regions are quite unlike humus soils – they are extremely hostile to life, and practically devoid of animals. Bacteria and protozoans (Fig. 161) are the sole inhabitants of such reducing sediments.

VI. The Significance of Animals in an Ecosystem

Animals are much more conspicuous in water than on land. On land they are surrounded by higher plants and are barely apparent, whereas on the bottoms of bodies of water they play a leading role. This is especially true of the ocean; the subdivision of marine bottom-dwelling animals has been done according to principles very like those used to subdivide terrestrial regions on the basis of plant communities. The large, well-known associations in the ocean are the pure animal communities of the coral reef; the benthic animal and plant communities (Fig. 131) hardly enter the macroscopic picture. On land, by contrast, animals in general contribute little to the flux of energy and the cycling of materials. Rarely do herbivores consume more than 10% of the energy fixed by the plants. For this reason, many ecosystem analyses have given only cursory treatment to the animals. Not a few researchers – particularly botanists interested in productivity and system analysts interested in the system as such – are inclined to deny the animals any appreciable role in the ecosystem.

But nothing could be further from reality. An animal like the roe deer, which preferentially consumes the buds of trees, can dramatically reduce the production of a forest; the rabbit, an equally selective feeder, in populations of normal density can keep the stands of many special plants down to the extent that they are no longer ordinarily encountered. After myxomatosis had drastically reduced the rabbit population in the British Isles certain wildflowers appeared in all sorts of places where people had been looking in vain for years. Beavers in an ecosystem can back up large lakes behind their dams; if the dams break extensive meadows appear in otherwise continuous woodland.

There is not much quantitative data available, and that has been taken from only a few animals. Indeed, effects such as those described can hardly be expressed quantitatively or included in treatments of energy flow or of the cycling of matter. Cycling and energy flow are but two facets in the intricate pattern of an ecosystem. We shall now present a few examples that illustrate the complex involvement of animals in ecosystems. These must be anecdotal and in some cases have not been completely worked out, but they reveal the difficulties inherent in this line of research. Leaf-eating beetles evidently cause no decline in production in a deciduous forest. Because the beetles eat holes in the upper leaves more light reaches those below, so that these can proceed with photosynthesis at a greater rate. Here loss and profit are in balance. In fact, there is an added benefit. The excrement of the beetles, which contains finely ground up leaves and in many cases is full of minerals, as well as the dead beetles, laden with nitrogen and phosphorus, fall to the ground and act like nuclei of crystallization – centers from which the decomposition of the leaf litter proceeds much more rapidly and efficiently than it would without these highly nutritive particles. The leaves shed from the trees contain very few nutrients, even for bacteria and fungi, and are excessively hard to digest. Only by way of the excreta and bodies of animals are they vulnerable to attack. With these aids, decomposition can occur rapidly; without them it would take so long that natural rejuvenation of the woods would be nearly impossible, for a thick layer of undecomposed leaves would pile up. The beetle is thus a fundamental element in the system. But its influence can be detected only by laborious and extremely meticulous research; without such research its effects might be perceived only when it was 80 or 100 years too late. Moreover, the effect varies on different types of soil. On chalky soils, where the decomposition of organic matter proceeds relatively rapidly, the contribution of the beetle corpses is less important than on acid sand and silicate soils (Herlitzius, 1977). Because plant-eating insects preferentially infest trees in which the sap flow is not completely intact, aging trees with lowered productivity, but which are still highly competitive, tend to be invaded and killed. Competition is thus reduced for the younger trees, and in the long term (as Mattson and Addy, 1975, have shown in the USA) productivity of the system is increased.

Aphids withdraw from their host plants large amounts of fluid containing sugar. These sucking insects also require nitrogen, present in much smaller concentrations in plant sap; by the time they have acquired enough nitrogen they have taken in far too much sugar, and this is excreted in the form of "honeydew." Are the plants seriously damaged by this loss? Experiments have shown that nitrogen-fixing bacteria can be brought to peak performance by aqueous solutions of sugar; after application of aqueous sugar solution the fixation of atmospheric nitrogen increases sharply. It is possible that the following system has developed by coevolution. The plants pass on the surplus carbohydrates they have formed to the sap-suckers, which pass most of these carbohydrates on to the soil and thus activate the bacteria to produce more of an element that is ordinarily a limiting factor for plant development. That is, the plant releases a surplus product in order to obtain adequate amounts of a factor in minimum. In such a case we would be dealing not with a plant and a pest, but with a system which – as a system – had been optimized (Owen and Wiegert, 1976).

Spittlebugs and their relatives can occupy tropical forests in such crowds that a continual fine rain drips from them onto the ground. They withdraw huge amounts of water from the plants. With what results? The soils of tropical rainforests contain almost no nutrients. Only the uppermost layer, the leaf litter, contains minerals that can be used by the plants, but in general this layer is too dry for the plant roots to penetrate. Because of the continual sprinkling by spittlebugs, some roots can invade even this top part of the soil and thus provide the plant with nutrients, whereas the greater part of the root mass, deeper in the soil, is responsible exclusively for the provision of water. In this case, too, plant and sucking insect together would represent an optimized system (Owen and Wiegert, 1976).

Studies of soil animals and litter decomposers have shown that they feed chiefly on bacteria, which carry out the actual decomposition. Without these animals the bacteria would be much more numerous. It would appear, then, that the soil animals inhibit decomposition of the litter. But the opposite is actually true. By their constant eating of bacteria, as has clearly been shown for Orchestia, these animals keep the population of bacteria in the exponential phase of growth. If the soil animals were not present the bacteria would rapidly increase to capacity and enter the stationary phase, in which they become practically inactive. By keeping the population density low the animals ensure the highest productivity of the bacteria – that is, the greatest effectiveness in breaking down the litter.

In the last analysis this example is simply another expression of the predator-prey relationship we found for the red grouse. In that case, too, the activity of the predators kept the rate of grouse reproduction as high as possible; without predation, the grouse reproduction rate falls off dramatically.

The lemmings in the arctic tundra show oscillations in population size, with mass reproduction bringing a peak every 3–5 years. The plants over large areas of tundra can be killed when the animals cut through their roots. But without the turnover of the soil that occurs at regular intervals by the large numbers of burrowing lemmings, breakdown of the ground litter in these arctic regions would take an exceedingly long time. The outbreaks of lemmings are necessary if organic material is not to accumulate to excess (Bliss, 1975).

The dung produced by the cattle introduced into the Australian steppes could not be naturally remineralized there. It accumulated on the ground, often drying into a hard crust, and destroyed the vegetation. Moreover, it provided a haven for stinging flies that had also been introduced by accident and proved severely annoying, and it gave the parasitic worms passed out by the cattle in their dung ample opportunity to reinfect them. In 1963 the Australian government began a massive effort to get rid of this nuisance. Many dung beetles were subjected to the most thorough tests, and finally – from April of 1967 on – about 275,000 beetles of four species were released. One species (Onthophagus gazella) reproduced and spread at a spectacular rate. Wherever this beetle settled in well, the cattle dung was worked into the soil in about two days (there the beetle larvae continued to feed on it). Two other African species have reproduced to a similar extent in other parts of Australia, with similar effects. The danger of parasites is very much less, and the troublesome flies have almost disappeared. But at the moment the additional release of mites is under consideration. All dung beetles normally carry mites with them; these move into the dung, reproduce there, and then attach to the next beetle to be carried to the next pile of dung. The European burying beetle Necrophorus is also associated with mites. Its larvae as a rule are subordinate in competition with fly maggots, but the mites of the beetle attack the fly's eggs – the competitors of the beetle larvae are eliminated. The relationship between the dung beetles and their mites is thought to be similar, and therefore experiments are currently underway to provide the dung beetles with their normal mites. No one pays much heed to dung beetles encountered in the countryside; but they can be vital to the functioning of the system, as the Australian example shows (Waterhouse, 1974).

For reasons obvious in these examples, Mattson and Addy (1975) do not regard the animals of an ecosystem in isolation, but rather consider the whole as a system for long-term optimization, with uniform operation over centuries. The example presented on p. 60, in which the ducks consumed the primary production almost completely and thus prevented a lake from reverting to land, is an argument in this direction.

But the principle of optimization becomes especially apparent when we consider the implications of animals as flower pollinators and seed-dispersing agents. Here they determine the structure of the plant community, although their function could not be adequately comprehended in the context of energy flow or the cycling of matter. Remarkably, these functions of animals – despite their outstanding significance to plant communities – are often neglected in botanically oriented analyses of ecosystems. This discrepancy has recently been pointed out, primarily by Heithaus (1974; cf. Bertsch, 1975; Hocking, 1968).

Plants growing in stands comprising a few species, or in natural monocultures, usually are wind-pollinated. The greater the number of species in a stand, and the more nearly random the distribution of the individuals of each species, the more the species are forced to adopt a better-aimed method of pollination, by animals. A logical consequence of pollination by animals is seed dispersal by animals. This connection has rarely been noted; Regal (1977), in particular, has called attention to it. When many species are distributed in a mosaic throughout a system, individuals have a good chance of survival only if the seeds can be transported over considerable distances. Thus, in quite general terms, we find that wind pollination prevails in the northern tundras and in the forests of northern and temperate latitudes, where the numbers of species are small, as well as in windswept steppes. Even in the temperate-zone forests pollination by insects and seed transport by birds gains in importance, and in subtropical and tropical forest regions animals are indispensable to most of the plants. It is true that many plants, whether pollinated by wind or animals, can exist for quite long times (sometimes for several generations, by parthenogenesis or self-pollination) without normal pollination. But in the long run an exchange of genes is necessary, as it is in animals. Just how necessary is evident in the complicated coevolution that has occurred between plant and insect. Examples have been presented in many research reports and handbooks. The enormous energy expenditure by the plants is understandable only in this context. Similar instances of coevolution to ensure seed transport are not known to such a great extent, and studies of this aspect have been fewer. Evidently real specialization, so common where pollination is concerned, is rare in the case of seed dispersal. But the fact that many plant seeds (mistletoe, tomato) germinate with difficulty or not at all unless they have passed through the body of an animal points in the direction of coevolution.

Under heavy grazing pressure most dicotyledonous plants disappear and are replaced by grasses and their relatives. When cattle are allowed into a wood to feed, it becomes overgrown with grass. It is quite certain that grasses, with their ability to grow continually, became so widespread only as a result of the activity of grazing mammals. Grazers can convert a plant community containing many species into a quite different, much less diverse community. A good example is given by the meadows in coastal regions which are sometimes flooded by the sea – the North Sea, for instance. Here many species grow naturally, but when sheep are kept on the meadows most of them disappear. Only meadow-grass (Puccinnellia), with its persistent growth, can withstand the pressure of grazing. The very dense grass vegetation that is produced under the influence of the sheep holds the soil together much more firmly than did the original salt-meadow vegetation. For this reason sheep are a valuable aid in converting these coastal fields into land that can be used for agriculture. The natural vegetation allows channels to open in the ground from time to time, so that conversion to dry land is much less likely; the system is maintained.

Another outstanding way in which plant-eating animals can affect ecosystems is by transmitting plant diseases (fungus and virus diseases). Bark beetles, long-horned beetles, and sap-sucking insects are disease carriers. Previously this fact has been viewed entirely as an injurious aspects of insects, so that at present nothing can be said about its significance in a balanced ecosystem. An ecologist would infer that such diseases do the most damage to already weakened plants, thus making room for other species that find conditions more favorable at that location. The spread of such pathogens would then be regarded as a mechanism by which turnover in the ecosystem is accelerated. But this is pure speculation, unsupported by any data.

The examples discussed so far have illustrated the significance of animals to the ecosystem as a whole. The importance of many animal species to the existence of other animals has been known for much longer, and indeed is self-evident. Woodpeckers that dig out holes provide breeding places for stock doves, owls, hoopoes, rollers, and tits, as well as for the dormouse and many insects. By their burrowing activities foxes in the forest, marmots and prairie dogs in the steppe, and rabbits in the savanna and at the edges of forests provide a continual supply of fresh soil; this is colonized by pioneer plants that otherwise could not exist in such mature stands. The rich fauna in the Florida Everglades results from the presence of alligators. These animals dig deep pits for themselves, where all sorts of other water animals – fish, turtles, insects – can shelter during the dry season, while at their margins even sensitive water plants can survive hot dry periods. If it were not for the alligators the water birds would find no food during the dry season, and the aquatic animals for which the Everglades are famous would die out (Fig. 162). In the mud flats of the North Sea the factor restricting colonization by sessile marine animals is lack of an adequate substrate. Only a few species are capable of living directly on the mud. One of these is the oyster. Where oysters appear, they are followed by a great number of sessile species which can now attach to the oyster shells. It was the observation of this relationship that led to the discovery by Möbius of the phenomenon of the biocenosis, and thus to the actual beginning of the science of ecology.

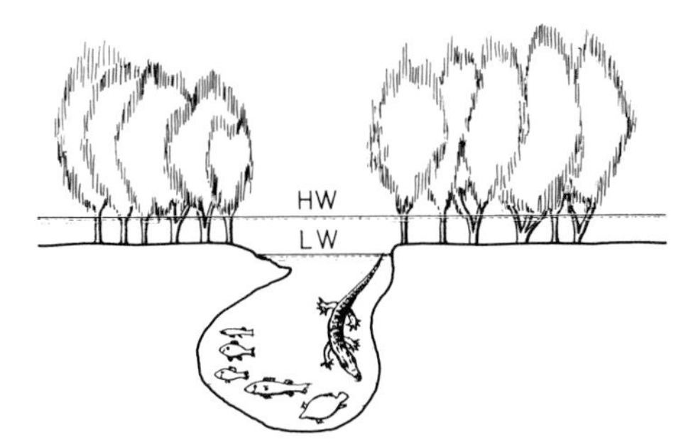

Fig. 162. Alligator holes in the Florida Everglades enable aquatic animals and plants to survive in this region. (George, 1972)

These examples should illustrate the point sufficiently, but let us mention just a few more of the many others that could be cited. The bug Chilaces typhae lives in cattails, but can enter the reeds only if the caterpillars of a moth have bored passages into them. Some parasitic wasps cannot find their host by themselves; they follow the trails of other species that prefer the same host and, having found their victims in this way, proceed to lay their eggs. Their larvae develop more rapidly than those of the first wasps, and thus win the competition against the actual discoverer of the host.

Given all these possibilities, it is quite inappropriate to limit an ecosystem analysis to energy flow and the cycling of matter. To do so would be like making a physiological study of an animal in which nothing but feeding and digestion were considered. The functions of animals described here determine the composition of the ecosystem, the level of primary production (of which they than consume 10%), and the form in which this primary production is delivered (i.e., by which species and which parts of the plant). The action of animals in an ecosystem can be compared with that of switches and amplifiers in an electronic system, or with the sense organs and nervous system of an individual organism. But these effects are very difficult to represent numerically. It is just as futile to try to express in numbers the way a system is affected by the removal of a pollinator or seed transporter as it would be to find a numerical expression for the loss of an individual's vision, hearing, or sense of smell (for further examples, see p. 197).

On land, too, animals can determine the entire structure of their system. Particularly striking examples are the islands colonized by sea birds, which build up large deposits of guano in arid regions and can thus be an economic factor of major importance. Sea-bird colonies situated in trees can kill the trees with their droppings; cormorants have been known to do this. In the water, animals dominate the structure of the habitat as the higher plants ordinarily do on land.

VII. Changeable and Constant Ecosystems

The climax concept implies that ecosystems under uniform climatic conditions (disregarding seasonal fluctuations) remain constant over long time spans. But what is meant by constant, and by long? We have encountered the cycles of many small-mammal populations in the northern hemisphere, and these themselves show that the concept must be qualified. Recently, moreover, ecological research has provided a great deal of evidence that wide fluctuations can occur in large overall systems, even over relatively long periods of time. Because ecological papers have been published only in recent years, there is much in the field that amounts to speculation – documented by more or less scanty evidence. But anyone interested in analyzing the effect of humans on ecosystems must know how constant the ecosystems would be without the influence of modern technology. For lack of anything more reliable, we shall therefore consider in detail a number of speculative suggestions.

In the region of the Neusiedler Lake, on the border between Austria and Hungary, there is very little precipitation. The lake exists only because water flows into it from the relatively low surrounding mountains, where the precipitation is somewhat heavier. It is surrounded by a broad belt of reeds, which is steadily expanding; the lake is very shallow and thus privides an excellent place for Phragmites communis to grow. The precipitation in the region of the lake is much less than the amount of water consumed by the vegetation. Now we know that at times in the past the lake has temporarily not existed; toward the end of the last century it was not there, and before that it appeared and vanished again several times. The following hypothesis suggests itself. At present the reeds are spreading, and using up more and more of the small amount of water that flows in. After a while the reeds will occupy almost the entire lake, and then they will use up considerably more water than can be supplied by precipitation and streams; they will "pump the lake dry." When that happens, the reeds will soon die off and disappear. Where once there was a lake, now there is a dry basin, which begins to fill up with the water that flows into it. The lake is reformed, reeds begin to invade it, and once again it is pumped dry. If this were the case, we would have a regular alternation based entirely on one key species – the reed. (Is this an alternation between two different ecosystems, or should we regard the whole thing as one fluctuating ecosystem?) If the weather conditions were constant over very long periods, the result would have to be a quite regular cycle. But it is certain that irregularities will be introduced by long-term changes in the climate. (Since this hypothesis was proposed for the Neusiedler Lake, doubts have been loudly voiced; but even the doubters firmly attest the likelihood of fluctuating lakes in arid regions.)

The situation in the North American taiga is similar; there stands of spruce and pine are intermingled to form a mosaic. Pines grow very rapidly, and allow a great deal of light to reach the ground. Beneath the pines spruce can grow, and as they do so they gradually suppress the pines, for they cast an almost unbroken shade on the ground. Eventually the pines are replaced by a very dense stand of spruce. Once this has happened, insect "pests" can spread and reproduce at a high rate. They destroy the entire stand of spruce, and pines grow again in its place. The ecosystem "taiga" here would be subdivisible into two temporal ecosystems, pine forest and spruce forest. In the taiga of northern Europe a similar alternation is achieved because the spruce are kept down by fire (Zackrisson, 1977). Calluna heaths seem unable to exist indefinitely. Older plants die off, and their place is taken by lichens of the genus Cladonia. Afterward either the heather returns or it is preceded by Arctostaphylos, which later is replaced by Calluna. One such cycle takes between 50 and 80 years.

It may be that the large mammals on the steppe of eastern Africa greatly overexploit their habitat. In the long run, such overusage results in the dying out of the essential food plants. When the food on which a species relies becomes scarce or unobtainable, the species will vanish from the steppe ecosystem. Once that has happened the plants can reestablish themselves. Petrides (1974) has suggested that in these regions there is a very long cycle in which grass steppe (with the typical large mammals) alternates with thorn-bush steppe (with other typical large mammals). And similar relationships may be found in humid regions (oases) in central Iceland, where rapid reproduction of cotton grass (Eriophorum) accompanied by whooper swans perhaps alternates in a regular cycle with sedge and pink-footed geese (Gardasson, personal communication).

Observations made on Spitsbergen make it seem likely that the reindeer subspecies living there, entirely free of predators or parasites, goes through regular population cycles, with a maximum every 50–100 years. The animals may reproduce until they reach a density of between 15 and 20 deer per square kilometer. In such large numbers they completely use up their chief winter fodder, lichens; at this point the very slow-growing lichens have nearly vanished from the system. Now the population collapses, the lichens spread again, and the reindeer can recover. But a number of other things are associated with these cycles. There is intense competition between lichens and vascular plants. If the lichens are overexploited there is extensive invasion of vascular plants. That is, the reindeer population collapses at the time when production of suitable summer food is greatest. While the reindeer are at their minimum and lichen production is high the sun's rays penetrate deeper into the soil than they do when it is covered by vascular plants. The surface of the permafrost thaws out during the "lichen season" and greater amounts of minerals become available to feed the plants. These also enable the vascular plants to flourish – provided that the reindeer remove the competition offered by the lichens.

There is rather more extensive documentation of the shorter-term lemming cycles in North America; these certainly amount to considerably more than the simple cycle of one small mammal. Here, again, the frozen soil probably thaws to a much deeper level after a lemming peak has destroyed the plant cover, so that more food is available to the plants and these in turn are more nutritious (especially with respect to phosphorus; cf. Figs. 163 and 164). The beaver evidently plays a fundamental role in boreal forest regions, for it can convert forest to lakes which ultimately, after the beaver dam has broken, become meadows and eventually are reforested.

A great deal of speculation is inherent in all these examples. But we must now accept it as probable that many habitats are not actually constant, but on the whole undergo extremely wide fluctuations. Only very long-term ecological studies, in which the alterations in plant communities are continually monitored, can provide information about them. The time for ecological "snapshots" is past. From now on we must start with the notion that long-term changes of systems are entirely possible and should be examined.

Whereas in the cases previously cited a single "key species" has subjected a whole system to dramatic oscillations, in other cases the inconstancy of individual species appears not to involve the whole system to such an extent. All our forest trees – oak, beech, maple, pine, spruce – produce seeds only at irregular intervals. This is interpreted as an adaptation to plant-eating animals (cf. p. 145 f., Fig. 106). In an acorn or beechnut year seed production can be so high that all the herbivores in the forest could live, in theory, for more than two years entirely on acorns or beechnuts. Naturally, the fauna changes markedly during a seed year. Some species – the crossbills – are precisely adjusted to seed years. Both male and female crossbills sing, and they can breed at any season; they wander throughout a very large region and breed when and where they find food. In North America the extinct passenger pigeon was presumably an ani-

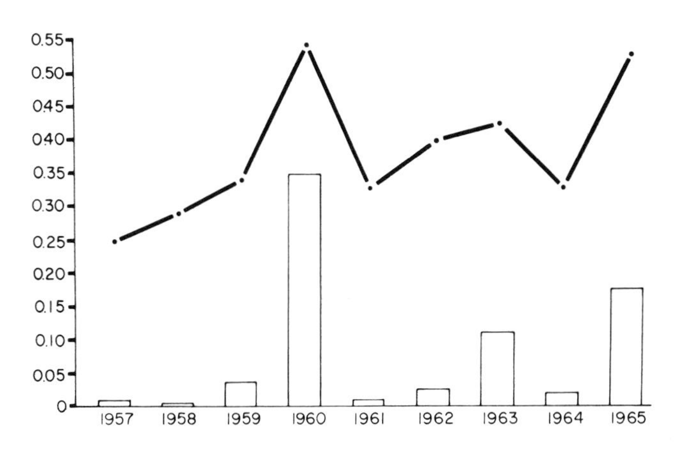

Fig. 163. Phosphorus content of plants *(curve)* and relative size of the lemming population *(bars)* in Point Barrow, Alaska. (Schultz, 1972)

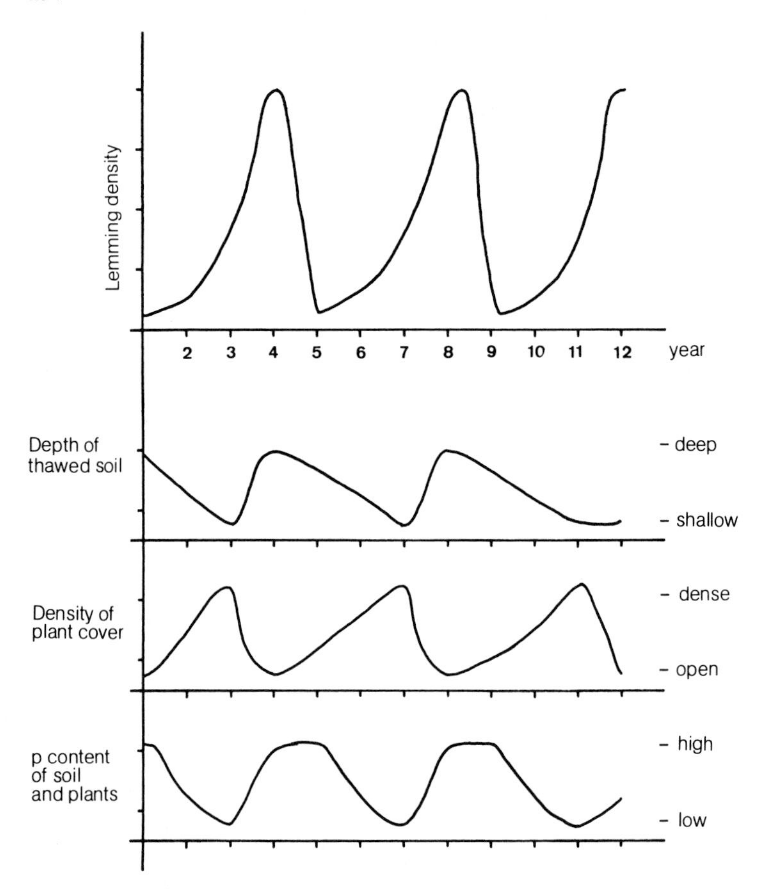

Fig. 164. Lemming cycles and their consequences, as postulated by Pitelka and Schultz in Alaska. High lemming densities tear open the closed plant cover, the permafrost thaws, nutrients are released from the soil so that the plants make better growth, and eventually the plant cover closes again. At the same time the plants contain more minerals and thus enable renewed growth of the lemming population. The hypothesis, in this strict form, could not be confirmed. (Remmert, 1973)

mal that also depended on years of abundant seed production by forest trees. In years without seeds the pigeon appears to have been not at all common, and to have lived in individual pairs. Only in seed years did they gather into the vast colonies that were so famous. Now their extinction is ascribed to the fact that humans reduced their numbers drastically after a series of prolific seed years, so that not enough birds were left to survive several years without seed production. For variations in seed production to occur special climatic conditions are required (certain temperatures during the autumn of the preceding year and spring of the current year), and the time elapsed since the preceding seed year is also a factor. It is uncommon – especially in the case of the oak – for two seed years to occur in succession. In a seed year the reserves deposited in the pith rays of the trees are largely exhausted, so that the trees are ill-prepared to produce seeds in the following year and do so only if the climatic conditions are especially favorable. Even a primeval forest in central Europe does not present a constant picture on the whole. Within it there is a mosaic of zones of rejuvenation, aging, maturation, and collapse (Fig. 165). In a forest that is uniform as a whole both shade and sun plants can thrive, and animals of the

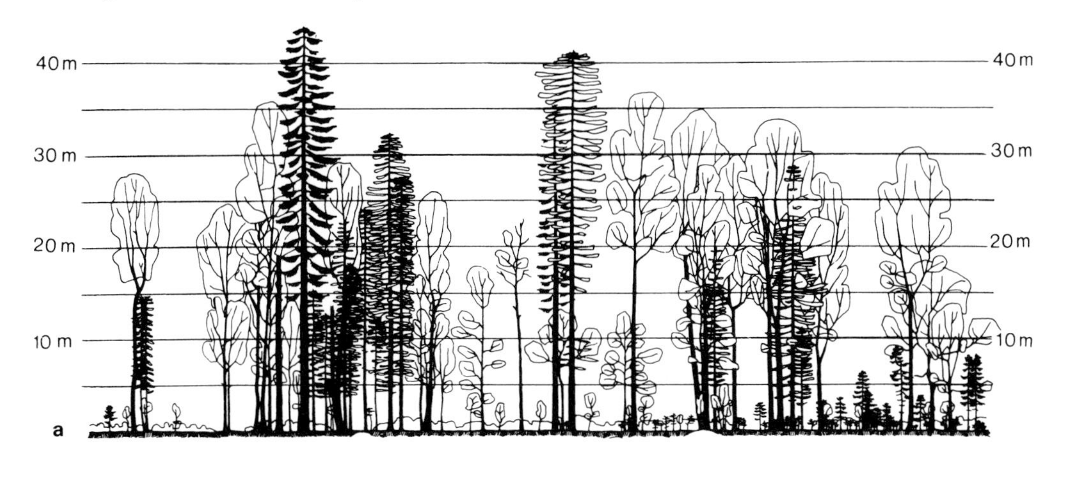

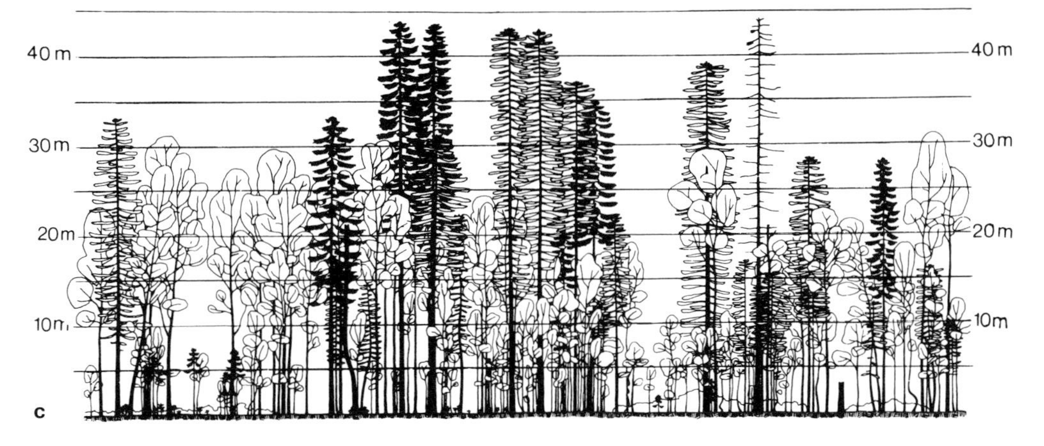

Fig. 165 a–e. The successive phases of a region of primeval forest in Austria (reconstructed from phases simultaneously present). **a** Beginning of rejuvenation, with scattered trees of extreme age. **b** Pronounced rejuvenation. **c** Highest growth rate. **d** Stage of maturity. **e** Aging phase. (Walter, 1973)

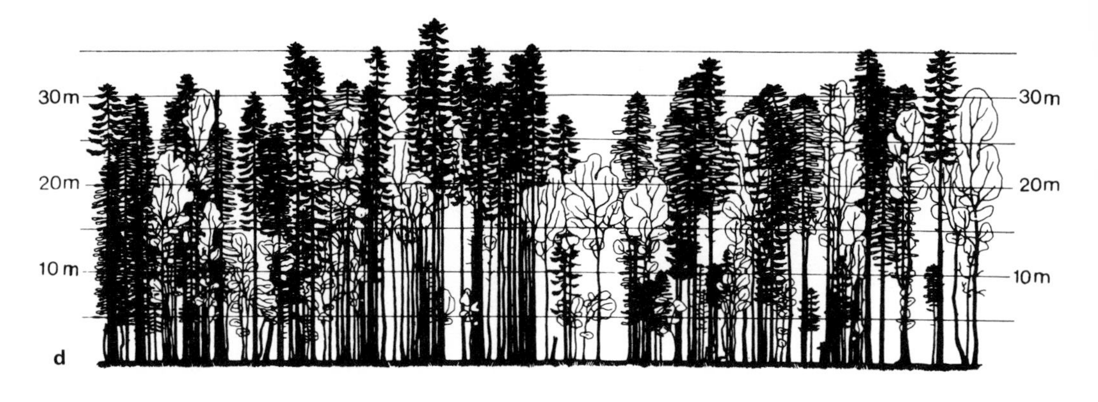

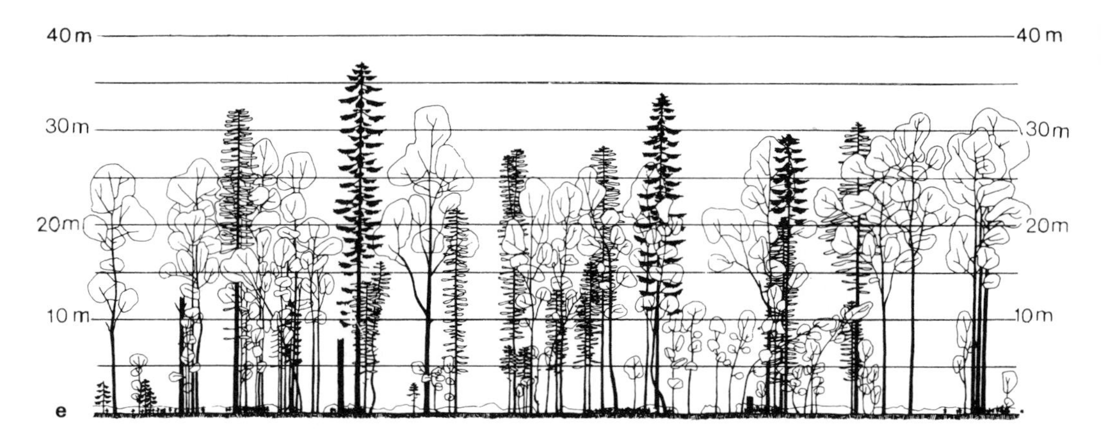

Fig. 165d, e: Legend see p. 235

herbaceous stratum live next to those of old and dead trees.

In the previous examples we have postulated an oscillation of the systems under uniform climatic conditions. In the following paragraphs we shall turn to examples of ecosystem oscillation that results from variations in climate. In reality, of course, the weather is never exactly the same from year to year. Accordingly, the flora and fauna are different in successive years. These changes are generally underestimated. Populations of both animals and plants in central Europe can be altered by a factor of 10 or more, simply under the control of climatic conditions. This is true, for example, of a dry grass region in southern Germany that was investigated over a long period (Figs. 120, 121, and 166–169). Not only was there a pronounced weather-dependent change in the species composition of the flora; the total production of above-ground material and the composition of the fauna were also changed. There were barely $^1/_{10}$ as many crickets and grasshoppers at a time when the fly population became very large. After a "favorable" summer the original proportions were largely restored. Corresponding fluctuations, based entirely on climatic conditions, have been shown to occur among birds and mammals (Figs. 170 and 171; cf. pp. 164 ff.).

Such variations in population size can be accompanied by large-scale transient or prolonged changes in the area occupied by the organism. For example, Bonelli's warbler is normally restricted to the region of the Alps and certain parts of the foothills. In 1947 and 1948, however, its range suddenly expanded to the northern edge of the central European mountains – into the

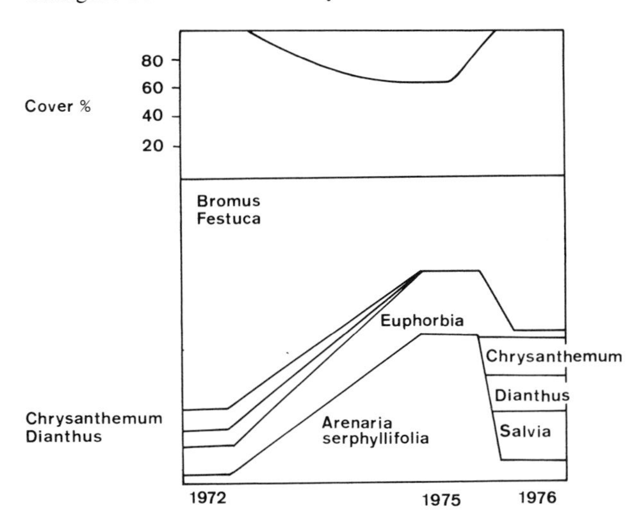

Fig. 166. Changes in amount of plant cover and in the composition of the plant community on a dry meadow in southern Germany

Harz, the Weserbergland, and as far as Deister, near Hannover. During these two years the bird was not at all uncommon in these forests, but then it vanished just as rapidly as it had appeared (Schlichtmann, 1951).

These completely "natural" oscillations in populations and systems make it infinitely difficult to demonstrate clear effects of newly introduced factors such as environmental poisons, tourism, and management procedures. And the legal system lays the burden of proof on the ecologist, rather than requiring the various manipulators to prove that their activities are not altering the system.

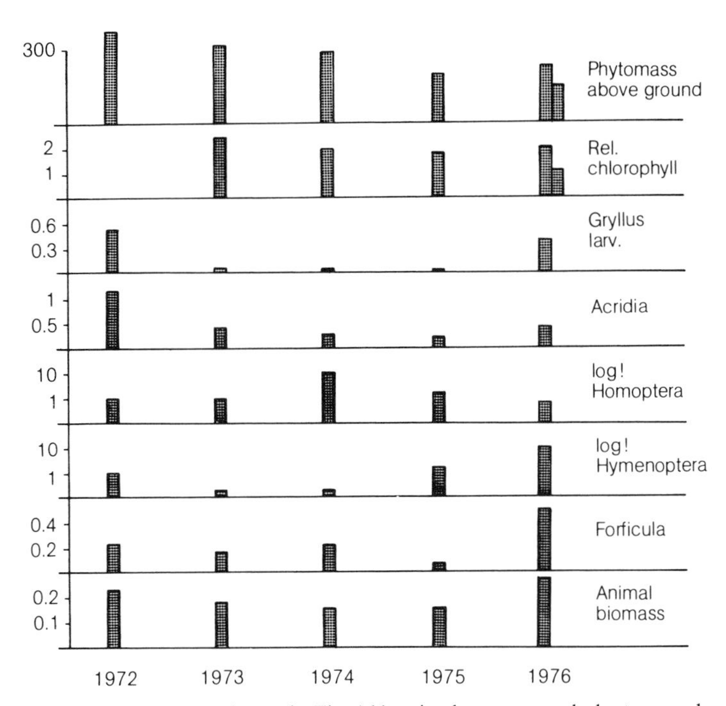

Fig. 167. The same region as in Fig. 166; animal captures and plant mass above ground (July)

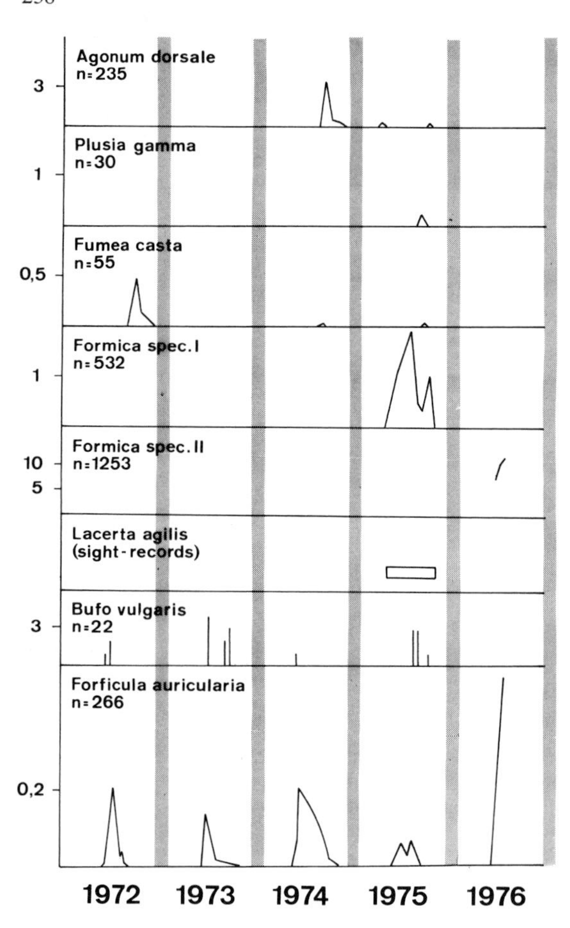

Fig. 168. The same region as in Fig. 166; capture frequency of particular species during the years of the study

There are systems which by their constancy contrast sharply with those just described. In general tropical rainforests and coral reefs are regarded as constant systems, and the meiofauna of the sea floor and the shores appears to be in this category as well. Fluctuations in these are thought to be slight and to affect only parts of the system, so that the system as a whole does not change. But as yet no studies have been done in which really solid data have been obtained for the same spot over many years, so that at present this claim cannot be supported by facts. Still, it will probably prove correct. At least in comparison with the changeable systems we have mentioned, these systems may well be considerably more constant, if not absolutely so. The mass outbreaks of toucans that have occasionally been said to occur in the jungle of central Brazil remind us of the need for caution. Very long-term studies of these systems are urgently needed.

It has often been assumed that small forms are typical of r systems and large forms, of K systems (cf. Table 7). In many cases this is correct, and it is probably generally valid to say that very large representatives of a particular systematic group tend under selection pressure to move toward the K part of the r-K continuum. But where small animals are concerned, the situation is much more complicated. In fact, the fauna of our soils – that of a forest floor or the meiofauna of the ocean bottom – appears to be composed of forms that have adapted far toward the K end of the continuum (cf. Fig. 3, p. 10). Some of these, despite their small size, have a very low rate of metabolism per unit weight (Zinkler, 1966), and

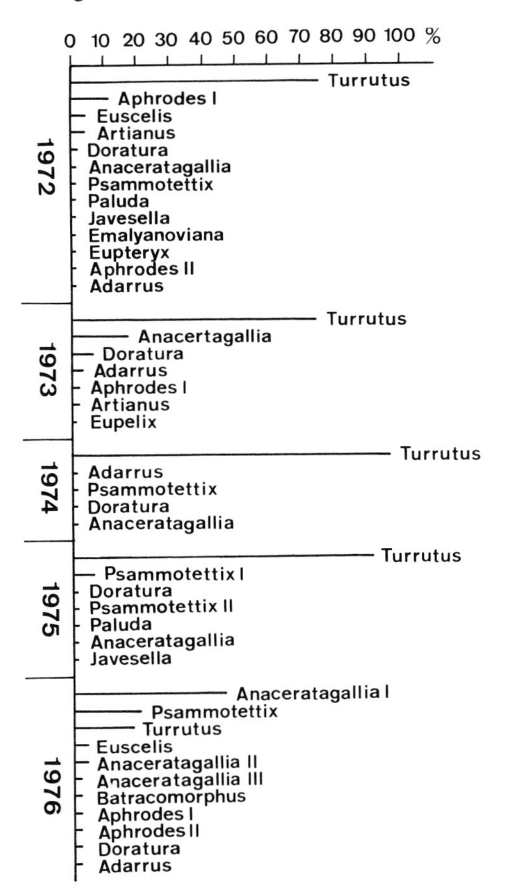

Fig. 169. The same region as in Fig. 166; the appearance of leafhoppers in the area studied (as in Fig. 167, considering only July of each year). Each bar represents one species. Diversity ranges from 0.22 (1974) to 1.73 (1976)

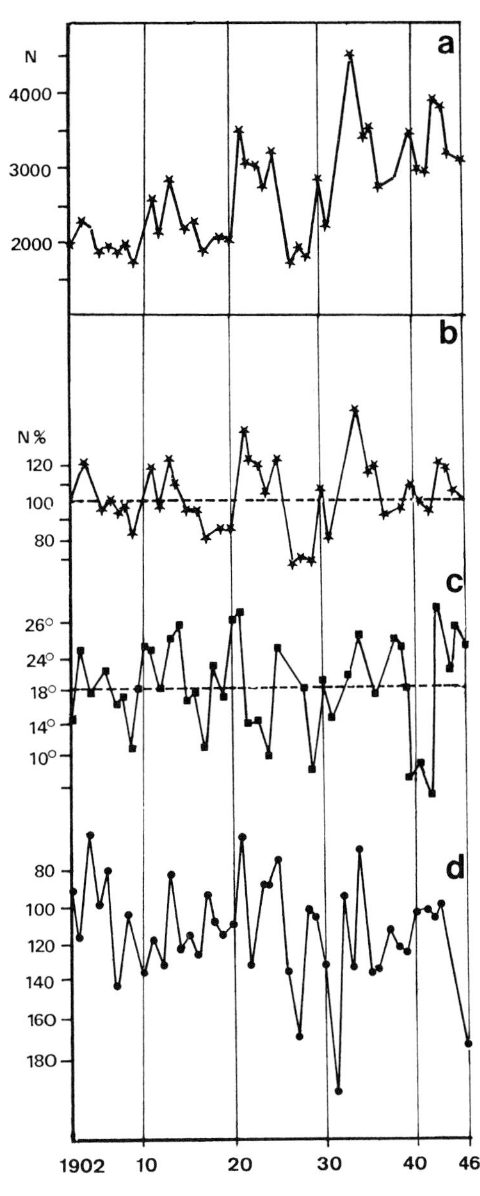

Fig. 170 a–d. Dependence of hare population size on weather factors (Denmark) **a** Number of hares shot by hunters. **b** The same curve with the rising trend eliminated. **c** Associated temperature curve (°C). **d** Precipitation curve. (Andersen in Huber, 1973)

most develop very slowly. Many mites and springtails in the soils of central Europe have only one or two generations per year (Schaller, 1963), and the same is true of many members of the meiofauna on the ocean bottom. Some turbellarians, polychaetes, archiannelids, and nematodes in this habitat have only a single generation per year, even though they are very small (Ax, 1968; von Thun, 1968). The rate of reproduction of all these animals is considerably lower than that of similar animals in more evanescent habitats. It is likely that the higher metazoans of the soil or ocean bottom never undergo parthenogenesis or asexual reproduction; on the contrary, they exhibit complicated mating behavior (Schaller, 1962; Ax, 1968) and the sperm is in many cases transferred by a spermatophore (cf. Fig. 3, upper right, Microhedyle). None of these features are integral consequences of their small size.

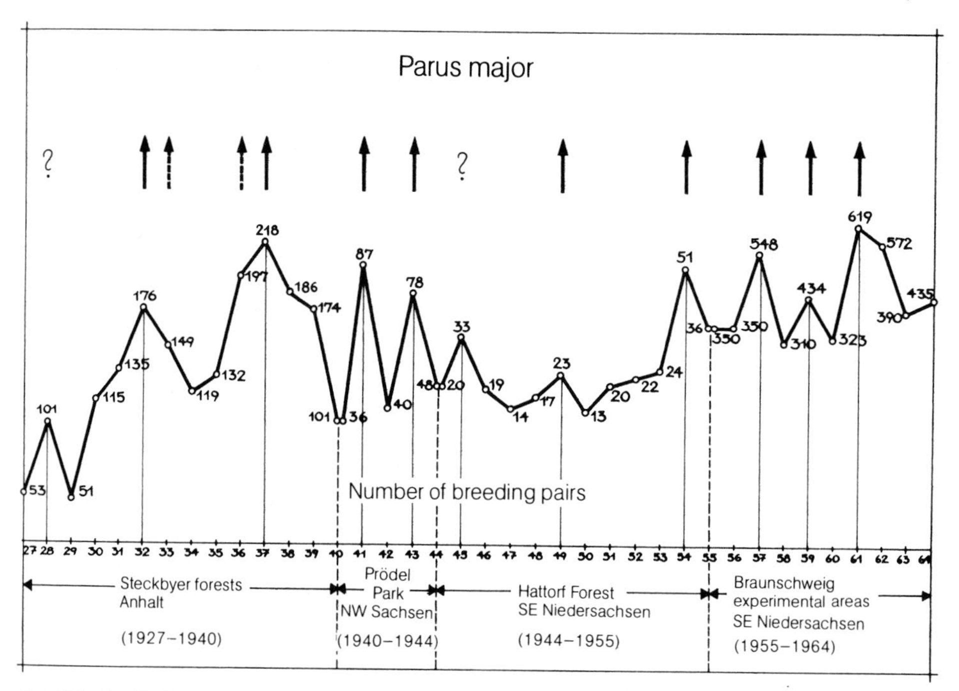

Fig. 171. Oscillations in the breeding population of the great tit (1927–1964). Because the studies could not be limited to a single region, different regions are included, as indicated at the bottom of the graph. (Berndt and Henss, 1967)

Other metazoans of similar size and systematic status have very high rates of development and reproduction (nematodes, mites, and springtails on animal corpses and in excrement; mass outbreaks of the mites and springtails that are "pests" in human grain fields and storehouses). But it is particularly notable that all these animals have large numbers of species, with which a high degree of food specialization is associated. Most soil mites of the family Oribatidae are strictly specialized to feed on either fungus hyphae or spores and bits of lichen, or on leaf litter and wood. There are also a few less specialized forms (Schuster, 1956). An examination of the digestive enzymes involved showed that the individual species differ in very characteristic ways; some can digest pectin with no difficulty, and others cannot digest it at all (Zinkler, 1971, 1972). In animals inhabiting beach detritus similar phenomena were found (Overgaard-Nielsen, 1966). This property is probably related to the observation that different forms which evidently require the same food need quite different proportions of animal protein. Finally, this system also includes highly specialized predators; some mites feed almost exclusively on fly eggs, and many nematodes are specific small-animal eaters (cf. p. 38 and Fig. 18). Quite a number of soil fungi have adapted to preying on small animals by developing specific mechanisms for their capture. On the whole, therefore, these communities of miniature animals meet the criteria of a K system (Table 7) more completely than any group so far analyzed.

VIII. Constancy and Stability

Since we have become aware that man's activities may profoundly affect ecosystems, we are more than ever concerned with understanding how stable they are. An extraordinary number of papers have been published that deal with this question, some of them in a purely speculative way. Only very recently have the available material and the conclusions drawn from it been subjected to critical scrutiny. At the outset it was necessary to establish a new terminology, in which the physical significance of terms used in everyday language was taken into account. In this terminology a constant system is a system in which under the prevailing climatic conditions only relatively small changes are observable. The term "inconstant" is applied to a system which exhibits more or less distinct differences when the same season is compared over many years. A stable system, when exposed to exogenous influences – the arrival of a new species of organism, unexpected climatic conditions, human intervention – is not fundamentally altered. Sensitive systems respond to such external influences by changing in a way that cannot be compensated. Elastic systems initially respond like sensitive systems, but relatively rapidly return to their original state (Orians, 1974). (These definitions differ greatly from those formerly in common use; the latter were imprecise and should no longer be used, for they equate constancy with stability.)

Once the concepts and terms had been clarified, it became clear that constancy and stability are no more equivalent than inconstancy and sensitivity. In the preceding section we discussed systems that were not subjected to any external influence. Now we shall consider what happens when sudden changes are imposed.

A critical analysis of such questions cannot be strictly limited to considerations within the realm of the natural sciences. In principle it is impossible to predict whether the addition of an animal to a system is more significant than the removal of an animal. We cannot say whether such biological factors are more or less important then human intervention by way of fire, bulldozers, insecticides or herbicides. Nor can we weigh the various forms of human intervention against one another. And in evaluating any single influence, we must take into account the season during which it acts. We know that fires have quite different effects in spring than in summer or in autumn, and that fires spreading through the terrain against the wind have much more severe consequences than those swept along with the wind. When we discuss these questions here, we must remember that it is inappropriate to compare even superficially similar effects, whether quantitatively or qualitatively.

The profound effects of intervention in existing systems are well illustrated by the islands on which early seafarers turned loose goats and pigs, in order to use them as provisions on later journeys. For example, when the island South Trinidad (20° S) was discovered in 1700, the goats and pigs that were introduced destroyed all the existing woodland. In the nineteenth century no trees were left alive. At that time the island was a gruesome sight to the sailors; from far away they could see the dead trees, projecting like skeletons into the tropical sky. Eventually they fell to the ground (Murphy, 1936). The collapse of an ecosystem does not always take such a dramatic form, or occur at a time so distant from its cause. On the originally forested island St. Helena goats were introduced in 1502 (some sources say 1513), and here too the forest and all the native fauna were destroyed. In 1731 the endemic tree species and the land snails that had lived in them were declared to have been destroyed. Darwin, on his journey (1836), found hardly any but English species of plants on the island (quoted by Murphy, 1936). On both of these islands a new system came into being. This system, too, can be relatively easily changed by the introduction of new species.

That the equilibrium state reached after intervention represents a system in itself, which must be considered no differently than the original system, is illustrated by the island Amrum. This island originally had no forests; after several small-scale plantations were established, from the

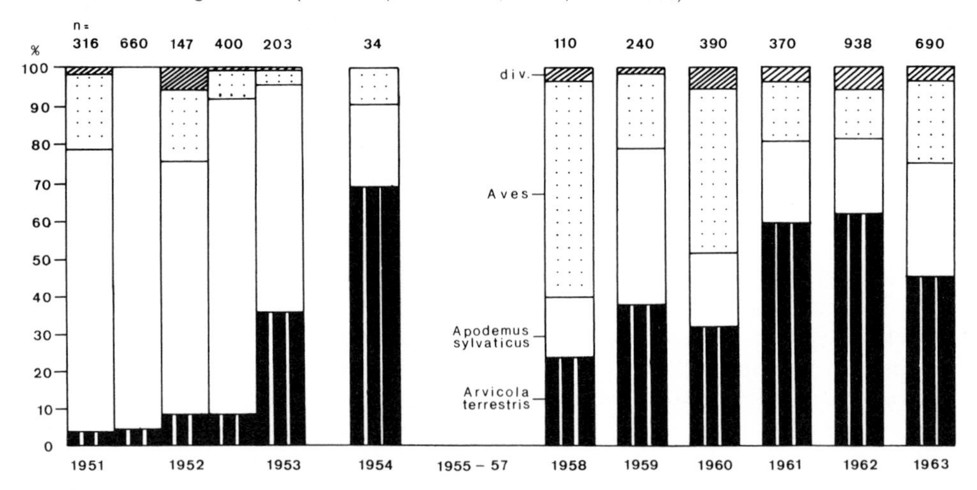

Fig. 172. Changes in the composition of the small-mammal population (as reflected in the changed composition of the prey taken by the long-eared owl) on the North Sea island Amrum, after large-scale forest planting was begun. (Remmert, 1964)

1950 s on the large-scale planting of mostly non-European trees was undertaken. The mouse Apodemus sylvaticus (which does not live in forests), formerly very common on Amrum, declined greatly in number, whereas the European water vole (Arvicola terrestris) became widespread. At present we cannot be sure whether the water vole had previously lived on Amrum, but if it did it was extremely rare. In the new plantations it found ideal conditions, and it damaged these plantations to such an extent that in several cases they had to be replanted. From the woodland the voles invaded the dikes that protected the island and undermined them extensively. During the next severe storms the dikes broke, and the sea flooded good pastureland. Because of the planting of trees on a primarily treeless island known to have been colonized and cultivated by man at least since the Bronze Age – an indefensible procedure, from the viewpoints of either ecology or forestry – this man-made system got out of control. Now a new equilibrium is likely to be reached, but it will have nothing in common with the previous equilibrium, reached under human influence. Instead of the previous central heath of the island (with Empetrum nigrum as the dominant plant), the center of the island will be invaded by a bushy woodland with an entirely different fauna (Fig. 172).

This sort of modification of a system has been known to occur in many places. The destruction of the forests in those parts of Ireland and western Scotland with abundant precipitation led to the formation of high moors; the destruction of tropical rainforest regions resulted in poor secondary forests on sandy soils with little precipitation and low productivity. In the Scandinavian birch forest mass outbreaks of the moth Oporinia autumnata (Fig. 151) destroyed the trees and pushed back the boundaries of the forest. In place of the forest we now find a typical tundra with typical tundra equilibria.

There are other systems that resist change even when exposed to strong external influences. One of these is the central European beech forest, which can tolerate regular logging, removal of the wood, and plowing and still, in general, regenerates. Naturally such regeneration does not proceed altogether smoothly. In logged-over or cultivated parts of a forest, for example, the fluctuations of mouse populations can

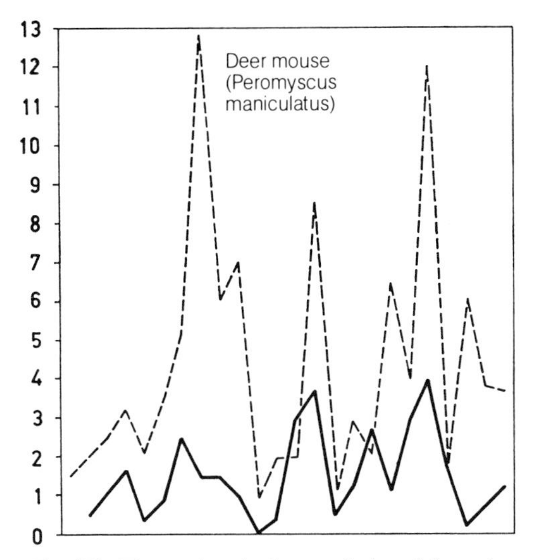

Fig. 173. Fluctuations in the population of deer mice in a clear-cut area *(dashed line)* and in the adjacent primeval forest *(solid line)*. (Tschumi, 1973)

be much greater than in the virgin forest (Fig. 173). The Phragmites stands on lake shores are also extraordinarily resistant to fluctuations in water level, to mowing, and to poisoning. The only thing that can cause large-scale damage to Phragmites is the force of large waves at a time when the young plants are breaking through the water line. Even then regeneration can be rapid, and as a rule occurs with no difficulty.

Because the terms "constancy" and "stability" have been confused, and on the basis of theoretical considerations, species diversity has often been regarded as the cause of stability. Constant systems are in fact characterized by a great variety of species. The use of the word "food web" has caused many to adopt the appealing notion that each organism, as part of a network, contributes to keeping the entire network together. On p. 210 it was mentioned that there has as yet been no quantitative confirmation that such networks exist. On the contrary, large streams of energy seem to flow in certain directions, with little quantitative importance attaching to the bridges between these major lines of flux. Moreover, the theoretical considerations are not compelling. Species diversity can also be likened to a framework with a very open structure, compared with which poverty of species resembles a simple wall. The delicate filigree can be much more easily destroyed than the solid masonry. Finally, remember that food – energy flow and the cycling of matter – is only one aspect of the system (p. 226 f.).

Today it is becoming more and more widely accepted that inconstancy and stability or elasticity of a system are related. Arguments favoring this view include the fact that the destruction of vegetation by mass outbreaks of an insect or lemming amounts in the end to the same thing as destruction by a human driving a bulldozer. Systems adapted to mass outbreaks of insects or lemmings can also recover after attack by a bulldozer. Systems like the reeds fringing lakes, which can withstand bombardment by ice fragments and waves in the spring, desiccation and flooding, and the formation of H_2S, can also survive exposure to poisons of human origin and other human machinations. In this view systems with many species, such as the tropical rainforest or a coral reef, are specially sensitive to exogenous influences – whether they take the form of removal or introduction of an organism or of human intervention. In this view, such regions should be handled with great care during the current population and technological explosion. In this view species diversity would be a mechanism to bring about biological constancy in the system – but an extremely vulnerable and thus hazardous mechanism, which could evolve only during long periods of constant abiotic conditions, largely free of marked exogenous fluctuations. [The degree of species diversity that some systems can reach is brought out by a glance at the Amazonian rainforest; within a test area of 2,000 m² Fittkau and Klinge (1972) found over 500 species of trees and palms over 1.5 m in height.]

But species diversity must once again be specially defined. A zoological or botanical garden contains a very great number of different species – much greater, in the case of certain plant and animal groups, than we would ever encounter in the field. A high degree of diversity (which implies an approximately uniform frequency of occurrence of the comparable species) can be a mechanism to ensure constancy only if it evolves in a comparatively constant habitat. In such a situation three things appear to produce constancy. First, very many species with largely overlapping ecological requirements coexist and can in part replace or represent one another; this phenomenon can be particularly effective when the population density of a prey object is variable, as we have seen on p. 144. This mechanism keeps all populations low and in a state of high productivity. Second, the fact that the individual organisms are in general likely to be randomly distributed in the biotope makes it harder to locate them. It does occasionally happen that there is a mass outbreak of an insect on a tree, but the remaining trees of the same species are not discovered and are spared. The factor described on p. 211 operates in the same way; that is, in the presence of various plant species of different form and with different smells the sense organs of an insect, for example, are not as successful in guiding it to the correct food plant. Third, organisms can kill one another off. This effect seems to operate primarily at the level of the roots of higher plants. Many compositaceous plants have in their roots substances that kill root nematodes and thus drastically reduce the populations of these worms in the soil. The consequence is that other plants in a mixed stand are also protected from infestation by nematodes (Gommers, 1973). But all these mechanisms are vulnerable to a change in environmental conditions and thus present a risk; they bring about constancy only as long as the conditions under which the system evolved remain unchanged.

This view marks a radical change from the way stability was interpreted only a few years ago. Whether the new interpretation is correct only the future can tell. But because the question is of such central importance to man in his environment, it is crucial to begin major empirical studies that can be continued for long periods. We have as yet no really reliable information as to whether the existence of "natural" islands within a cultivated landscape automatically ensures a degree of buffering of the cultivated areas against calamitous invasion by undesirable organisms, or not. It is almost too late for such studies, but they should be undertaken as quickly as possible.

No one denies that strips of vegetation to protect against wind and water are important in regions where wind or water erosion is a problem. But it is not clear whether such small areas of woodland among cultivated fields also exert a benign influence on the fields by way of their flora and fauna, or whether any such influence can counteract the disadvantages – these woods make the operation of machinery more awkward and take up space that could otherwise be used to increase the harvest. On the basis of general ecological experience, the expert is a priori inclined to favor the woods. But such woods do offer a refuge from which weeds and undesirable animals can spread out into the fields – the woods are not inhabited only by the species the farmer would prefer! Only very long-term studies can be of help in resolving this problem.

It also happens in many cases that the introduction of new organisms to a system has no particular effect. The American gray squirrel has crowded out the native squirrel in many parts of England, but in all other respects the system appears to be unchanged. The introduction of reindeer to central Iceland or onto South Georgia seems to have been tolerated by those systems with no difficulty. The same is true of the introduction of the rainbow trout to central European streams, whereas the armored catfish, a spawn-eater, decimated the populations of other fish species. Many systems can also respond elastically. When Elodea canadensis was accidentally introduced to central Europe, the plant multiplied in all bodies of water that did not flow too rapidly, to the extent that it began to be called "water plague." There was a real worry that all the ponds, lakes and rivers would be clogged. But the system was flexible; after some time the plague plant became less common, and today it is an inconspicuous feature of European lakes and streams. The reasons for this recession are not entirely clear; some ascribe it to a fungus and others to a nematode. Nor is it at all certain whether the two species suspected were already present in Europe or came from America and became established only after the plant was widespread.

The new view of elasticity and stability raises more questions than it provides answers. How do natural monocultures such as a beech forest or a stand of reeds come to be elastic or stable? Why are our artificially created monocultures so sensitive? It cannot be due to the quality of the plants as food, for beech leaves and Phragmites are also highly nutritious foods. Is it entirely a matter of soil conditions, which when the turnover is low always provide substances to support new growth? Has it to do with special "key species" that are capable of such resilience (but what makes them "capable"?)? Here is one of the great challenges to zoology, about which not even speculation is possible. Using the r-K formalism explained in previous sections, one could say something like the following. A cultivated plant, under our conditions of cultivation, is shifted along the K-r continuum, moving progressively away from the wild form toward the r end. It is true that all our crop plants are within the r range at the outset, but breeding enhances this tendency; the plants become less resistant to competitors, channel much of the substance produced into the reproductive organs, and are short-lived. On the other hand, the "pests" that live on these crop plants shift from their natural r position into the K range; their food is always abundant. These animals, which had always been selected for numerous progeny and which had evolved no intrinsic mechanisms for population control, are suddenly confronted with an inexhaustible, ever-present source of food. This is but a preliminary formulation, but perhaps it deserves further development.

It will also be necessary to find a way of treating the influences to which a system responds so that they can be compared and quantified. No ecologist doubts that the large number of species in highly diverse habitats represents a method of keeping the frequencies of occurrence of organisms constant under constant, very favorable conditions. Here the introduction or removal of a single species can perhaps be compensated more easily than in other habitats where there are fewer species. However, the different species have different value ratings. To give an extreme example – the removal of the beech from a beechwood certainly has an effect different from the removal of a particular species of springtail.

The difficulty of analyzing the effects of intervention in systems comprising many species is quite evident in the example of the Antarctic Ocean. Here whale fishing has brought about a drastic change in the predator spectrum of krill (a planktonic crustacean, Euphausia superba). Whereas the whales once dominated this spectrum, now seals play the most important role. The stock of baleen whales has fallen from about 43 million tons to 7 million tons. It can be calculated that this amounts to a difference of 150 million t in the annual consumption of krill, under the assumption that the whales used to eat the same amount of krill per ton body weight as they do today. In fact, however, it appears that as the competition for food decreased the relative consumption has risen, which has led to an increase in the rates of growth, maturation, and reproduction in blue whales, rorquals, and sei whales. There are some indications that other krill consumers have also profited from the reduction in the whale population; crab-eater seals (Lobodon) now in some locations become

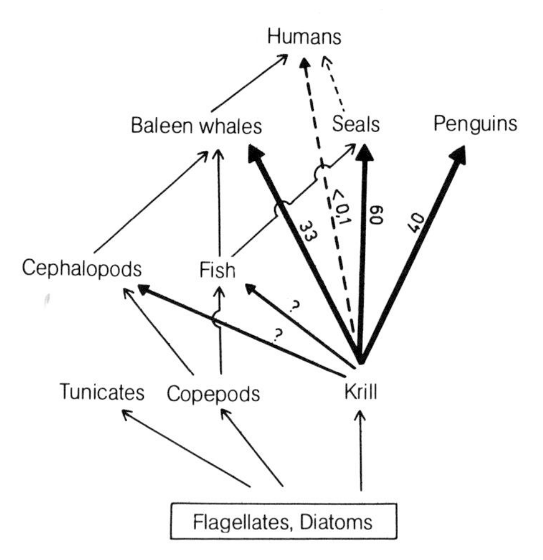

Fig. 174. The central position of krill in the antarctic food system. The numbers indicate the present-day krill consumption, in millions of tons per year. (Hempel, 1977)

sexually mature at the age of 2.5 rather than 4 years, and the penguin colonies seem to have increased, although the primary limiting factor in their case is thought to be the availability of nesting places and winter food. The flows of energy we observe today are illustrated by the diagram of a food web in the Antarctic Ocean (Fig. 174). Before whale fishing was begun in the Antarctic, the baleen whales consumed not 33 million tons of krill annually, as they do today, but about 180 million tons; much of the remainder is evidently now consumed by seals, penguins, and other organisms. But the calculation does not work out exactly; the uncertainty to which all such estimates are subject is apparent (Hempel, 1977). It is evident that the constancy of ecosystems and their members is in general considerably overestimated. This conclusion can be drawn from many of the diagrams presented here, but must be emphasized again. The researcher studying a particular case may, if he is not on guard, take years that are optimal with respect to his special interest for normal years. Fisher (1954) has pointed out this danger. A primeval forest exhibits very long cycles in which the emphasis on the various parts of the mosaic gradually shifts (Fig. 165). But very brief fluctuations can probably be observed in all animal populations. They are found among tits, for example (Fig. 171), and fluctuations of the same form are likely to affect all small birds. Even marine faunas – the prime example of constancy – can be subject to pronounced fluctuations. That of the Baltic Sea is particularly well known. The population of guillemots and razor-billed auks can fall almost to zero after extremely cold winters, and decades are required for it to return to a size of about 50,000 (which is evidently the maximum for the Baltic Sea; cf. Remmert, 1955). Extraordinarily cold winters also have a marked influence on the actual marine fauna. The shore fauna (barnacles, mussels) is almost completely destroyed. Mussels can release their byssus threads and drop into the depths, from which they return to the surface in the spring (Tiedtke, 1964). The situation is similar in the North Sea (Ziegelmaier, 1964, 1970). In attempting to set up an energy budget for the clam Scrobicularia plana on the British coast, Hughes (1970 a, b) encountered tremendous difficulties. Models can be used only if the population is in equilibrium, but Hughes found not a single population in that state. The fluctuations of the bottom fauna in the North Sea can be derived from the stomach contents of flatfish; in different years those occupying a given region consume quite different kinds of food (Fig. 175; Thorson, 1957). In order to determine the influence of environmental chemicals on the fauna of the North Sea bottom, a program has been underway for years in which regular collections are quantitatively evaluated. So far no deleterious effects have been detectable, but there have been very dramatic changes in the stocks of all species. The number of species varies between 80 and less than 20, and the number of individuals per square meter in the macrofauna ranges from about 300 to 1500 (Fig. 176; Rachor and Gerlach, 1976). A now famous phenomenon is "El Nino," a displacement of the food-rich, cold Hum-

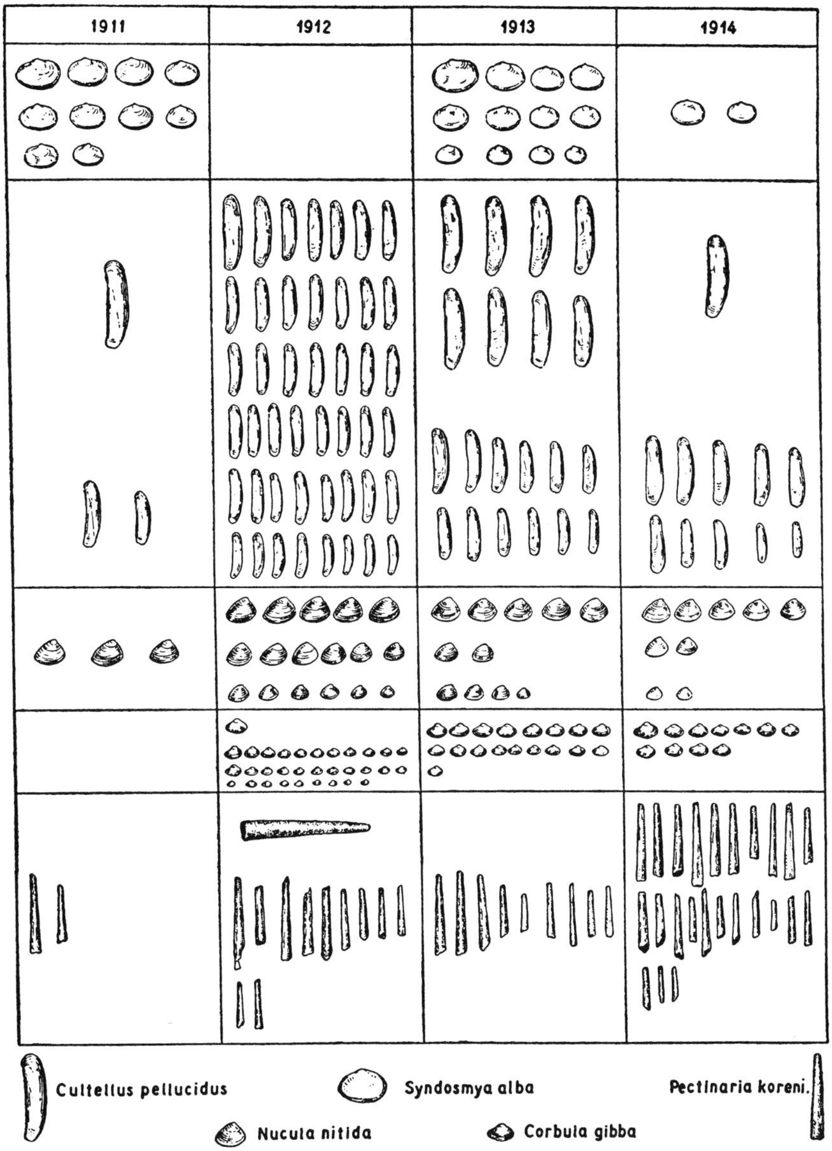

Fig. 175. Changes in the food spectrum of the flatfish in the western Lim Fjord, Denmark. (Thorson, 1957)

boldt current by water in which food is scarce, which leads to collapse of the population of guano birds on the Peruvian coast as a result of collapse of the fish population. When the Humboldt current, carrying its abundance of food, is pushed into the depths, the high primary production that normally distinguishes this region is interrupted. The primary consumers display a correspondingly reduced rate of reproduction, and the fish population collapses. When the currents interact in this way, as happens at irregular intervals of between 1 and 13 years, the event is reflected in the prices of fish and fodder supplies throughout the world (Idyll, 1973).

As yet no long-term quantitative studies of this sort have been made in warm tropical regions, so that it is hard to say whether the constancy that has been claimed for them

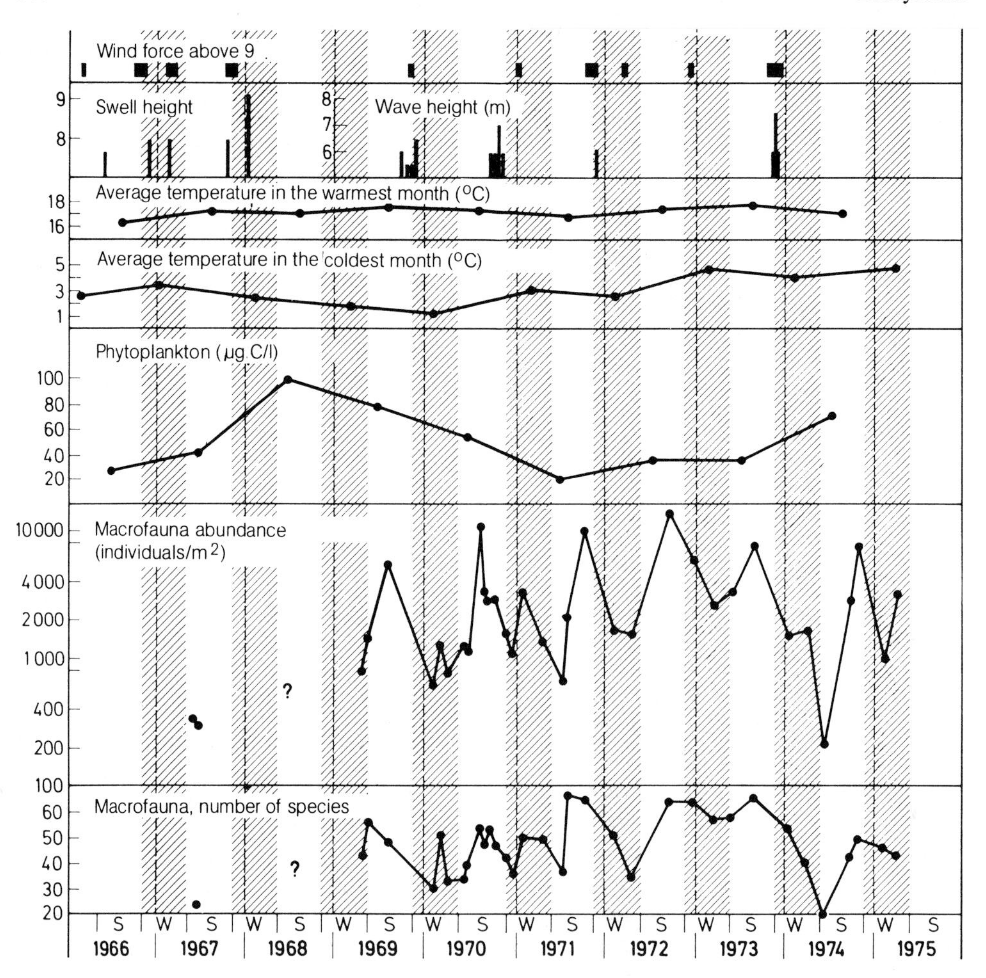

Fig. 176. Changes in the macrofauna on the bottom of the North Sea near Helgoland, in successive years. (Gerlach, 1976)

will be found at a quantitative level. Not only must these studies be continued over a really long period, but even then it must be admitted that constancy is always relative. There is no place in the present and future literature for exhaustive analyses of the ecological situation at a particular moment, if they are not embedded in a well-defined broad concept. Of what use is even the most meticulous determination of bird population density, with all the conceivably related parameters, if it reflects but a single year? It can only cause confusion, for it is entirely likely that the population in the next year can differ by a factor of ten. Only really long-term programs are a useful contribution.

That populations of organisms fluctuate, for reasons usually not immediately discernible, presents a problem of considerable practical significance. We have seen that such fluctuations may be quite irregular; many species, such as the field cricket, in central Europe pass several years under suboptimal conditions, growing fewer in number each year, and then suddenly become very numerous in the rare year when conditions are optimal. Because the basic biochemical events are the same in all organisms, it has seemed a good idea to use organisms as bioindicators, observing the responses of those that are more rapidly affected by environmental poisons than man and thus give advance warning of potentially dangerous situations, before they actually become hazardous to the humans who have brought them about. There are many examples of this sort. The danger of DDT would have gone unnoticed if it were not for the catastrophic decline of the peregrine falcon; the creeping mercury pollution in the Swedish lakes and the Baltic Sea would not have been noted except for the reduced numbers of sea eagles and ospreys in these regions. Indeed, if it had not been shown that the feathers grown by the osprey while in Scandinavia and the Baltic area contain mercury, whereas those grown in Africa contain none, the mercury problem would never have been followed up. Only when the lakes in a part of southern Norway covering an area the size of Switzerland became practically devoid of fish did we become aware how dangerous precipitation containing sulfur can be; this element accumulates in Scandinavia in the snow cover, and following the spring thaw causes a very sudden dramatic drop in the pH of the lake water. The disappearance of lichens from the trees in our cities is yet another example (for more examples see Ehrlich et al., 1975).

These warnings have all come from chance observations. There has been no lack of effort to develop a more systematic approach. The earliest such attempt is the well-known saprobe system, applied to bodies of water. A number of species of organisms are diagnostic of water quality, and can be used for relatively simple and quick evaluations. More recently a system has been established in which lichens serve as indicators of urban air pollution. The success of this system has provided an incentive for a greatly intensified search for bioindicators in recent years (cf. Kunick and Kutscher, 1976; Miyawaki and Tüxen, 1977).

On the other hand, one must have serious reservations about such a search. In fact, lichens appear to respond chiefly to SO_2, so that they tell us nothing about other substances in the air. The saprobe system can give only very approximate evidence; one does not know exactly what the organisms are indicating. Furthermore, the term bioindicator is used quite differently by different authors. In some cases organisms are meant which exhibit changes in frequency of occurrence associated with environmental changes. But in view of the large fluctuations that occur normally, for entirely different reasons, this procedure is problematic – consider the chance parallelism between the decline in the human birth rate and the decline of the stork in central Europe. The term bioindicator is also used for a species that accumulates certain pollutants in specific tissues, so that chemical analysis of the organism is a promising tool in the search for pollutants. Because the possible applications vary so widely, there are great differences of opinion as to the usefulness of bioindicators. It is indisputable that once a harmful substance has been recognized a technical device can indicate its presence more rapidly and reliably than an organism can. On the other hand, it is equally indisputable that technical devices cannot be built to react to substances of which we are so far unaware. Early warning about such pollutants can be obtained only by way of organisms. And we cannot know in advance that any particular species will respond; there is no organism on earth that is not a possible indicator. Many people reject proposals for widespread monitoring of organisms as too complicated, but these pessimists forget that general observations have proved extremely useful in the cases of DDT and mercury. Toward this end, most of the ornithological groups in the European countries have agreed to make regular counts of the bird population in the largest possible area. The entire British Isles, for example, is included in this "Common Bird Census" (cf. Svensson, 1970). When so many species are counted, over so large a region, it seems a reasonable assumption that effects of even previously unknown substances potentially harmful to man will be detected. But the example of the ornithologists should be widely followed, with observations extended to higher plants, lichens, certain insect groups (such as butterflies and moths), and – as has already been begun – to marine fauna (cf. Fig. 176). If all these organisms were monitored as bioindicators we could really expect to receive timely warning of any new noxious substance, before human life is endangered. Having been warned, we ought to be able to devise techniques of pollution control. This, again, will require a large-scale program of unlimited duration which, as in the case of bird observation, will depend on the cooperation of volunteers. However the program is organized, a program there must be – we cannot, as in the past, leave matters to chance.

IX. Case Studies of Ecosystems

1. Lake Nakuru (Kenya)

In the high plateau of eastern Africa, at an altitude of almost 200 m, lies Lake Nakuru, a shallow, very alkaline soda lake in the center of the Rift Valley. The high sodium carbonate content of the water, and the pH of 10.5, prohibit all but a few organisms. Those species that can exist in the lake are present in enormous numbers. The only primary producer is the blue-green alga Spirulina platensis. The populations of this plant can be so dense that the whole lake is like pea soup, and visibility extends to barely 10 cm. For every square meter of lake surface there is about 450 g dry weight of alga. Their spirally coiled trichomes are microscopically small (spiral diameter 10–30 μm). Because the alga mass is so dense, most of it does not receive enough light for photosynthesis, so that production is relatively low. The high winds that regularly churn the lake, mixing the water through its entire depth, are the reason that such dense algal masses can extend down to the lake bottom. As a result of the water turbulence caused by the wind, all the algae are at regular intervals brought close enough to the surface that they can perform photosynthesis.

Originally there was only one significant consumer of the phytomass generated by

Fig. 177 a, b. The filtration apparatus in the beak of the lesser flamingo retains blue-green algae **a**, whereas the much smaller diatoms and green algae slip through. (Vareschi, 1977 a)

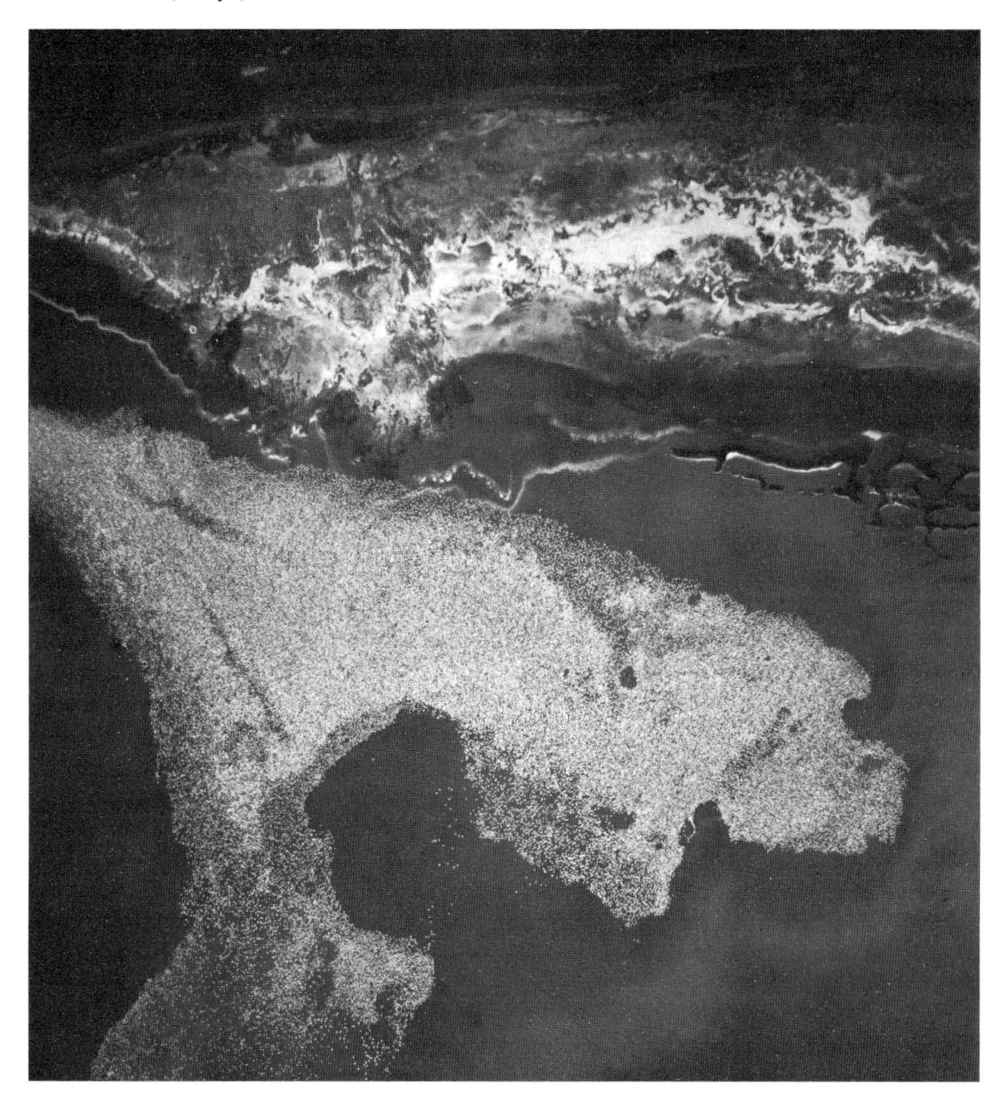

Fig. 178. Lake Nakuru, in eastern Africa. The band of flamingos extends along the shore like a white veil. (Original photo, E. Vareschi)

the single producer – the lesser flamingo (Phoenicopterus minor). In its beak is a highly specialized system of lamellae (Fig. 177). When the bird dips its head into the water it makes pulsating movements with the tongue, which press out the water and leave the algae behind, a green pulp in beak and throat. The system of lamellae makes the beak into a sort of filter. Aerial photographs made in 1972–1974 allowed the stocks to be determined quite precisely for those years; on the average, there were 915,000 flamingos, with peak values reaching 1,500,000 (Fig. 178). These flamingos do not breed at the lake, but only feed there. Their breeding grounds are restricted to soda lakes some distance away, even more hostile to life. So far the relationships between the breeding colonies and the birds at Lake Nakuru have not been entirely clarified; we cannot say whether the latter site is occupied only by the young, sexually immature birds. One reason to suspect this is that "play" nests

are regularly built – an activity typical of young flamingos. On the other hand it is very likely that a considerable part of the breeding population also feeds at Lake Nakuru. In any case, it is quite obvious that the young birds in the breeding colonies are not fed with food from Lake Nakuru, a fact which further simplifies analysis of the system. Not only are only one producer and one consumer present in the lake, but the consumer is always the same size; growth processes do not have to be included in the calculation.
Cage birds were used to determine the amount of water filtered by the flamingos. 29.4 l of water were filtered per hour by each bird; a free-living flamingo spends about 12–13 h per day feeding. Given the alga density in the lake, one can calculate a food intake of 65 g dry weight of Spirulina, or 281 kcal per day. Other researchers computed the energy requirement of free-living flamingos, and arrived at values between 154 and 391 kcal per day. In view of the difficulties we have discussed in determining the food requirements of free-living animals, the agreement between the two estimates appears good.
On average, then, flamingos withdraw from the lake 2.5 g dry weight of Spirulina per square meter per day. This is about half of the primary production. But in arriving at such percentages we encounter problems; if we extracted still more from the lake, more light would reach the lower water levels, and primary production would be increased accordingly. So when we say that half of the primary production is taken away per day, this is simply a theoretical, calculated value. In actuality consumption of the algae by the flamingos presumably increases production sufficiently to compensate for that consumption.
A few years ago, man introduced a fish into this very simple system, in order to improve the nutrition of the local human population. This fish, Tilapia, also feeds on Spirulina and multiplied rapidly. In its train, the lake has been invaded by fish-eating birds such as the pelican, eagle, and heron, and fish-eating mammals like the otter. Here we can give only a partial estimate of the quantitative relationships.
The distribution of the schools of fish in the lake is quite irregular and evidently random, to a great extent; superimposed on this distribution is a diurnal migration that concentrates the fish near the shores at midday, whereas at night they remain in central parts of the lake. The animals are usually to be found just below the surface. Vareschi (1979) estimates that the total stock in 1972 was about 90 t, and that it grew to about 400 t (dry weight) in 1973. The chief consumer of these fish is the white pelican (Pelecanus onocrotalus roseus), which does not breed in the immediate vicinity of the lake. In 1967–1969, 10,000–35,000 pelicans are said to have lived on the lake, in 1971/1972 there were about 2,500, and in 1974 the number again increased to about 10,000. Since 1968 there has been a breeding colony about 14 km away from Lake Nakuru, with about 4,000 pairs. In general two eggs are laid per pair, and the breeding success is ca. 50%. Each adult pelican needs about 1.33 kg of fish per day; the total consumption of a young bird from hatching to fledging averages 770 g (fresh weight). The breeding pelicans thus consumed 16,000–20,000 kg fish from Lake Nakuru every day. This corresponds to 72 kg P/PO_4 and 486 kg nitrogen per day (Vareschi, 1979). The data for all inhabitants of the lake can be assembled to give the flow diagram of Fig. 179 (Halbach, 1977).
Unexpectedly, in the spring of 1974, there was a sudden profound change in the lake. The blue-green algae almost entirely vanished, the water became clear and transparent, and the flamingos also disappeared. By contrast, a small green alga that previously had played an utterly insignificant role became somewhat more common. This alga is the food of a copepod that in turn serves as food for the greater flamingo (P. ruber). Normally these flamingos are present only in small numbers.

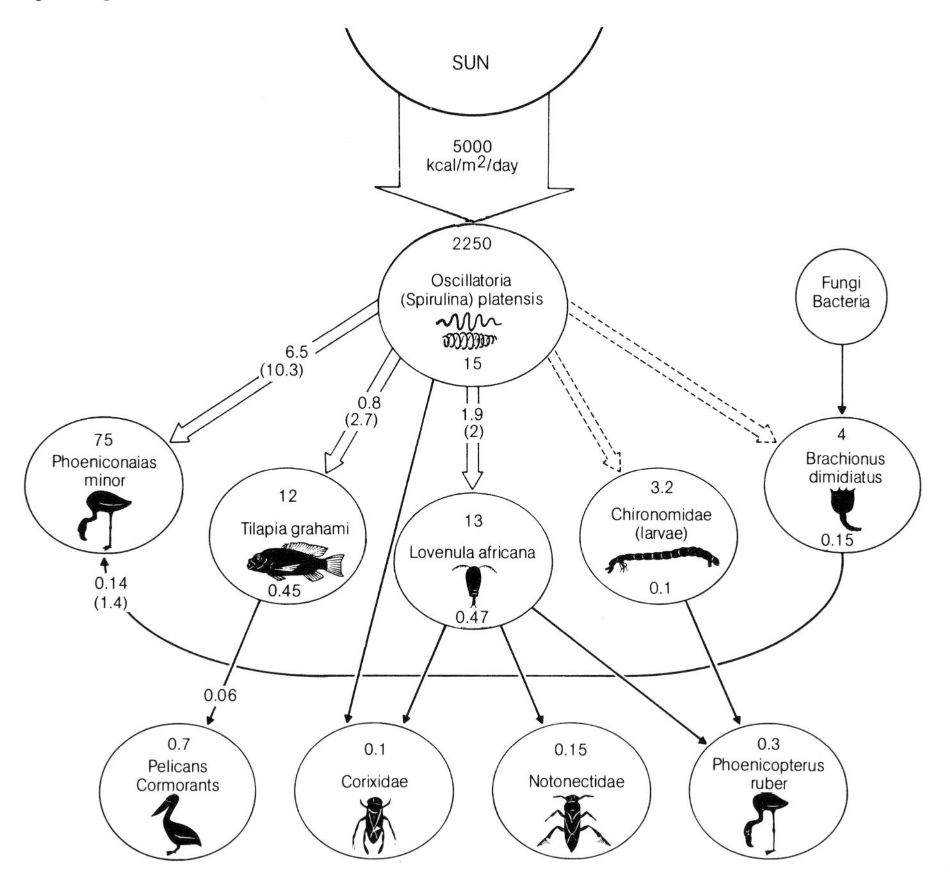

Fig. 179. Energy flow through the ecosystem Lake Nakuru, Kenya. The circles and ellipses represent species or groups of organisms, the upper number indicating biomass (kcal/m²) and the lower, production (kcal/m² × day). The arrows represent energy flow; the upper number on each arrow gives the mean rate of consumption (kcal/m² × day) and the lower, in parentheses, the maximal consumption rate (previously unpublished data of Vareschi). The numbers were obtained by preliminary calculations and under certain circumstances may change. (Vareschi in Halbach, 1977)

At the moment we can but speculate on the reasons for this sudden alteration. Subsequent research has shown that similar phenomena occasionally occurred in the past. Until the exact physiological requirements of the blue-green alga have been analyzed, together with the changes in the lake that caused their disappearance, little can be said. Here is a major problem complex for ecosystem research (Vareschi, 1977 a, b, 1979).

2. Spitsbergen

Spitsbergen, located between 77° and 80° north latitude, has long been one of the best-known regions, in biological terms, of the far Arctic. Since the beginning of this century faunistic and floristic, as well as ecological, studies have been carried out there in large numbers. This abundance of information, together with the seclusion of the island, appeared to justify an attempt at a total ecosystem analysis (Figs. 180–182).

Far arctic and alpine tundra regions are regarded as identical in most scientific papers and books. Before the ecosystem analysis was begun, this old assumption was reexamined. The invertebrate fauna of far arctic regions is characterized by a distinct reduction in number of systematic groups,

Fig. 180. Tundra on Spitsbergen

Fig. 181. A group of reindeer on Spitsbergen

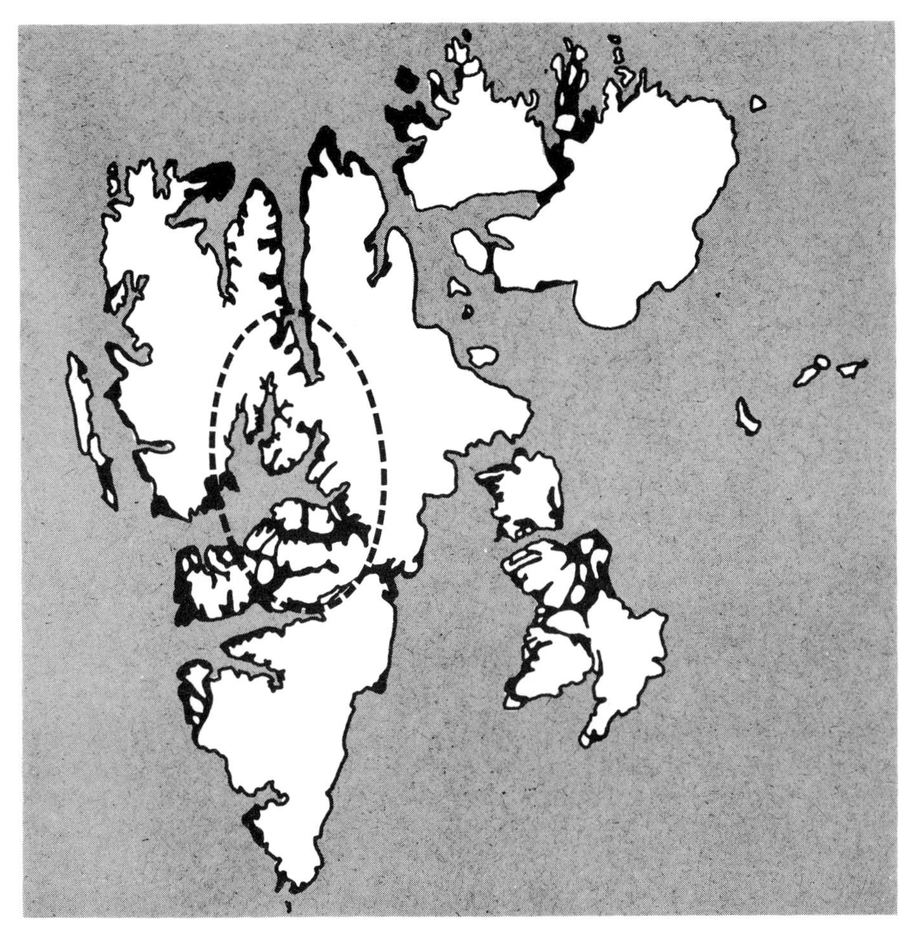

Fig. 182. Spitsbergen. The area circled is the warm inner fjord zone. *Black:* grazing grounds of the reindeer. (Remmert, 1966)

as compared with high alpine habitats. Cicadas, bugs, grasshoppers, beetles, and lepidopterans are almost entirely absent; in any case, they are so rare that they can hardly play a role in the system. The same is true of isopods, diplopods, and gastropods. When formalin traps are set out on Spitsbergen, the proportion of the catch represented by these groups is less than 1% of the total biomass, as compared with more than 40% in the Austrian Alps (Fig. 183). The groups that are less common or absent in the arctic, then, can be characterized as relatively large (over 8 mm in length) ectothermic animals and as ectothermic herbivores. By contrast, there is a relatively large number of small ectothermic animals (the nematoceran families Sciaridae and Mycetophilidae, parasitic Hymenoptera, and the spider families Linyphidae and Micryphantidae). These are forms that play a role in decomposing ground litter, as well as predators and parasites. When such regularities are found, they can only be based on ecological factors that are currently effective. What can these be?

Regions in the far north, such as Spitsbergen, are characterized by permanent light during the entire growing season; from the beginning of April to the middle of August the sun never sets. Correspondingly, there are no marked fluctuations in temperature over the course of the day during the grow-

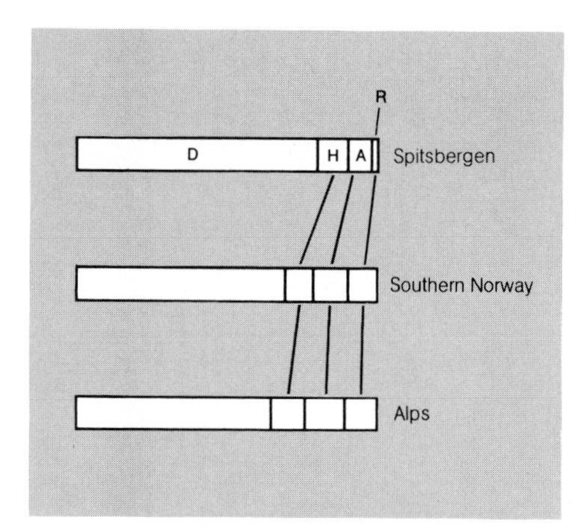

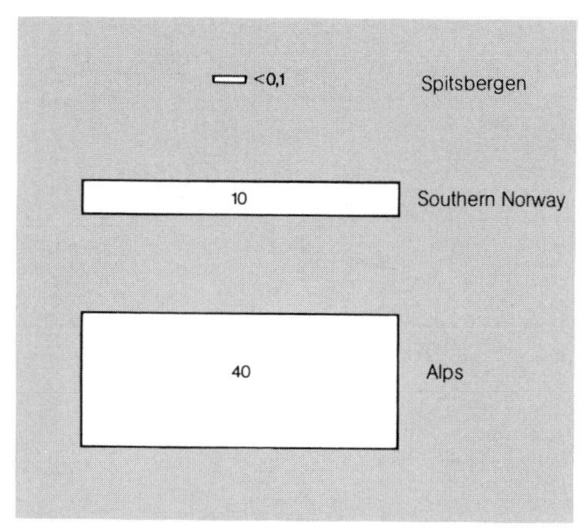

Fig. 183. *Left:* relative number of winged insects and spiders caught in formalin traps on Spitsbergen, in southern Norway, and at high altitudes in the Alps. *D* Diptera; *H* parasitic Hymenoptera; *A* Araneae; *R* remaining groups. *Right:* fraction of the biomass of the total catches in the three locations represented by the remaining groups *(R)*. (Remmert, 1972)

ing season. In this respect, the light and temperature situation contrasts sharply with that in alpine regions. What is the significance of such factors?

In the far north, as elsewhere, the activity of insects and birds is synchronized with the earth's rotation (Fig. 184). The diurnal changes in light intensity during the growing season are negligible, especially in view of the changes in intensity that occur when cloud masses pass in front of the sun. Some authors have described marked fluctuations in light intensity as a function of time of day, but this claim is misleading. Such observations have probably been made with meters having faulty cosine-correction, so that the elevation of the sun appears to have a much greater effect than is actually the case. In the laboratory, birds maintained under light conditions corresponding to those of the arctic summer are permanently active. Two possible timing signals have been tested in the laboratory (Krüll, 1976 a, b) – the azimuthal position of the sun in combination with landmarks, and the spectral composition of the light (more pronounced red components during the "night," measured integratively as color temperature; Fig. 185). Each of these factors acts as a weak timing signal, even

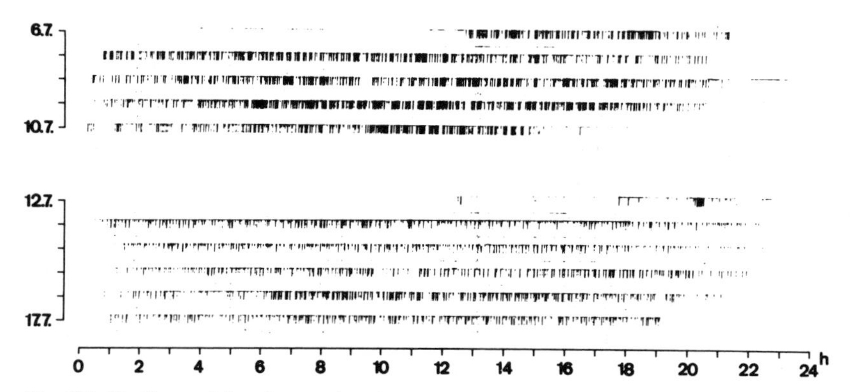

Fig. 184. Feeding activity of two pairs of snow buntings on Spitsbergen. The birds are synchronized with the earth's rotation; they make a regular pause between about 10 p.m. and 1 a.m. (Krüll, 1976)

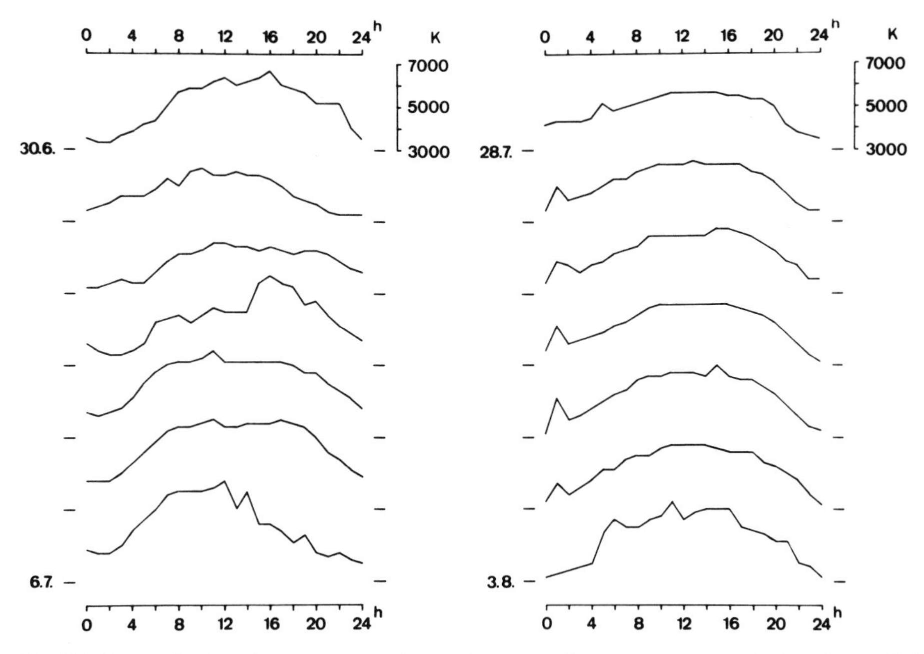

Fig. 185. Changes in the color temperature during the course of a day, over two weeks (one in June/July, the other in July/August) on Spitsbergen. The marks at either end of the curves indicate 3,000 Kelvin. (Krüll, 1976a)

when the light intensity (watts × cm²; lux) is kept constant. Birds are most readily synchronized during the reproductive season; birds during the refractory period, and castrated birds, normally do not respond to these signals. In the Longyeardal, a very narrow valley leading off the Adventdal on Spitsbergen, sunshine penetrates only at midnight. At all other times of day the floor of the valley is in the shade. If the activity of the birds of Spitsbergen were dependent on light intensity, they ought to be active around midnight in the Longyeardal. But the snow buntings that breed in the valley are asleep then, just like their conspecifics in the open tundra; they sleep through the only time when the sun is shining on their nest site.

Both factors, then, act as timing signals. Responses of both northern and southern species can be demonstrated. As far as diurnal rhythm is concerned there is no difference in principle between north and south. The apparent differences in daily light patterns can therefore be ruled out as an effective parameter.

What can we say about the other chief difference, the temperature? What are the effects of constant low temperature, as compared with temperatures oscillating about the same mean? Recall that there is no true adaptation to constant low temperatures. Animals and bacteria growing under such conditions require a very long time for a single generation. When the temperature fluctuates about the null point for development of a species, its rate of development is dramatically increased. Accordingly, the development of bacteria, fungi, and ectothermic animals occurs much more rapidly under the conditions of alpine tundra regions than in the far north. To pass through a single generation an insect on Spitsbergen needs much longer than in southern Norway or in the Alps, with identical mean temperatures. Moreover, because mortality is in part a function of time, the difference between north and south is further enhanced. Finally, the rate of reproduction differs depending on whether the temperature is constant or fluctuating (p. 32).

These considerations offer a simple physiological explanation of the absence of the larger ectothermic animals from the tundra on Spitsbergen. On the other hand, it is clear that in the relatively high humidity (which results from the constant low temperature) small insects have an advantage over large ones.

The lack of ectothermic herbivores in the far north is a side effect of the thermal situation. The digestibility of plant matter is drastically reduced at lower temperatures; below 10 °C plants can hardly be digested at all. At such temperatures, only organisms that grow very slowly can still use plant material as food (Schramm, 1972).

To summarize: Physiological experiments provide an explanation of the special composition of the Spitsbergen fauna. With this understanding, we have a solid basis for attempting an ecosystem analysis. Assuming that the climatic conditions remain the same, we can be sure that no sudden outbreak of a herbivorous insect will upset all our predictions. Predictability – and thus reproducibility – are the bases of research in the natural sciences, and both are represented here.

From a number of publications, we know quite a lot about the composition and density of the invertebrate fauna. On the average, in central Spitsbergen, 5,000 individuals of the groups Mycetophilidae and Sciaridae will be present per square meter of soil, or emerge during the course of a year. This number, together with springtails, mites, fly larvae, and enchytraeids, most probably is sufficient to decompose the litter layer (to the extent that animals are involved at all; fungi are probably the chief agents of biochemical decomposition). In any case, there is evidently no accumulation of the litter over a period of years. A major feature of moister, cooler regions is the presence of terrestrial chironomids; their larvae can be subdivided into four size classes, so that they evidently require four years to develop (Sendstad et al., 1977). There is a clear relationship between the productivity of the tundra and the number of animals caught per day. It holds regardless of the factors affecting productivity – altitude or fertilization (Hinz).

The description of the system so far has without doubt touched on but a small part of the role that animals play. They are sure to be important contributors to flower pollination. 79 plant species are originally insect-pollinated. Many of these can manage for long periods without pollination by insects, but in the very long term pollination by insects may well be necessary for survival of these plants in the far north. All of the numerous Sciaridae, Mycetophilidae, flies (Empididae, Syrphidae, Anthomyidae), and parasitic Hymenoptera are quite regular flower visitors. It is essentially impossible to really quantify these effects.

So far we have left the warm-blooded animals out of the discussion. They play a distinct and extensive role in the system, and because their numbers are relatively well known this role can in some cases be quantitatively estimated. The land birds are evidently of the least significance. The distances between their nests are relatively large. However, if a bird is very selective in the buds it eats, as the ptarmigan is known to be, its effect can be considerably more pronounced than one would at first expect. We shall return to this point. The marine birds, by contrast, play a leading role. Some of them breed relatively far away from the coast, on inland rocks. They transport enormous quantities of nutrients from the ocean to the land, and thus fertilize the tundra. Quite a lot is known about the amount of food they need and thus the amount of nutrients transported, as well as about the number of animals that breed on Spitsbergen. But a critical aspect, as far as the system is concerned, is precisely where they breed; a distinction must be made between inland and coastal colonies. So far no such data are available.

It is obviously easier to evaluate the warm-blooded herbivores – the Spitsbergen reindeer, the musk ox, the ptarmigan, the pink-footed goose, the barnacle goose, and the

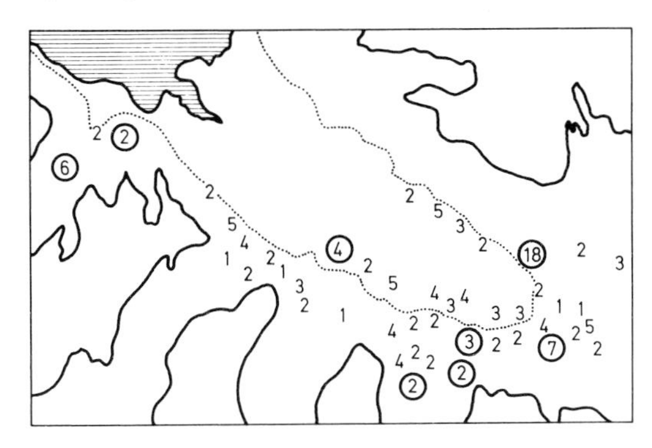

Fig. 186. The Adventdal in central Spitsbergen and the distribution of the reindeer there on July 2, 1974. The numbers of single reindeer present in each area are encircled. (Pöhlmann, 1975)

Brent goose. The number of representatives of these species is relatively well known. There are about 7,000–8,000 reindeer, 20,000 geese, and between 20,000 and 200,000 ptarmigans; there were about 50 (introduced) musk oxen, but these have probably died out since the count was made. What effect do these animals have on the tundra?

The reindeer occupies a special position from the human point of view, and therefore has received the most attention. Their dung has been collected over large areas and analyzed so as to obtain a measure of consumption by reindeer. For such calculations to succeed, a number of conditions must be met.

1. Laboratory studies must have clarified the relationship between the *amount of dung* produced per day and the *food intake*. A rather large number of such studies have been made, but the results obtained are in some cases contradictory.

2. The *remineralization* of the dung must be *slow*. This is in fact the case. From the number of reindeer and the weight of dung per unit area we conclude that about 30 years are required for remineralization of reindeer dung on Spitsbergen.

3. The *reindeer* must be relatively *uniformly distributed* over the tundra. If they traveled in large herds like those so characteristic of the caribou in North America and the Scandinavian reindeer, studies of this sort would be impossible. In fact, the behavior of the Spitsbergen reindeer is quite different from that of the other reindeer forms. Small groups (2–3 individuals) of adult stags, subadult animals, and females with a calf (and very often a yearling, probably the calf of the preceding year) are scattered quite uniformly through the tundra. Only very wet regions near rivers and the high Fjell plateau have a lower population. The groups are not constant; their composition changes often (Figs. 186 and 187, Pöhlmann, 1976).

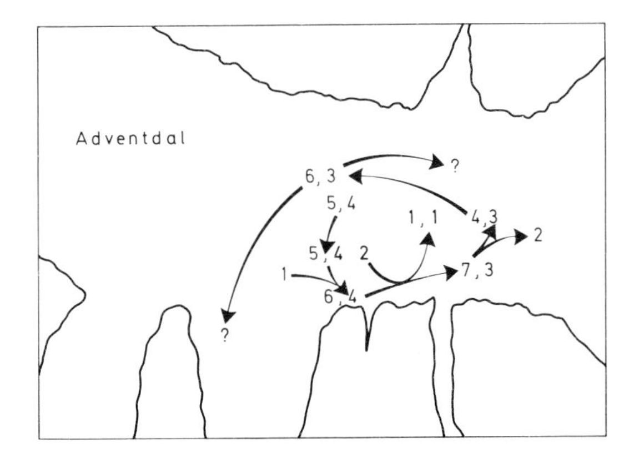

Fig. 187. The history of a reindeer group in the Adventdal during the course of a week. n = number of adults, number of calves. (Pöhlmann, 1975)

The implication of these findings is that collection of dung in test areas from which all the dung was removed a year before can accurately indicate the production of dung per year. Because the number of reindeer is known, the dung production per day per animal can be calculated.

These calculations can be related to primary production (Brzoska, 1976). The highest annual above-ground production is that of meadows with Dupontia fisheri (235 g/m²), Eriophorum scheuchzeri (408 g/m²), Alopecurus alpinus (227 g/m²), or Poa arctica vivipara (421 g/m²). Because these are very wet, they are not much utilized by the reindeer, at least in summer. The animals tend rather to concentrate in the dry areas where Cassiope, Salix, and Dryas grow. Here the productivity is much lower. The biomass above ground amounts to almost 300 g/m², but only about 10% of this can be regarded as annual production. Moreover, one must bear in mind that large areas have no vegetation at all, and that the different plant communities cover different fractions of the total area of the island. The conclusion that has been drawn is that the reindeer, with a current population density of about 10 animals per square kilometer, graze off about 10%–15% of the annual above-ground production. This is a very high value, and one may ask whether it is normal.

Around 1900 Spitsbergen housed large numbers of reindeer. Then the population collapsed, probably as a result of overhunting, and the Spitsbergen reindeer nearly became extinct. Recovery of the population was very slow, but gradually it rose to the present size (Fig. 188).

The question is: was the population uniformly large prior to 1900? Evidently it was not. Examination of the soil reveals a subfossil layer of dung with sharp upper and lower limits. This indicates a short-term massive increase in numbers of reindeer. The layer can be dated by carbon particles, showing that the mass reproduction must have occurred around 1900. Probably the reindeer population goes

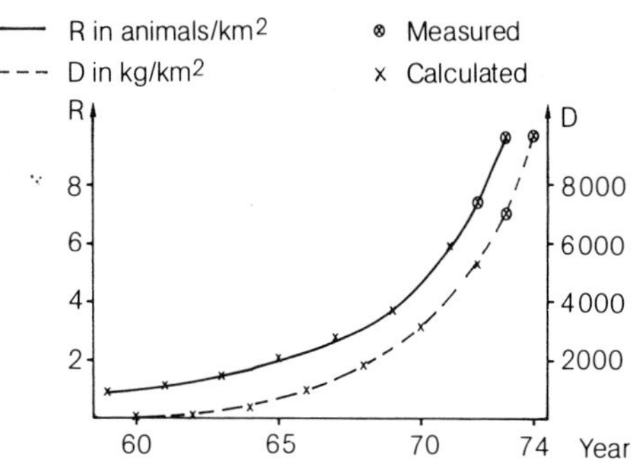

Fig. 188. Increase in the reindeer population in the Adventdal, Spitsbergen *(R)* and accumulation of reindeer dung *(D)* in the same region. The total population of reindeer on Spitsbergen rose correspondingly. (Pöhlmann, 1975)

through large-amplitude cycles, with a period of between 50 and 100 years. The way in which the system as a whole is affected by such cycles has been discussed on p. 233 (cf. Fig. 189). The most noteworthy aspect is that according to that hypothesis the reindeer population collapses when the production of vascular plants is greatest, for then the essential winter fodder, lichens, is used up. But here, again, we face the difficulties of ecosystem research – all our studies are terminated too soon. Moreover, there appear to be profound differences between the various reindeer populations, which include their food requirements. The reindeer of northern Greenland and the northern islands of the Canadian Archipelago probably – in contrast to all the other reindeer so far investigated – do not require lichens as food in winter. Because the Spitsbergen reindeer is very closely related to the Greenland form, and because as yet no analyses of the winter feeding habits of the Spitsbergen reindeer have been done, caution is imperative.

But all these interpretations, as presented on p. 234 f., are still hypothetical. The apparently simple ecosystem Spitsbergen is in fact a very complicated system, and only very patient, long-term analysis can give results. Although Spitsbergen is now a national park, there is no guarantee that such a long-term study can actually be carried out without disturbance of the system

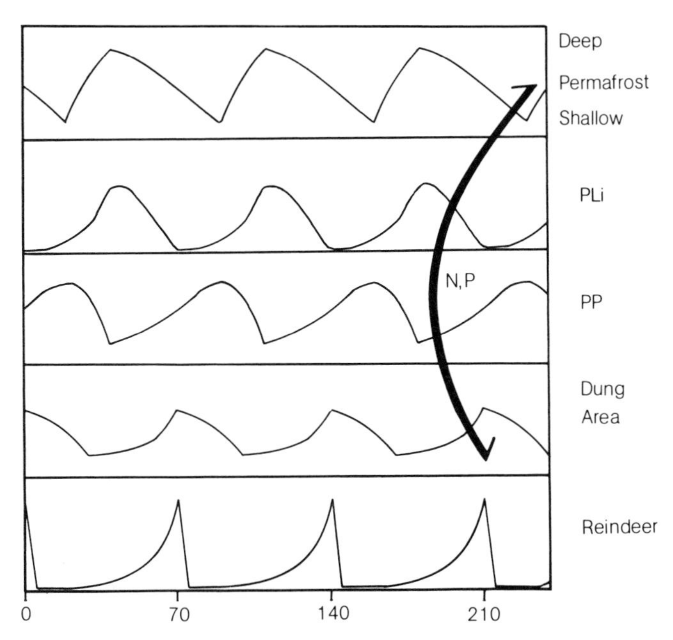

Fig. 189. Conceptual model of the action of a factor. When a reindeer population fluctuates widely, as that on Spitsbergen probably does, there are effects on the thawing of the permafrost in summer *(top)*, on the productivity of the lichens *(PLi)*, on the above-ground production by green plants *(PP)*, on the amount of undecomposed dung per unit area, and on the composition of the plant communities. The amount of feces in an area and the depth of the thawed soil together affect the availability of important minerals *(N, P)* to the plants. As yet the validity of this model has not been confirmed, but a number of studies suggest that it is very probably correct

by man. But an analysis of this system would be of great value to people in the future, and if ecologists are to carry it out, they must be allowed the necessary time and given access to the necessary areas.

3. Central Europe

When the glaciers began to melt after the last Ice Age, and a tundra landscape began to spread through central Europe, man was already in residence. Human activities affected the development of the tundra and the subsequent forest communities in a variety of ways. There never was an ecosystem in central Europe that did not include man. His influence was especially pronounced in regions through which he could move easily – less in the wet jungle regions than along the rivers, around the lakes and on the coast, less in large swamps than on dry flatland, which offered the special advantage that it could be kept free of forest with relatively little effort. The Lüneburger Heide began to be formed even before the Bronze Age. Some Bronze-Age graves are covered by soil strata that clearly reveal the influence of humans as destroyers of the forest. When the Romans expanded their empire into central Europe, they entered terrain quite different from a solid primeval forest. Even then, there must have been extensive unforested areas where herds of domestic animals grazed. We can begin to see a really comprehensive picture of the region once the central European peoples had been converted to Christianity; from that time on regular chronicles were kept, and there was continuity in geographical names. This period marks the beginning of an upsurge in civilization, accompanied by a great hunger for land. There were waves of effort to clear the terrain, which pushed back the forest over large areas; large-scale deforestation was characteristic of the years between 950 and 1300. (German place names ending in *-rode* and *-reuth* come from this time; *-roden* means to clear the land.) The virgin soil thus created gave high yields when cultivated, and the people saw no more need to worry about fertilization than they did during the early years after clearing of the forests in North America. There was plenty of both field and forest. In these circumstances, especially during the Gothic period (approximately 1250–1500 in central Europe), the economy flourished to an extent almost unknown today, and power became concentrated in the cities. The guilds instituted a weekly day of rest ("blue Monday"), and the per capita consumption of meat was extraordinarily high, even by present-day standards. The profound effects that this economic development had had by the end of the Gothic period have been described by Jansen (1913). In Holstein, for example, a free workman in the fifteenth century in one day could earn half a bushel of rye, three-quarters of a bushel of oats, or a whole bushel of root vegetables; he could earn a lamb ready for slaughter in 3–4 days, a sheep in 6–7 days at most, and a fat cow in 22 days. On the lower Rhein, in Klevischen, between 1470 and 1510, a day-laborer working for both board and wages received for six work days, on the average, ¼ bushel of rye, 10 lb pork or 12 lb veal, 6 large pitchers of milk, and 2 bundles of wood, and in 4–5 weeks he had accumulated enough cash to buy a work smock, 6 ells of linen, and a pair of shoes. It is known that in Aachen at the end of the fourteenth century a day-laborer earned a sheep in 5 days, a wether in 7 days, a pig in 8 days, and in a single day almost two geese. An ordinance by the Saxon dukes Ernst and Albrecht, in the year 1482, decrees: "The workers and mowers shall be satisfied when, apart from their wages, they receive twice a day, at noon and evening, four foods: soup, two meat dishes and one vegetable; on festival days, however, five foods: soup, two kinds of fish and two side vegetables. In the morning and evening, between the mealtimes, one should give them no more than cheese and bread, but no cooked food." The general restriction of meat-eating during the middle of the sixteenth century was one of the most important signs of the sad transformation that occurred in the agricultural and social status of Germany; in the case of the working class, it is explicable entirely by the fact that the daily wage had fallen to only half that between 1450 and 1500. Meat, once the food of the poor, became increasingly the prerogative of the rich.

The system by which the land was employed can be described roughly as follows. At a relatively long distance from the settlements, the chief activity was the cutting of wood for buildings, primarily pine and spruce. Closer to the settlements, and to some extent within them as a protection against fire, beeches and oaks, in particular, were encouraged to grow. The distances separating the oaks were large. Under these conditions acorns were produced almost every year, whereas in a dense stand of oaks they appear only at relatively long and, above all, irregular intervals. Acorns and beechnuts were the basis for nearly the entire meat production (they fed the herds of pigs). Wood for burning was obtained from bushy coppices situated, wherever possible, fairly close to the villages. Because saws were unheard of at that time, and all wood had to be felled and trimmed with the axe, such coppices were particularly useful providers of fuel. The rest of the land was cultivated, the predominant aim being to obtain sugar and grain. At that time sugar, to be used for preserving and sweetening food, was known only in the form of honey, so that plants that provided honey were favored crops. The spread of large areas of heather (Calluna) and nectar-bearing trees such as the linden was of great importance to the human population. Finally, wood was used extensively for the production of glass, for extracting salt from salt water, and for making tar and all metal goods. The consumption of wood was enormous. The soil could not tolerate this exploitation for very long. The social upheavals beginning at the end of the Gothic (around 1500), which reached their peak with the Plague, the Peasant Wars, the Hussite Wars, and the Thirty-Years War, to a great extent resulted from the economic emergency. On the fields no more than "the second corn" could be harvested; that is, only twice as much grain was harvested as was sown. The forest was almost used up, but it seemed that the requirements of each person could not be cut back correspondingly. Moreover, the population had become dense, which encouraged the spread of diseases. The result of this crisis is general knowledge; the population declined by one-third, many settlements were given up and left derelict, and cities, villages, and the nobility were poverty-stricken. The cities and the minor nobles, which during the Gothic had become more or less independent entities in central Europe, no longer had the economic power to resist consolidation under the rule of a few princes. The soil could regenerate in the deserted villages, precipitation and weathering from below brought renewed nutrients, and the forest expanded again. The composition of this new forest was dictated, of course, by the way the land had been used and by the impoverished soil. The newly formed large states had technical advantages that permitted the cultivation of land that had been untouched before, such as moors and large regions of swamp. But as early as 1700 there began to be an increased effort to reclaim the derelict settlements. The forms of land use had not changed in principle, so that larger yields were obtained for only a short time. The general distress was alleviated to some extent by the introduction of the undemanding potato, and by the discovery that fields could be fertilized with leaf and needle litter from the forest. Once the farmers had begun systematically removing the litter, the nutrient cycles in the forests (chiefly coniferous stands and heaths) were interrupted. From the Nürnberger Reichswald, a region covering 30,000 ha, about 100,000 cart-loads of leaf and needle litter were carried away per year (about 16 m^3/ha/yr), and this continued over many decades (Sperber, 1968). This was the origin of the situation that existed in central Europe between 1800 and 1850. Not only was the formation of humus, so necessary for the retention of water and minerals , brought to a halt, but the minerals were actually removed from circulation in the forest.

The cities sank deeper into poverty; indeed, they became bankrupt. The attempt by the city of Nürnberg in 1796, for example, to join the Prussian state failed, because the Prussian king was unwilling to take over the city's debts. In the villages, a not inconsiderable part of the population starved during the winters. Central Europe at that time was largely unforested, as we can see in the paintings of the Romantics. One must understand that forests act as "nutrient pumps." Nutrients are brought up from the depths of the soil and deposited on the surface. Here they are fixed; in addition, nitrogen fixation is performed by the blue-green algae on the trees and the root bacteria in the superficial layers of soil. As a result, the nutrient content of the top soil stratum in the forest rises, slowly but steadily.

The situation was not changed until artificial fertilization was achieved, by bringing in nitrogen-containing fertilizers such as Chile salpeter and guano from South America. Once the fields were fertilized, the cultivation of sugar beet reduced the importance of bee-keeping. The sugar beet is the ideal crop in this case; it "improves" the soil in that all its components, minerals in particular, are returned to the field. Only the sugar itself, synthesized from carbon dioxide and water, is taken away ("only the sunshine is sold"). At the same time, large-scale cultivation of the potato, a crop that will grow almost anywhere, reduced the threat of starvation – although the population had to revise radically its eating habits to take advantage of this new food. Higher yields made reforestation possible, though on the depleted soils it was rarely possible for trees other than pines and spruce to grow. The result was the present central European landscape. The present-day forests were the only tree stands that could be raised on the poor soil. Such forest types – pine forest, above all – were naturally highly vulnerable to mass outbreaks of plant-eaters. Great calamities occurred; all plantations in the Nürnberger Reichswald, for example, were utterly destroyed toward the end of the nineteenth century. The fertilization of poor forest regions that is often done today achieves nothing more than to restore the forest soil to its original fertility. It is really an attempt to cancel out the influence of man. The poor pine forests on sandy soil with ericaceous undergrowth, which are so similar to Nordic pine forests, may well be entirely anthropogenic. Anyone who wants to preserve them should realize that he is protecting a type of vegetation created by man, which otherwise, with all its animal inhabitants, would not be encountered in central Europe. In this brief survey, of course, a great deal has been much simplified and to some extent may seem exaggerated. It is obvious that not all the forests and all the soils became exhausted simultaneously, and that the process was halted before a state of total depletion was reached (when plague reduced the human population). Nevertheless, the general course of events was as has been described here.

The history of the central European habitat has been followed in the context of the forests. It would also have been possible to take the wet regions as an example of the changes that have occurred. As recently as the last century, the tidal zone extended from the North Sea to Rendsburg in Schleswig-Holstein – about 100 km away from the coast. Here hundreds of square kilometers were regularly flooded and left dry, in the rhythm of the spring and neap tides. We have no idea how this habitat may have actually looked; all that is apparently certain is that a kind of plover, the dotterel, had its nesting grounds there. Our rivers became deep channels for ship traffic at the end of the Middle Ages. Originally many of these were shallow waterways, forming a highly branched network like a spiderweb over the flatland; these were made into canals by banking and deepening their beds. Place names based on the word "ford" reflect this situation; at the two German towns called Frankfurt, one could pass on foot through the Oder or Main rivers. We can hardly imagine now how these rivers looked then; they must have flowed very slowly, and their broad valleys must have been a proper amphibian wilderness.

The history of the habitats in central Europe, then, is a history of the eternal striving of mankind. There was never a balance between the human population and the environment. Always man took away more than he gave back. Always man eventually destroyed whatever habitat he occupied; only now is he beginning to return what he has been removing from the outset – perhaps over millenia. But this repayment goes hand in hand with a neglect of other stresses that are imposed on the habitat. We are still no closer to achieving equilibrium. Everywhere energy is being used up, in agriculture and forestry (in the form of fossil energy), and still more in urban life. It is high time we gave some thought to the population density and the standard of living that will make a balance possible.

And the situation in central Europe is only an example. We could say the same of very many regions. Northern Sweden, too, was deforested over broad expanses around 1800 (Zackrisson, 1977). The great wave of emigration from Europe to the New World was partially due to this fact. That the Viking settlements on Iceland and Greenland flourished is based in part on the vast amounts of driftwood that had accumulated there over the centuries. This wood was used up, and soon the Vikings had to travel back to Norway in search of more (but found that the wood there was also severely depleted); descriptions of the famous journeys from Greenland to the North American coast tell us that they were undertaken in order to find wood. Though climatic changes – the minor Ice Age – and many other factors were also involved, the exhaustion of the wood supply (by shipbuilding, for example) was one of the causes of the downfall of the settlements in Iceland and Greenland.

E. Outlook

E. Outlook

The attentive reader will no doubt have felt a certain frustration at several places in this book. Didn't I just read that this phenomenon has these particular effects, and these particular causes? And here at another place it says something completely different!

These many facets of a single phenomenon are the difficulty and the charm of ecology. Each individual has evolved toward a high rate of reproduction, toward the avoidance of a predator, toward the ability to find a favorable biotope. The population comprising these individuals has evolved toward maintenance of optimal density, and toward the sacrifice of individuals less suited to the conditions of the moment without disadvantage – or rather, to the advantage – of the population. The populations of which the system is composed have evolved toward maintenance of the system, toward optimization of system function. To express these relationships anthropomorphically: the interests of the population do not necessarily coincide with those of the overall system. Evolution and coevolution thus have in the last analysis preprogrammed a "schizophrenia" in the individual, which one must see and recognize if one wants to pursue ecology. This apparent schizophrenia is resolved by long-term evaluation of the overall system. In evolution, too, we encounter this preprogrammed schizophrenia. Coevolution should lead to harmony in ecosystems, and as a rule this is just what it does. The evolution of "new technologies" by plants and animals, however, in the course of the earth's history has led to the rapid reconstruction of large ecosystems. The sharks, with their high mobility and their highly developed sense organs, replaced the ammonites that had previously drifted through the seas. The teleosts, which because of their swim bladder consumed relatively little energy while swimming in open water, were superior to the sharks. For the echolocation systems of the dolphins, the teleost swim bladders were an ideal target.

Similar complexity confronts us when we consider the many relationships, causes, and effects within a system of ecological interactions. The beginner may ask what is the actual significance of cycles of mass reproduction – those of lemmings, to take but one example. The answer to this is that the question was incorrectly phrased. The significance of such cycles to the individual lies in the fact that the individual must adjust to different population densities at different phases of the cycle, and thus adopt different forms of social behavior. Individuals that cannot do this are at a selective disadvantage. The significance to the population lies, among other things, in the opportunity for gene exchange with neighboring populations during the periods of increasing population density, and in the attainment of a mean population level which, if it were to be kept constant, would result in a horde of predators. For the predators, the significance of mass reproduction lies in the possibility of raising many progeny in some years, whereas in others few or none at all can be produced. That the predators of lemmings tend not to stay in one particular location is related to the cycles of their prey; the predators gather wherever the lemming population happens to be high. The significance of synchronization among the lemming maxima in different regions, and of the synchronization of these maxima with those of other small mammals, hares or grouse, is that predator pressure is thus kept as low as possible during the peak period. The significance of mass reproduction to the system "tundra," which has evolved along with the lemming, is that turnover is accelerated in the detritus stratum.

And as many answers as there are to the question of the effects of the lemming cycles can be given to the question of their cause. No answer is correct in itself. That apparently simple ecological questions can be simply answered is a fallacy. The ecologist who is expected to predict the effect of one or another alteration in the maze of factors cannot fall back on generalized models. Such models exist, and I have tried to describe some of them. But every system is different.

All these generalized models fail at just the point that is the sole measure of achievement in the natural sciences – in their application to the situation in nature. Though they become more and more complex – so much so that major computing installations can be occupied for hours in dealing with them – it is only in certain subareas of autecology that models have reached the degree of perfection necessary to give reliable predictions. Such a degree has been reached, for example, in the meticulous analysis of photosynthetic performance by some plants. But all models of population ecology or ecosystem research are either so specialized that they exactly match things past and finished, or so general that they permit no predictions. Just as there is no generalized model of the organism most subjected to biological analysis, the human being, many ecologists today feel that there will probably never be a model of populations and habitats that is general and allows predictions. The extremely important article by Hedgepeth (1977) on "models and muddles" – very stimulating to read, but far too little known – and the illustration that follows represents the situation well.

The discrepancies between different systems are far greater than the variation among humans in a population. But just as the physician must analyze each individual case before deciding on the therapy, so must the ecologist investigate the physiological and biochemical causes underlying each case before he can make a reliable prediction. For such an exhaustive study, there is almost never enough time. A person's experience, and his intrinsic computer – the human brain – can help him here. Computer experts again and again find their machines outclassed by the achievements of the human brain, while biologists are amazed by the performance of electronic computers. The experience that an ecologist has had over many years with phenomena that appear to stand in mutual contradiction, and which his brain has stored, generally serves him better in evaluating a situation than would a single analysis leading to a general model – however great the technical effort expended. I have tried here to give the reader some insight into ecological thinking, and in the process to rely on the capabilities of the human brain – without falling into the trap of simplification, as so often happens. The ecologist needs a solid grounding in the fields of systematics, genetics, physiology, and biochemistry. He needs to be schooled in dealing with the additive and multiplicative effects of factors, and he needs to know that no generalized solution will help him in his task, but only detailed examination of the individual case.

There are laws to protect us from medical quacks. But no one guards us from ecological quacks, and their number and influence are increasing alarmingly. Ecology is a biological science. Without a solid biological substructure rural and urban planning, ecological recommendations, and suggestions for measures to protect the environment can only be quackery – dangerous quackery, because the prospect of quick and simple solutions is so seductive.

Co-operative environmental research
Subproject research design
Data bank
Process models
Integrated site models
Comprehensive biome models
Real-world application

References

Alexander RD, Moore TE (1962) The evolutionary relationships of 17-year and 13-year cicadas, and three new species (Homoptera, Cicadidae, Magicicada). Miscellaneous Publications Museum of Zoology, University of Michigan 121:5–59

Ali KE et al. (1977) Harnstoff-„Recycling" beim Eiweißstoffwechsel der Lamas. Umschau in Wissenschaft und Technik, 338–339

Andersen FS (1956) Effects of crowding in Endrosis sarcitrella. Oikos 7/2:215–226

Andersen FS (1957) The effect of density on the birth and death rate. Annual report 1954–1955, statens skadedyrlaboratorium government pest infestation laboratory, Springforbi, Danmark, pp 5–27

Andersen FS (1961)Effect of density on animal sex ratio. Oikos 12:1–16

Andersen KP, Ursin E (1977) A multispecies extension to the Beverton and Holt theory of fishing, with accounts of phosphorus circulation and primary production. Medd fra Dan Fisk Havunders NS 7:319–435

Arntz WE, Hempel G (1972) Biomasse und Produktion des Makrobenthos in der Kieler Bucht und seine Verwertung durch Nutzfische. Verh dtsch Zool Ges Helgoland, pp 32–36, Stuttgart

Aschoff J, Günter B, Kramer K (1971) Energiehaushalt und Temperaturregulation. In: Gauer, Kramer, Jung (eds) Physiologie des Menschen, Vol 2. Urban & Schwarzenberg, München Berlin Wien

Ashkenazi IE, Hartman H, Strulovitz B, Dar O (1975) Activity rhythms of enzymes in human red blood cell suspensions. J interdiscipl Cycle Res No 4, 6:291–301

Aumann GD, Emlen JT (1965) Relation of population density to sodium availability and sodium selection by microtine rodents. Nature 208:198–199

Ax P (1966) Die Bedeutung der interstitiellen Sandfauna für allgemeine Probleme der Systematik, Ökologie und Biologie. Veröffentlichungen des Instituts für Meeresforschung in Bremerhaven. Sonderband II, pp 15–66

Ax P (1968) Populationsdynamik, Lebenszyklen und Fortpflanzungsbiologie der Mikrofauna des Meeressandes. Verh. d. Deutschen Zool. Gesellsch. in Innsbruck, pp 66–113

Baeumer K (1971) Allgemeiner Pflanzenbau, UTB Uni-Taschenbuch, Stuttgart, pp 264

Ballester A, Albo JM, Vieitez E (1977) The allelopathic potential of Erica scoparia. Oecologia (Berl.) 30:55–61

Beck L (1971) Bodenzoologische Gliederung und Charakterisierung des amazonischen Regenwaldes. Amazoniana 3:69–132

Beck L (1972) Der Einfluß der jahresperiodischen Überflutungen auf den Massenwechsel der Bodenarthropoden im zentral-amazonischen Regenwaldgebiet. Pedobiologa 12:133–148

Bejer-Petersen B (1975) Length of development and survival of Hylobius abietis as influenced by silvicultural exposure to sunlight, Kgl. Vet.- og Landbohojsk. Arsskr., Kopenhagen, pp 111–120

Bell HV (1971) A grazing ecosystem in the Serengeti. Sci Amer 225/1:86–93

Benson AA, Lee RF (1975) The role of wax in oceanic food chains. Sci Amer 232/3:90–101

Berndt R, Henß M (1967) Die Kohlmeise, Parus major, als Invasionsvogel. Die Vogelwarte 24:17–37

Bertsch A (1975) Blüten – lockende Signale. Ravensburg 143 pp

Bezzel E, Ranftl H (1974) Vogelwelt und Landschaftsplanung. Tier und Umwelt 11/12:92

Bezzel E, Reichholf J (1974) Die Diversität als Kriterium zur Bewertung der Reichhaltigkeit von Wasservogel-Lebensräumen. J Ornithol 115:50–61

Blair-West JR et al. (1968) Physiological, morphological and behavioural adaptation to a sodium deficient environment by wild native Australian and introduced species of animals. Nature 217:922–928

Bliss LC (1975) Devon Island, Canada. In: Rosswall, Teal (eds) Structure and function of tundra ecosystems, Ecol. Bulletins NFR 20, Stockholm, pp 17–60

Boeckh J (1967a) Reaktionsschwelle, Arbeitsbereich und Spezifität eines Geruchsrezeptors auf der Heuschreckenantenne. Z vergl Physiol 55:378–406

Boeckh J (1967b) Inhibition and excitation of single insect olfactory receptors, and their role as a primary sensory code. Olfaction and taste II, Proc of the 2nd Int Symp Tokyo, Pergamon Press, Oxford, pp 721–735

Botkin DB, Jordan PA, Dominski AS, Lowendorf HS, Hutchinson GE (1973) Sodium dynamics in a northern ecosystem. Proc Nat Acad Sci (Wash.) 70/10:2745–2748

Braekke F (1976) Impact of acid precipitation on forest and freshwater ecosystems in Norway. In: Research report Fagrapport FR, vol 6, As, Oslo

Brown CE (1909–1966) A cartographic representation of spruce budworm Choristoneura fumiferana (Clem.) Infestation in Eastern Canada. Publication

No. 1263 du service canadien des forèts ministère des pèches et des forèts, Ottawa
Brüll H, Lindner A, von Luterotti L, Scherzinger W (1977) Die Waldhühner, Parey, Hamburg Berlin
Brzoska W (1976) Produktivität und Energiegehalte von Gefäßpflanzen im Adventdalen (Spitzbergen). Oecologia (Berl.) 22:387–398
Bückmann A (1963) Das Problem der optimalen Befischung. Eine Darstellung zur Methodik der Fischereibiologie. Arch Fischerwiss 14/Beiheft 1:1–107
Bulla LA (ed) (1973) Regulation of insect populations by microorganisms. Ann NY Sci 217:243
Bulnheim HP, Siebers D (1976) Salzgehaltsabhängigkeit der Aufnahme gelöster Aminosäuren bei dem Oligochaeten Enchytraeus albidus. Verh Dtsch Zool Ges, p 212
Burdon JJ, Chilvers GA (1976a) Controlled environment experiments on epidemics of Barlew Mildew in different density host stands. Oecologia (Berl.) 26:61–72
Burdon JJ, Chilver GA (1976b) The effect of planting patterns on epidemics of damping-off disease in cress seedlings. Oecologia (Berl.) 23:17–29
Burschel P (1966) Untersuchungen in Buchenmastjahren. Forstwissenschaftliches Zentralblatt 7/8:193–256
Caswell H, Reed F Plant-Herbivore interactions: The indigestibility of C_4 – bundle sheath cells by grasshoppers. Oecologia (Berl.)
Caswell H, Reed F, Stephenson SN, Werner PA (1973) Photosynthetic pathways and selective herbivory: A hypothesis. Amer Natural 107:465–480
Ceska V (1974) Experimentelle Untersuchungen über den Nahrungsbedarf und den Jahreszyklus der Schnee-Eule (Nyctea Scandiaca). Verh. dtsch. Ges. Ökolog, pp 199–201
Chapman RF, Bernays EA (ed) (1978) Insect and host plant. Proc 4th Int Symp 1978 Entomol Exp appl 24:200–766
Christensen NL (1977) Fire and soil-plant nutrient relations in a pine-wire-grass savanna on the Coastal Plain of North Carolina. Oecologia (Berl.)
Christiansen M (1948) Epidemiaktikt Sygdomsudbred blandt Ederfugle ved Bornholm, foraasaget af dyriske snyltere. Dansk orn Foren Tidsskrift, p 42
Conver JL, Howick, GL, Corzette MH, Kramer SL, Fritzgibbon S, Landesberg R (1978) Visual predation by planktivores. Oikos 31:27–37
Crisp DJ (1965) Surface chemistry, a factor in the settlement of marine invertebrate larvae. Proc. 5th Marine Biol Symp, Göteborg, pp 51–65
Cummins KW, Wuycheck JC (1971) Caloric equivalents for investigations in ecological energetics. Mitt Int Vereinig Limnolog 18:1–158
Curio E (1976) The ethology of predation. In: Zoophysiology and ecology, vol 7. Springer, Berlin Heidelberg New York
Danin A (1976) Plant species diversity under desert conditions. Oecologia 22:251–259
Dawkins R (1978) Das egoistische Gen. Springer, Berlin Heidelberg New York, pp 1–246
Denison WC (1973) Life in tall trees. Sci Amer 228/6:74–81
Dobler E (1977) Correlation between feeding time of the Pike Esox lucius and the dispersion of a school of Leucaspius delineatus. Oecologia (Berl.) 27:93–96
Dobzhansky Th (1947) Adaptive changes induced by natural selection in wild populations of Drosophila. Evolution 1:1–16
Dogiel A (1963) reworked and completed by Poljanski GI, Cheissin EM: Allgemeine Parasitologie. Parasitolog Schr-Reihe 16
Dolinger PM, Ehrlich PH, Fitch WL, Breedlove DE (1973) Alkaloid and predation patterns in Colorado lupine populations. Oecologia (Berl.) 13:191–204
Drift van der J, Janson E (1977) Grazing of springtails on hyphal mats and its influence on fungal growth and respiration. Soil organisms as components of ecosystems. Ecol Bull 25:203–209
Drosopolous S (1977) Biosystematic studies on the Muellerianella complex (Delphacidae, Homoptera, Auchenorrhyncha). Medd Landbouwhogeschool Wageningen 77/14:1–133
Efford IE, Tsumura K (1973) Uptake of dissolved glucose and glycine by Pisidium, a freshwater bivalve. Can J Zool Vol 51:825–832
Ehleringer JR (1978) Implications of quantum yield differences on the distributions of C_3 and C_4 grasses, Oecologia (Berl.) 31:225–267
Ehrlich PR (1975) The population biology of coral reef fishes. Ann Rev Ecol Syst 6:212–247
Ehrlich PR, Ehrlich A, Holdren JP (1975) Humanökologie: Der Mensch im Zentrum einer neuen Wissenschaft. Springer, Berlin Heidelberg New York
Ehrlich PR, White RR, Singer MC, McKechnie, Stephen W, Gilbert, Lawrence E (1975) Checkerspot Butterflies: A historical perspective. Science 188:221–228
Eisfeld D (1975) Der Eiweiß- und Energiebedarf des Rehes (Capreolus C. Capreolus L.), diskutiert anhand von Laborversuchen. Verh dtsch Ges Ökol, pp 129–139
Ellenberg Ch u. Heinz (1969) „Kal“ – Das Kahlwerden von Kulturwiesen Islands als ökologisches Problem. Bericht aus der Forschungsstelle Neôri Ås, Hveragerôi (Island) No 3, pp 3–47
Ellenberg Heinz (1971a) Zur Kartenübersicht der Kahlschäden an den Kulturwiesen Islands im Jahre 1969. Bericht aus der Forschungsstelle Neôri Ås, Hvergerôi (Island) No 7, pp 2–22
Ellenberg Heinz (ed) (1971b) Integrated experimental ecology methods and results of ecosystem research in the German Solling project. In: Ecological studies, vol 2, Springer, Berlin Heidelberg New York
Ellenberg Heinz (1974) Zeigerwerte der Gefäßpflanzen Mitteleuropas. Scripta Geobotanica IX/9:5–97
Ellenberg Hermann (1974) Beiträge zur Ökologie des Rehes (Capreolus capreolus L. 1758). Daten aus

dem Stammhamer Versuchsgehege. Dissertation, Kiel

Ellenberg Hermann (1978) Zur Populationsökologie des Rehes (Capreolus capreolus L., Cervidae) in Mitteleuropa. Spixiana (München), Suppl. 2:1–211

Elster H-J (1974) Das Ökosystem Bodensee in Vergangenheit, Gegenwart und Zukunft. Schriftenreihe VG Bodensee 92:233–250

Emeis W (1950) Einführung in das Pflanzen- und Tierleben Schleswig-Holsteins. Möller, Rendsburg

Enders F (1975) The influence of hunting manner on prey size, particularly in spiders with long attack distances (Araneidae, Linyphiidae, and Salticidae). The Amer Natural 109/970:737

Enright JT (1976) Climate and population regulation. The biogeographer's dilemma. Oecologia (Berl.) 24:295–310

Ernst W (1974) Mechanismen der Schwermetallresistenz. Verh dtsch Ges Ökol, pp 189–197

Erz W (1964) Populationsökologische Untersuchungen an der Avifauna zweier nordwestdeutscher Großstädte (unter besonderer Berücksichtigung der populationsdynamischen Verhältnisse bei der Amsel, Turdus merula merula L.). Z wiss Zool 170:1–111

Farb P (1965) Die Ökologie. Time-Life (Netherland)

Fisher J (1952) The fulmar. Collins, London

Fisher J (1954) Birds as animals. I. A history of birds. In: Biological sciences. Hutchinsons University Library

Fittkau EJ (1973 a) Artenmannigfaltigkeit amazonischer Lebensräume aus ökolologischer Sicht. Amazoniana 4:321–340

Fittkau EJ (1973 b) Crocodiles and the nutrient metabolism of Amazonian Waters. Amazoniana 6/1:103–133

Fittkau EJ (1974) Zur ökologischen Gliederung Amazoniens I. Die erdgeschichtliche Entwicklung Amazoniens. Amazoniana V 1:77–134

Fittkau EJ, Klinge H (1972) Filterfunktionen im Ökosystem des zentralamazonischen Regenwaldes. Mitt Dtsch Bodenkundl Ges 16:130–135

Fittkau EJ, Klinge H (1973) On biomass and trophic structure of the central. Amazonian rain forest ecosystem. Biotropica 5:2–14

Fittkau EJ, Irmler U, Junk WJ, Reiss F, Schmidt GW (1975 a) Ecological division of the Amazon region. In: Golley FB, Medina E (eds) Tropical ecological systems, trends in terrestrial and aquatic research. Springer, Berlin Heidelberg New York, pp 289–311

Fittkau EJ, Klinge, H, Rodrigues WA, Brunig E (1975 b) Biomass and structure in a central amazonian rain forest. Tropical ecological systems. Trends in terrestrial and aquatic research. Golley FB, Medina E (eds). Springer, Berlin Heidelberg New York, pp 115–122

Florey E (1970) Lehrbuch der Tierphysiologie. Eine Einführung in die allgemeine und vergleichende Physiologie der Tiere. Thieme, Stuttgart

Ford MJ (1977 a) Metabolic costs of the predation strategy of the spider Pardosa amentata. Oecologia (Berl.) 28:333–340

Ford MJ (1977 b) Energy costs of the predation strategy of the web-spinning Spider Lepthyphantes zimmermannae. Oecologia (Berl.) 28:341–350

Foster WL, Tate J (1966) The activities and coactions of animals at sapsucker trees. Living Bird 5:87–113

Franz JM, Krieg A (1972) Biologische Schädlingsbekämpfung. Parey, Hamburg

Fraser J (1965) Treibende Welt. In: Verständliche Wissenschaft, Vol 85. Springer, Berlin Heidelberg New York

Freeland WJ, Winter JW (1975) Evolutionary consequences of eating: Trichosurus vulpecula (Marsupialia) and the genus Eucalyptus. J Chem Ecol 1:439–455

Frisch K v (1965) Tanzsprache und Orientierung der Bienen. Springer, Berlin Heidelberg New York, p 578

Funke W (1973) Food and energy turnover of leafeating insects and their influence on primary production. Ecological Studies 2:89–93

Funke W, Weidemann G (1973) Food and energy turnover of phytophagous and predatory arthropods. Ecological Studies 2:100–109

Gardasson A, Sigurdson JB (1972) Research on the pink-footed goose 1971. Reykjavik (isländ.)

Gardasson A, Sigurdson JB (1974) Studies on the breeding biology of the pink footed goose; studies on plant production and grazing by pink footed goose. Progress Report. Reykjavik (isländ.)

Gates DM (1965) Heat, radiant and sensible, Ch. 1, Radiant energy, its receipt and disposal. Meteorological Monographs 6/28:1–26

Gates M, Schmerl RB (1975) Perspectives of biophysical ecology. Springer, Berlin Heidelberg New York

Geisler G (1971) Pflanzenbau in Stichworten. II. Die Ertragsbildung. Hirt, Kiel

Geist V, Walther F (ed) (1974) The behaviour of Ungulates and its relation to management, vol 1, vol 2, IUCN Publications, new series No. 24, Morges, Switzerland, pp 11–511, pp 512–940

George JC (1972) Everglades wildguide. Natural History Series, US Dept. of the Interior

Gerlach SA (1958) Die Mangroveregion tropischer Küsten als Lebensraum. Z Morph Ökol Tiere 46:636–730

Gerlach SA (1959) Über das tropische Korallenriff als Lebensraum. Verh dtsch Zool Ges, pp 356–363

Gerlach SA, Schrage M (1969) Freilebende Nematoden als Nahrung der Sandgarnele Crangon crangon. Oecologia (Berl.) 2:362–375

Gerlach SA (1971 a) On the importance of marine meiofauna for benthos communities. Oecologia (Berl.) 6:176–190

Gerlach SA (1971 b) Produktionsbiologie des Meeres. Biologie in unserer Zeit 1:1–34

Gerlach SA (1976) Meeresverschmutzung. Diagnose und Therapie. Springer, Berlin Heidelberg New York

Gessner F (1957) Meer und Strand. VEB Deutscher Verlag der Wissenschaften, Berlin, pp 1–426

Gillmor R Ecological isolation in birds. Blackwell, Oxford-Edinburgh, pp 1–404

Glück E (1979) Abhängigkeit des Bruterfolges von der Lichtmenge am Neststandort. J Orn 120:215–220

Golley FB (1960) Energy dynamics of a food chain of an old-field community. Ecological Monographs 30:187–206

Gommers FJ (1973) Nematicidal principles in compositae. Meded Landbouwhogesch. Wageningen Nederland 73(17):1–71

Görner P, Andrews P (1969) Trichobothrien, ein Ferntastsinnesorgan bei Webespinnen. Z vergl Phys 64:301–317

Goss-Custard JD (1977) The energetics of prey selection by redshank, tringa totanus (1.), in relation to prey density. J Anim Ecol 46:1–19

Gyllenberg G (1969) The energy flow through a Chorthippus parallelus (Zett.) (Orthoptera) population on a meadow in Tvärminne, Finland. Acta Zool Fennica, p 123

Gyllenberg G (1974) A simulation model for testing the dynamics of a grasshopper population. Ecology 55:645–650

Haarlov N (1960) Microarthropods from Danish soils: ecology, phenology. Oikos, Suppl. 3:11–176

Hahn J, Aehnelt E (1972) Die Fruchtbarkeit der Tiere als biologischer Indikator für Umweltbelastungen. Tagung Gießen, pp 49–54

Halbach U (1969) Das Zusammenwirken von Konkurrenz und Räuber-Beute-Beziehungen bei Rädertieren. Zool Anz, Suppl 33; Verh Zool Ges, pp 72–79

Halbach U (1973) Life table data and population dynamics of the Rotifer Brachionus calyciflorus Pallas as influenced by periodically oscillating temperature. In: Wieser W (ed) Effects of temperature on ectothermic organisms. Springer, Berlin Heidelberg New York, pp 217–228

Halbach U (1977) Probleme der Ökosystemforschung am Beispiel der Limnologie. Verh Dtsch Zool Ges, pp 41–66

Harder W (1965a) Elektrische Fische. Umschau 15:467–473

Harder W (1965b) Elektrische Fische. Umschau 16:492–496

Hashimoto H (1957) Peculiar mode of emergence in the marine Chironomid Clunio (Dipt. Chironomidae). Sci Rep Tokyo Kyoiku Daigaku B 8:217–226

Haukioja E, Niemälä P (1977) Retarded growth of a geometrid larva after mechanical damage to leaves of its host tree. Ann Zool Fenn 14:48–52

Hedgpeth JW (ed) (1957) Treatise on marine ecology and paleoecology. The Geological Society of America Memoir 67, vol 1, pp 1–1296

Heithaus ER (1974) The role of plant-pollinator interactions in determining community structure. Ann Missouri Bot Gard 61/3:675–691

Helfferich B, Gütte JO (1972) Tierernährung in Stichworten. Hirt, Kiel

Hempel G (1977) Biologische Probleme der Befischung mariner Ökosysteme. Naturwissenschaften 64:200–206

Hendrichs H (1977) Untersuchungen zur Säugetierfauna in einem paläotropischen und einem neotropischen Bergregenwaldgebiet. Säugetierkundliche Mitteilungen, BLV Verlagsgesellschaft mbH München 40, 25. Jg. Heft 3, pp 213–225

Hendrichs H (1978) Die soziale Organisation von Säugetierpopulationen. Säugetierkundliche Mitteilungen, BLV Verlagsgesellschaft mbH München 40, 26. Jg. Heft 2, pp 81–116

Henrikson L, Oscarson G (1978) Fish predation limiting abundance and distribution of Glaenocorisa p. propinqua. Oikos 31:102–105

Herlitzius R, Herlitzius H (1977) Streuabbau in Laubwäldern. Oecologia (Berl.) 30:147–173

Hewson R (1976) A population study of mountain hares (Lepus timidus) in north-east Scotland from 1956–1969. J Anim Ecol 45:395–414

Hinz W Zur Ökologie der Tundra Zentralspitzbergens. Norsk Polarinstitutt Skifter No 163, pp 4–47

Hochachka PW, Somero GN (1973) Strategies of Biochemical Adaptation. Saunders, Philadelphia, pp 1–358

Hocking B (1968) Insect-flower associations in the high Arctic with special reference to nectar. Oikos 19:359–388

Hoffmann K-H (1973) Der Einfluß der Temperatur auf die chemische Zusammensetzung von Grillen (Gryllus, Orthopt.). Oecologia (Berl.) 13:147–175

Hoffmann K-H (1974) Wirkung von konstanten und tagesperiodisch alternierenden Temperaturen auf Lebensdauer; Nahrungsverwertung und Fertilität adulter Gryllus bimaculatus. Oecologia (Berl.) 17:39–54

Hoffmann K-H (1976) Organic body constituents of Protophormia terrae-novae (Dipt.) from Spitzbergen compared with flies from a laboratory stock. Oecologia (Berl.) 23:13–16

Hölldobler B Chemische Verständigung bei sozialen Insekten. Grzimeks Tierleben, Ergänzungsband Verhaltensforschung, Kindler Verlag, pp 486–494

Hölldobler B, Haskins CP (1977) Sexual calling behavior in primitive ants. Science 195:793–794

Holling CS (1966) The functional response of invertebrate predators to prey density. Memoirs of the Entomological Society of Canada 48:3–86

Holst D v (1969) Sozialer Streß bei Tupajas (Tupaia belangeri). Die Aktivierung des sympathischen Nervensystems und ihre Beziehung zu hormonal ausgelösten ethologischen und physiologischen Veränderungen. Z vergl Physiol 63:1–58

Holst D v (1972) Renal failure as the cause of death in Tupaia belangeri exposed to persistent social Stress. J Comp Physiol 78:236–273

Hörnfeld B (1978) Synchronous population fluctuations in voles, small game, owls and Tularaemia in northern Sweden. Oecologia (Berl.) 32

Huber W (1973) Biologie und Ökologie des Feldhasen. Wald und Wild, Beiheft No 52, to Zeitschr des Schweiz Forstvereins Inst f Waldbau, ETH-Zürich, pp 223–237

Hughes RN (1970a) Population dynamics of the bivalve Scrobicularia plana (Da Costa) on an intertidal mud-flat in North Wales. J Anim Ecol 39:333–356

Hughes RN (1970b) An energy budget for a tidal-flat population of the bivalve Scrobicularia plana (da costa). J Anim Ecol 39:357–381

Hylleberg J (1976) Resource partitioning on basis of hydrolytic enzymes in deposit feeding mud snails (Hydrobiidae). Oecologia (Berl.) 23:115–125

Ibrahim I (1973) Vergleichende Untersuchungen zur Thermophilie von Bakterien der Gattung Bacillus. Zbl Bakt, Abt II 128:269–273

Idyll CP (1973) The anchovy crisis. Sci Amer 228/6:22–29

Immelmann K (1962) Besiedlungsgeschichte und Bastardierung von Lonchura castaneothorax und Lonchura flaviprymna in Nordaustralien. J Orn 103:344–357

Ingolfsson A (1970) Hybridization of glaucous gulls Larus hyperboreus and herring gulls L. argentatus in Iceland. IBIS 112:340–362

Jakovlev V (1956) Wasserdampfabgabe der Acrididen und Mikroklima ihrer Biotope. Verh dtsch zool Ges, p 136

Jakovlev V, Krüger F (1954) Untersuchungen über die Vorzugstemperatur einiger Acrididen. Biologisches Zentralblatt 73:633–650

Jaksic Fabian M, Montenegro G (1979) Resource allocation of Chilean herbs in response to climatic and microclimatic factors. Oecologia (Berl.) pp 2–9

Janssen J (1913) Geschichte des Deutschen Volkes. Herder, Freiburg

Jansson A (1967) The food-web of the Cladophora-belt fauna. Helgoländer wissenschaftliche Meeresuntersuchungen 15:574–588

Jansson BO (1972) Ecosystem approach to the Baltic problem. Bull Ecol Res Committee/NFR, No 16, pp 1–82

Jenkins D, Watson A, Miller GR (1964) Predation and red grouse populations. J appl Ecology 1:183–195

Jones DA (1973) Co-evolution and cyanogenesis. Taxonomy and Ecology, pp 213–242

Joosse Els NG, Testerink GJ (1977) The role of food in the population dynamics of Orchesella cincta (Linne) (Collembola). Oecologia (Berl.) 29:189–204

Kaiser H (1974) Verhaltensgefüge und Temporalverhalten der Libelle Aeschna cyanea (Odonata). Z Tierpsychol 34:398–429

Kändler R (1962) Die Fischereierträge der Meere als Ausdruck ihrer unterschiedlichen Produktionsleistungen. Kieler Meeresforschungen 18:121–127

Kaufmann O (1932) Einige Bemerkungen über den Einfluß von Temperaturschwankungen auf die Entwicklungsdauer und Streuung bei Insekten und seine graphische Darstellung durch Kettenlinie und Hyperbel. Z Morphol Ökol Tiere 25:353–361

Kenagy GJ (1973) Adaptations for leaf eating in the great basin kangaroo rat, Dipodomys microps. Oecologia (Berl.) 12:383–412

Kinzelbach RK (1976) Beiträge zur Kenntnis der Symbiose von Pagurus prideauxi und Adamsia paliata (Crustacea: Paguroidea; Coelenterata: Actiniaria). Verh dtsch Zool Ges, p 213

Kirschbaum U (1972) Flechtenkartierungen in der Region Untermain zur Erfassung von Immissionsbelastungen. Tagungsbericht der Gesellschaft für Ökologie, Gießen, pp 133–140

Kleiber M (1967) Der Energiehaushalt von Mensch und Haustier. Parey, Hamburg

Klopffleisch U (1976) Ökolog. Untersuchung an der Feldgrille Gryllus campestris L am natürlichen Standort. Staatsex Arb (Köln, unpubl.)

Kluge M, Lange OL, Eichmann M v, Schmid R (1973) Diurnaler Säurerhythmus bei Tillandsia usneoides: Untersuchungen über den Weg des Kohlenstoffs sowie die Abhängigkeit des CO_2-Gaswechsels von Lichtintensität, Temperatur und Wassergehalt der Pflanze. Planta (Berl.) 112:357–372

Koch J, Heinig S (1977) Daphnis nerii – ein Labortier? (Lep., Sphingidae). Entomol Z, pp 57–62

Kremer P (1978) Giftalgen und Algengifte. Biologie in unserer Zeit 8:97–103

Krishnamurthy L (1978) Production, dominance, diversity and stability in some semi-arid grazing lands. Intecol Bull 6:14–20

Krüll F (1976a) The position of the sun is a possible zeitgeber for arctic animals. Oecologia (Berl.) 24:141–148

Krüll F (1976b) Zeitgebers for animals in the continuous daylight of high arctic summer. Oecologia (Berl.) 24:149–157

Kruuk H (1978) Foraging and spatial organisation of the European badger, Meles meles L. Behav Ecol Sociobiol 4:75–89

Kunick W, Kutscher G (Hrsg) (1976) Vorträge der Tagung über „Umweltforschung" der Universität Hohenheim. Daten und Dokumente zum Umweltschutz No 19, Hohenheim, pp 5–220

Lagerspetz K (1963) Humidity reactions of three aquatic amphipods, Gammarus duebeni, G. oceanicus and Pontoporeia affinis in the air. Exp Biol 40:105–110

Lahti S, Tast J, Uotila H (1976) Fluctuations in small rodent populations in the Kilvisjärvi area in 1950–1975. Luonnon Tutkija 80:97–107

Lampe RP (1977) Aspects of the predatory strategy of the North American badger. Diss Abstr Int 37:12

Lampert W (1976) Die „kritische" Futterkonzentration als mögliche Ursache für Assoziationen und Sukzessionen von Zooplankton. Verh dtsch Zool Ges, pp 214
Lamprey HF (1964) Estimation of the large mammal densities, biomass and energy exchange in the Tarangire game reserve and the masai steppe in Tanganyika. East Afr Wildlife J 2:1–59
Lange OL, Schulze E-D, Koch W (1970) Experimentell-ökologische Untersuchungen an Flechten der Negev-Wüste. Flora 159:38–62 (1970)
Larcher W (1973) Ökologie der Pflanzen. Uni Taschenbuch 232. Ulmer, Stuttgart
Leuthold W (1977) African Ungulates, a comparative review of their ethology and behavioral ecology. Springer, Berlin Heidelberg New York, pp 1–307
Lieth H, Whittaker RH (1975) Primary productivity of the biosphere. Ecological Studies, p 14
Linsenmair KE, Jander R (1963) Das „Entspannungsschwimmen" von Velia und Stenus. Die Naturwissenschaften 50:231
Lockley RM (1953) Puffins. J.M. Dent, London
Lohmann M (1974) Ökofibel. Deutscher Naturschutzring, Bundesverband für Umweltschutz e.V. Bonn-Oberkassel
Lopez GR, Levinton JS, Slobodkin LB (1977) The effect of grazing by the detritivore Orchestia grillus on Spartina litter and its associated microbial community. Oecologia (Berl.) 30:111–128
Lundberg A (1979) Residency, migration and a compromise: adaptations to nest-site scarcity and food specialization in three Fennoscandian owl species. Oecologia 38:1–13
MacArthur RH, Connell JH (1961) Biologie der Populationen. München: BLV Verlagsgesellschaft 1970
Macfadyen A (1961) Metabolism of soil invertebrates in relation to soil fertility. Ann appl Biol 49:215–218
Mackinnon J (1974) In search of the red ape. Ballantine, New York
Magaard L, Rheinheimer G (ed) (1974) Meereskunde der Ostsee. Springer, Berlin Heidelberg New York, pp 1–269
Markl H (1972) Neue Entwicklungen in der Bioakustik der wirbellosen Tiere. J Ornithol 113/1:91–104
Markl H, Hauff J (1973) Die Schwellenkurve des durch Vibration ausgelösten Fluchttauchens von Mückenlarven. Die Naturwissenschaften 60. Jg 9:432–433
Markl H, Tautz J (1978) Caterpillars detect flying wasps by hairs sensitive to airborne vibration. Behav Ecol Sociobiol 4:101–110
Markl H, Lang H, Wiese K (1973) Die Genauigkeit der Ortung eines Wellenzentrum durch den Rückenschwimmer. Notonecta glauca L. J comp Physiol 86:359–364
Marler P (1977) Sound transmission and its significance for animal vocalization. Temperate habitats. Behav Ecol Sociobiol 2:271–290
Marler P, Marten K, Quine D (1977) Sound transmission and its significance for animal vocalisation. Tropical forest habitats. Behav Ecol Sociobiol 2:291–302
Martin PS (1973) Discovery of America. Science 179:969–974
Mattson WJ, Addy ND (1975) Phytophagous insects as regulators of forest primary production. Science 190:515–522
Maurer E (1964) Buchen- und Eichensamenjahre in Unterfranken während der letzten 100 Jahre. Allgemeine Forstzeitschrift 31:469–470
McNaughton SJ (1967) Relationships among functional properties of California grassland. Nature 216:168–169
McNaughton SJ (1968) Definition and quantification in ecology. Nature 219:180
Mech DL (1966) The wolves of Isle Royale. Fauna of the national parks of the U.S., Ser. 7. Washington D.C.
Meinertzhagen R (1954) Bulletin of the British Ornithologists' Club 74/9:97–112
Merkel E (1957) Die ökologischen Ursachen der Massenvermehrung des großen Fichtenborkäfers in Südwestdeutschland. Part 1. Freiburg
Merkel G (1977) The effects of temperature and food quality on the larval development of Gryllus bimaculatus (Orthoptera, Gryllidae). Oecologia (Berl.) 30:129–140
Miyawaki A, Fujiwara K (1970) Vegetationskundliche Untersuchungen im Ozegahara-Moor, Mittel-Japan. The National Parks Association of Japan, Tokyo
Miyawaki A, Tüxen R (1977) Vegetation science and environmental protection, Maruzen, Tokyo
Möglich M, Maschwitz U, Hölldobler B (1974) Tandem calling: a new kind of signal in ant communication. Science 186:1046–1047
Mohr H (1969) Lehrbuch der Pflanzenphysiologie. Springer, Berlin Heidelberg New York
Moss R (1969) A comparison of red grouse (Lagopus l. scoticus) stocks with the production and nutritive value of heather (Calluna vulgaris). J Anim Ecol 43:103–122
Moss R, Watson A, Parr R (1973) The role of nutrition in the population dynamics of some game birds (Tetraonidae). XI. Int Congr of Game Biologists, Stockholm, pp 193–201
Mühlenberg M (1976) Freilandökologie. Quelle & Meyer, Heidelberg
Muller CH (1967) Die Bedeutung der Allelopathie für die Zusammensetzung der Vegetation. Z Pflanzenkrankh (Pflanzenpath) Pflanzenschutz 74:332–346
Müller FJ (1974) Territorialverhalten und Siedlungsstruktur einer Mitteleuropäischen Population des Auerhuhns Tetrao Urogallus Major C.L., Brehm. Dissertation, Marburg
Müller P (1974) Aspects of zoogeography. Dr. W. Junk, Den Haag

Müller W (1969) Auslösung der Metamorphose durch Bakterien bei den Larven von Hydractina echinata. Zool Jb Anat 86:84–95

Murphy RC (1936) Oceanic birds of South America I. Macmillan, New York

Myers J, Krebs CJ (1974) Population cycles in rodents. Sci Amer, pp 38–46

Myrberget S (1973) Geographical synchronism of cycles of small rodents in Norway. Oikos 24:220–224

Nagy KA, Milton K (1979a) Aspects of dietary quality, nutrient assimilation and water balance in wild howler monkeys (Alouatta palliata). Oecologia 39:1–13

Nagy KA, Milton K (1979b) Energy metabolism and food consumption by wild howler monkeys (Alouatta palliata). Ecology 60/2:6

Nellen W (1978) Probleme der wirtschaftlichen Nutzung mariner Ökosysteme. Verh Ges Ökol, 7. Jahresvers., pp 67–76

Neumann D (1961) Ernährungsbiologie einer rhipidoglossen Kiemenschnecke. Hydrobiologia 17:133–151

Neumann D (1962) Über die Steuerung der lunaren Schwärmperiodik der Mücke Clunio marinus. Verh dtsch Zool Ges, pp 275–285

Neumann D (1965) Die intraspezifische Variabilität der lunaren und täglichen Schlüpfzeiten von Clunio marinus (Diptera: Chironomidae). Verh dtsch Zool Ges, pp 223–233

Neumann D (1966) Die lunare und tägliche Schlüpfperiodik der Mücke Clunio: Steuerung und Abstimmung auf die Gezeitenperiodik. Z vergl Phys 53:1–61

Neumann D (1968) Die Steuerung einer semilunaren Schlüpfperiodik mit Hilfe eines künstlichen Gezeitenzyklus. Z vergl Phys 60:63–78

Neumann D (1969) Die Kombination verschiedener endogener Rhythmen bei der zeitlichen Programmierung von Entwicklung und Verhalten. Oecologia (Berl.) 3:166–183

Neumann D (1971) Eine nicht-reziproke Kreuzungssterilität zwischen ökologischen Rassen der Mücke Clunio marinus. Oecologia (Berl.) 8:1–20

Neumann D (1976) Mechanismen für die zeitliche Anpassung von Verhaltens- und Entwicklungsleistungen an den Gezeitenzyklus. Verh Dtsch Zool Ges, pp 9–28

Neuweiler G (1978) Die Echoortung der Fledermäuse. Rheinisch-Westfälische Akademie der Wissenschaften, Vorträge. 271:58–82

Newell RC, Johnson LG, Kofoed LH (1977) Adjustment of the components of energy balance in response to temperature change in Ostrea edulis. Oecologia (Berl.) 30:97–110

Nielsen OB (1966) Carbohydrases of some wrack invertebrates. Natura Jutlandica 12:191–194

Nuorteva P (1963) The influence of Oporinia autumnata (Bkh.) (Lep., Geometridae) on the timber-line in subarctic conditions. Ann Ent Fenn 29:270–277

Odum EP (1967) Ökologie. Moderne Biologie. BLV München

Ohnesorge B (1976) Tiere als Pflanzenschädlinge. Thieme, Stuttgart

D'Oleire-Oltmanns W (1977) Combustion heat in ecological energetics. What sort of information can be obtained? In: Applications of calorimetry in life sciences. De Gruyter, Berlin New York, pp 315–324

Orians GH (1974) Diversity, stability and maturity in natural ecosystems. Proc Inst int Congr Ecology. Pudoc, Wageningen, pp 64–65

Owen DF, Wiegert RG (1976) Do consumers maximize plant fitness? Oikos 27:488–492

Pallmann H, Eichenberger E, Hassler A (1940) Eine neue Methode der Temperaturmessung bei ökologischen oder bodenkundlichen Untersuchungen. Ber Schweiz bot Ges 50:377–362

Pardi L (1960) Innate components in the solar orientation of littoral amphipods. Biological clocks, vol XXV. The Biological Laboratory. Cold Spring Harbor, L.I., New York, pp 395–401

Pearson TH (1968) The feeding ecology of sea-bird species breeding on the Farne Islands, Northumberland. J Anim Ecol 37:521–552 (1968)

Persson L (1974) Endoparasitism causing heavy mortality in eider ducks in Sweden. XI. Int Congr of Game Biologists, Stockholm, pp 255–258

Petrides CA (1974) The overgrazing cycle as a characteristic of tropical savannas and Grasslands in Africa. Proc. 1st int. Congr. Ecology. Pudoc, Wageningen, pp 86–91

Petrusewicz K, Macfadyen A (1970) Productivity of terrestrial animals. Blackwell, Oxford-Edinburgh

Pflüger W, Neumann D (1971) Die Steuerung einer gezeitenparallelen Schlüpfrhythmik nach dem Sanduhr-Prinzip. Oecologia (Berl.) 7:262–266

Phillipson J (1966) Ecological Energetics. Arnold, London

Pielou EC (1974) Population and community ecology: principles and methods. Gordon and Breach, New York

Pöhlmann H (1975) Ökologische Untersuchungen an Rentieren in Spitzbergen. Verh Ges Ökolog, pp 89–92

Pöhlmann H (1976) The food requirements of reindeer and geese, and their importance for the Arctic tundra of the Adventalen, Spitzbergen, Dissertation, Erlangen, pp 3–106

Porter WP, Gates DM (1969) Thermodynamic equilibria of animals with environment. Ecological Monographs 39:227–244

Precht H, Christophersen J, Hensel H (1955) Temperatur und Leben. Springer, Berlin Göttingen Heidelberg

Pschorn-Walcher H, Zwölfer H (1968) Konkurrenzerscheinungen in Parasitenkomplexen als Problem der Biologischen Schädlingsbekämpfung. Anzeiger für Schädlingskunde 5:71–76

Quednau W (1957) Über den Einfluß von Temperatur und Luftfeuchtigkeit auf den Eiparasiten Trichogramma cacoeciae Marchal. (Eine biometrische Studie.) Mitteilungen aus der Biologischen Bundesanstalt für Land- und Forstwirtschaft 90:5–63

Rachor E, Gerlach AS Variations in macrobenthos in the German Bight. Int. Council for the Exploration of the Sea. Symp. on the changes in the north sea fish stocks and their Causes No. 11, pp 1–16

Regal PhJ (1977) Ecology and evolution of flowering plant dominance. Science 196:622–629

Reichholf H u J (1973) „Honigtau" der Bracaatinga-Schildlaus als Winternahrung von Kolibris (Trochilidae) in Süd-Brasilien. Bonn Zool Beitr 24:7–14

Reichholf J (1975 a) Biogeographie und Ökologie der Wasservögel im subtropisch-tropischen Südamerika. Anzeiger der Ornithologischen Gesellschaft in Bayern 14/1:2–69

Reichholf J (1975 b) Die quantitative Bedeutung der Wasservögel für das Ökosystem eines Innstausees. Verh Ges Ökolog, pp 247–254

Reichholf J (1977) Die Ökostruktur der Innstauseen. Bild der Wissenschaft 8:36–41

Reichle DE (1973) Analysis of temperate forest ecosystems. Ecological Studies 1:1–304

Reise K (1976) Feindruck auf der Wattfauna der Nordsee. Dissertation, Göttingen, pp 1–141

Reiss F (1973) Zur Hydrographie und Makrobenthosfauna tropischer Lagunen in den Savannen des Território de Roraima, Nordbrasilien. Amazoniana IV/4:367–378

Reiss F (1974) Vier neue Chironomus-Arten (Chironomidae, Diptera) und ihre ökologische Bedeutung für die Benthosfauna zentralamazonischer Seen und Überschwemmungswälder. Amazoniana V/1:3–23

Reiss F (1976 a) Charakterisierung zentralamazonischer Seen aufgrund ihrer Makrobenthosfauna. Amazoniana VI/1:123–134

Reiss F (1976 b) Die Benthoszoozönosen zentralamazonischer Varzeaseen und ihre Anpassungen an die jahresperiodischen Wasserstandsschwankungen. Biogeographica 7:125–135

Reiss F (1977) Qualitative and quantitative investigations on the macrobenthic fauna of Central Amazon lakes. I. Lago Tupé, a black water lake on the lower Rio Negro. Amazoniana VI/2:203–235

Remane A (1940) Die Tierwelt der Nord- und Ostsee. I: Ökologie. Akademische Verlagsgesellschaft, Leipzig

Remane A (1955) Die Brackwasser-Submergenz und die Umkomposition der Coenosen in Belt- und Ostsee, Kieler Meeresforschungen 11:59–73

Remane A, Schlieper C (1971) Biology of brackish water. Die Binnengewässer, pp 372, Stuttgart

Remane R, Koch J (1977) Merkmalsverschiebungen im Bau der Genitalarmatur der zentraliberischen Populationen des Euscelis-incisus Kb.-alsius Rib.-Formenkreises – ein Indiz für Introgressionsphänomene? Zool Beitr 23:133–167

Remmert H (1955) Substratbeschaffenheit und Salzgehalt als ökologische Faktoren für Dipteren. Zool Jb (Systematik) 83:453–474

Remmert H, Aves (1957) Tierwelt der Nord- und Ostsee 38:1–102

Remmert H (1960) Über tagesperiodische Änderungen des Licht- und Temperaturpräferendums bei Insekten (Untersuchungen an Cicindela campestris und Gryllus domesticus). Biologisches Zentralblatt 79:577–584

Remmert H (1962) Der Schlüpfrhythmus der Insekten. Steiner, Wiesbaden

Remmert H (1964) Änderungen der Landschaft und ökologischen Folgen, dargestellt am Beispiel der Insel Amrum. Veröffentlichungen des Instituts f. Meeresforschung in Bremerhaven 9:100–108

Remmert H (1965 a) Biologische Periodik. In: Handbuch der Biologie, Vol 5. Akademische Verlagsgesellschaft Athenaion, Frankfurt, pp 335–441

Remmert H (1965 b) Distribution and the ecological factors controlling distribution of the European wrackfauna. Botanica Gothoburgensia 3:179–184

Remmert H (1965 c) Über den Tagesrhythmus arktischer Tiere. Z Morphol Ökol Tiere 55:142–160

Remmert H (1966) Zur Ökologie der küstennahen Tundra Westspitzbergens. Z Morphol Ökol Tiere 58:162–172

Remmert H (1967) Physiologische-ökologische Experimente an Ligia oceanica (Isopoda). Z Morphol Ökol Tiere 59:33–41

Remmert H (1968 a) Die Littorina-Arten: Kein Modell für die Entstehung der Landschnecken. Oecologia (Berl.) 2:1–6

Remmert H (1968 b) Über die Besiedlung des Brackwasserbeckens der Ostsee durch Meerestiere unterschiedlicher ökologischer Herkunft. Oecologia (Berl.) 1:296–303

Remmert H (1969 a) Der Wasserhaushalt der Tiere im Spiegel ihrer ökologischen Geschichte. Naturwissenschaften 56:120–124

Remmert H (1969 b) Tageszeitliche Verzahnung der Aktivität verschiedener Organismen. Oecologia (Berl.) 3:214–226

Remmert H (1972) Die Tundra Spitzbergens als terrestrisches Ökosystem. Umschau in Wissenschaft und Technik 2:41–44

Remmert H (1973) Über die Bedeutung warmblütiger Pflanzenfresser für den Energiefluß in terrestrischen Ökosystemen. J Orn 114:227–249

Remmert H (1976 a) Die Bedeutung der Tiere in terrestrischen Ökosystemen. Bayerisches Landwirtschaftliches Jahrbuch 53:96–101

Remmert H (1976 b) Gibt es eine tageszeitliche ökologische Nische? Verh Dtsch Zool Ges, pp 29–45

Remmert H (1980) Arctic animal ecology. Berlin-Heidelberg-New York

Remmert H, Ohm P (1955) Etudes sur les rockpools des pyrenees-orientales. Vie et Milieu 6:194–209

Remmert H, Wünderling K (1970) Temperature differences between Arctic and Alpine meadows and their ecological significance. Oecologia (Berl.) 4:208–210

Ribaut JP (1964) Dynamique d'une population de Merles noirs, Tudus merula L. Revue Suisse de Zoologie 71/42:815–902

Riess W (1975) Kontrolliertes Brennen – eine Methode der Landschaftspflege. Mittl flor-soz Arbeitsgem. NF 18:265–271

Riess W (1976a) Die Wirkungen kontrollierten Feuers auf den Boden und die Mikroorganismen. Umweltschutz, Umwelterhaltung, Immissionsschutz, Pflanzenschutz und Schädlingsbekämpfung. Ergänzende Nachfolge „Natur, Kultur und Jagd", pp 259–262

Riess W (1976b) Der Feuereinsatz und seine Technik in der Landschaftspflege. Natur und Landschaft 51:284–287

Riess W, Tüxen R (1976c) Bibliographie der Arbeiten über Einfluß des Feuers auf die Vegetation. Excerpta Botanica, Section B, 15:277–310

Risch SJ (1979) A comparison, by sweep sampling of the insect fauna from corn and sweet potato monocultures and dicultures from Costa Rica. Oecologia (Berl.) 42:195–211

Roeder KR (1968) Neurale Grundlagen des Verhaltens. Beispiele aus der Insektenwelt. Physiologisches Kolloquium, Vol IV. Huber, Bern Stuttgart

Rohmeder E (1967) Beziehungen zwischen Frucht- bzw. Samenerzeugung und Holzerzeugung der Waldbäume. Mitteilungen aus der Staatsforstverwaltung Bayerns 36:1–23

Rüppell G, Gösswein E (1972) Die Schwärme von Leucaspius delineatus (Cyprinidae, Teleostei) bei Gefahr im Hellen und im Dunkeln. Z vergl Physiol 76:333–340

Ryszkowski L (1975) The ecosystem role of small mammals. Ecological Bulletins/NFR, Biocontrol of Rodents 19:139–145

Salomonsen F (1935) Aves. In: Spark R (ed) Zoology of the faroes, vol III

Sarkissian IV (1974) Regulation by salt of activity of citrate synthetases from osmoregulators and osmoconformers. Trans NY Acad Sci Ser II/36:775–782

Schaller F (1963) Die Unterwelt des Tierreiches. Springer, Berlin Heidelberg New York, pp 1–125

Schauermann J (1973) Zum Energieumsatz phytophager Insekten im Buchenwald II. Die produktionsbiologische Stellung der Rüsselkäfer (Curculionidae) mit rhizophagen Larvenstadien. Oecologia (Berl.) 13:313–350

Scherzinger W (1976) Rauhfuß-Hühner, Nationalpark Bayerischer Wald, Arbeiten No 2. Bayerisches Staatsministerium für Ernährung, Landwirtschaft und Forsten

Schildknecht H (1970) Die Wehrchemie von Land- und Wasserkäfern. Angewandte Chemie, pp 17–25

Schlichter D (1973) Meerwasser als Nahrungsquelle: Aufnahme gelöster organischer Verbindungen. Verh Ges Ökolog, pp 25–38

Schlichter D (1974) Aufnahme in Meerwasser gelöster Aminosäuren durch Anemonia sulcata Pennant. Z Morphol Tiere 79:65–74

Schlichter D (1975) The importance of dissolved organic compounds in sea water for the nutrition of Anemonia sulcata Pennant (Coelenterata). Proc 9th Europ mar biol Symp, pp 395–405

Schlichter D (1976) Umweltanpassung im makromolekularen Bereich. Verh Dtsch Zool Ges, p 215

Schlichtmann W (1951) Bemerkungen zur Ornis Niedersachsens. Beitr Natkde Nieds 4:36–44

Schmidt-Koenig K (1972) Animal orientation and navigation. Scientific and Technical Information office. National Aeronautics and Space Administration, Washington D.C., pp 1–605

Schmidt-Nielsen K (1953) The desert rat. Sci Amer. pp 189–191

Schmidt-Nielsen K (1965) Physiology of salt glands. In: Sekretion und Exkretion. Springer, Berlin Heidelberg New York, pp 269–288

Schmidt-Nielsen K (1975) Animal physiology. Adaptation and environment. Cambridge University Press, pp 1–699

Schneider D (1974) Chemical communication in danaid butterflies. In: Olfaction and taste, vol V. Academic Press, New York, p 327

Schneider D et al. (1975) A pheromone precursor and its uptake in male danaid butterflies. J comp Phys 97:245–256

Schnitzler H-U (1978) Die Detektion von Bewegungen durch Echoortung bei Fledermäusen. Verh Dtsch Zool Ges, pp 16–33

Schöne HK (1977) Die Vermehrungsrate mariner Planktondiatomeen als Parameter in der Ökosystemanalyse. Habilitationsschrift der Mathematisch-Naturwissenschaftlichen Fakultät der RWTH Aachen, Aachen, pp 1–323

Schorger AW (1955) The passenger pigeon. Its natural history and exstinction. Univ. Wisconsin Press, 424 pp

Schramm U (1972) Temperature-food interaction in herbivorous insects. Oecologia (Berl.) 9:399–402

Schröder D (1972) Bodenkunde in Stichworten. Hirt, Kiel

Schröder W (1974) Warum rotten Raubtiere ihre Beute nicht aus? Die Pirsch 8:380–385

Schultz AM (1972) A study of an Ecosystem: The Arctic Tundra. In: Cycles of essential elements

Schulze E-D (1972) Die Wirkung von Licht und Temperatur auf den CO_2-Gaswechsel verschiedener Lebensformen aus der Krautschicht eines montanen Buchenwaldes. Oecologia (Berl.) 9:235–258

Schulze ED, Lange OL, Koch W (1972) Ökophysiologische Untersuchungen an Wild- und Kulturpflanzen der Negev-Wüste. Oecologia (Berl.) 9:317–340

Schwerdtfeger F (1968) Ökologie der Tiere, Vol II: Demökologie. Parey, Hamburg, pp 11–448

Schwerdtfeger F (1975) Ökologie der Tiere, Vol III: Synökologie. Parey, Hamburg, pp 11–451

Seelemann U (1968) Zur Überwindung der biologischen Grenze Meer-Land durch Mollusken. Oecologia (Berl.) 1:130–154

Seeley MK (1979) Irregular fog as a water source for desert dune beetles. Oecologia (Berl.) 42:213–227

Seitz A (1977) Selektiver Fischfraß stabilisiert die Koexistenz dreier Daphnia-Arten. 70. Jahrestagung der DZG Erlangen

Sendstad E, Solem JO, Aagard K (1977) Studies of terrestrial chironomids (Diptera) from Spitsbergen. Norsk Ent Tidsskr 24:91–98

Sengbusch P v (1977) Einführung in die Allgemeine Biologie. Springer, Berlin Heidelberg New York

Sharp MA, Parks R, Ehrlich R (1974) Plant resources and butterfly habitat selection. Ecology 55/4:870–875

Siebers D, Bulnheim HP (1976) Salzgehaltsabhängigkeit der Aufnahme gelöster Aminosäuren bei dem Oligochaeten Enchytraens albidus. Verh dtsch Zool Ges 1976, 212, Stuttgart

Siegl H (1953) Untersuchungen über den Samenertrag der Fichte im Herbst 1951. Forstw Cbl 11/12:369–379

Slijper EJ (1966) Riesen des Meeres. Eine Biologie der Wale und Delphine. Verständliche Wissenschaft. Springer, Berlin Heidelberg New York, pp 1–115

Slobodkin L, Richman S (1956) The effect of removal of fixed percentages of the newborn on size and variability in populations of Daphnia pulicaria (Forbes). Limnology and Oceanography 1:209–237

Sondheimer E, Simeone JB (1970) Chemical ecology. Academic Press, New York

Sperber G (1968) Der Reichswald bei Nürnberg. Aus der Geschichte des ältesten Kunstforstes. Mitteilungen aus der Staatsforstverwaltung Bayerns, p 37

Sperlich D (1973) Populationsgenetik. Grundlagen der modernen Genetik, Vol 8. Fischer, Stuttgart

Spindler K-D, Müller WA (1972) Induction of metamorphosis by bacteria and by a lithium-pulse in the larvae of Hydractinia echinata (Hydrozoa). Wilhelm Roux' Arch 169:271–281

Staiger H (1954) Der Chromosomendimorphismus beim Prosobranchier Purpura lapillus in Beziehung zur Ökologie der Art. Chromosoma 6:419–478

Stein W (1960a) Biozönologische Untersuchungen über den Einfluß verstärkter Vogelansiedlung auf die Insektenfauna eines Eichen-Hainbuchen-Waldes I. Z angew Entomol 46:345–370

Stein W (1960b) Biozönologische Untersuchungen über den Einfluß verstärkter Vogelansiedlung auf die Insektenfauna eines Eichen-Hainbuchen-Waldes II. Z angew Entomol 47:196–230

Stein W, Franz J (1960) Die Leistungsfähigkeit von Eiparasiten der Gattung Trichogramma (Hym., Trichogrammatidae) nach Aufzucht unter verschiedenen Bedingungen. Naturwissenschaften 47:262–263

Steinhausen G (1913) Die Geschichte der deutschen Kultur, 2nd edn. Leipzig u. Wien

Stenson JAE (1978) Differential predation by fish on two species of Chaoborus (Diptera, Chaoboridae). Oikos 31:98–101

Stern K, Tigerstedt PMA (1974) Ökologische Genetik. Fischer, Stuttgart

Strenzke K (1951) Grundfragen der Autökologie. Acta biotheoretica 9/4:163–184

Strenzke K (1953) Ökologie der Wassertiere. In: Handbuch der Biologie, Vol III, No 5/6, Akademische Verlagsgesellschaft Atheneion, Frankfurt, pp 115–192

Strenzke K (1954) Nematalycus nematoides n.g.n.sp. aus dem Grundwasser der algerischen Küste. Vie et Milieu 4:638–647

Stresemann E, Timofeef-Ressovxky NW (1947) Artentstehung in geographischen Formenkreisen. I. Der Formenkreis Larus argentatus-cachinnansfuscus. Biologisches Zentralblatt 66:57–75

Svärdsson G (1957) The "invasion" type of bird migration. British Birds 50:314–343

Svensson S (1970) Bird census work and environmental monitoring. Bull Ecol Res Commun 9:5–52

Tahvanainen JO, Root RB (1972) The influence of vegetational diversity on the population ecology of a specialized herbivore, Phyllotreta cruciferae (Coleoptera: Chrysomelidae). Oecologia 10/4:321–346

Tast J, Kalela J u. O (1971) Comparisons between rodent cycles and plant production in finish Lapland. Ann Acad Scient Fenn A, IV 186:1–14

Tautz, J, Markl H (1978) Caterpillars detect flying wasps by hairs sensitive to airborne vibration. Behav Ecol Sociobiol 4:101–110

Raylor RJ (1977) The value of clumping to prey: experiments with a mammalian predator. Oecologia 30:1–10

Thauer RK, Jungermann K, Decker K (1977) Energy conservation in chemotrophic anaaerobic bacteria. Bact Rev 41:100–180

Thiele HU (1968) Bodentiere und Bodenfruchtbarkeit. Organischer Landbau, Internationale Fachzeitschrift für Biologie und Technik im Landbau 1+2:6–8, 29–31

Thiele HU (1971) Über die Facettenaugen von land- und wasserbewohnenden Crustaceen. Z Morphol 69:9–22

Thorson G (1957) Bottom communities (sublittoral or shallow shelf). Geol Soc Amer, Memoir 67/1:461–534

Thun von W (1968) Autökologische Untersuchungen an freilebenden Nematoden des Brackwassers. Dissertation Kiel 1–72

Tiedtke B (1964) Über die ökologische Bedeutung eines extrem kalten Winters für die eulitorale Hartbodenfauna der Kieler Förde. Schr-Naturw Ver Schlesw-Holst 35:33–60

Tischler W (1955) Synökologie der Landtiere. Fischer, Stuttgart

Tischler W (1965) Agrarökologie. Fischer, Jena

Tischler W (1975) Ökologie mit besonderer Berücksichtigung der Parasitologie. In: Wörterbuch der Biologie. Fischer, Jena

Tischler W (1977) Einführung in die Ökologie. Fischer, Stuttgart

Tokuyama T, Kuraishi H, Aida K, Uemura T (1973) Hydrogen sulfide evolution due to pantothenic

acid deficiency in the yeast requiring this vitamin; with special reference to the effect of adenosine triphosphate on yeast cysteine desulfhydrase. J Gen Appl Microbiol 19:439–466
Tschumi P (1973) Die Bedeutung des Raubwildes in Tiergemeinschaften. Wild und Wald. Beih Z Schweiz. Forstverein 52:137–157
Utida S (1955) Fluctuations in the interacting populations of host and parasite in relation to the biotic potential of the host. Ecology 36:202–206
Utida S (1957) Cyclic fluctuations of population density intrinsic to the hostparasite system. Ecology 38:442–449
Utida S (1958) On fluctuations in population density of the rice stem borer Chilo suppressalis. Ecology 39:587–599
Utschick H (1976) Die Wasservögel als Indikatoren für den ökologischen Zustand von Seen. Verh orn Ges Bayern 22:395–438
Vaartaja O, Salisbury PJ (1965) Mutual effects in vitro of micro-organisms isolated from tree seedlings, nursery soil, and forests. Forest Science 11/2:160–168
Vareschi E (1977a) Biomasse und Freßrate der Zwergflamingos im Lake Nakuru (Kenia). Verh dtsch Zool Ges Abstract
Vareschi E (1977b) The ecology of Lake Nakuru (Kenya). I. Abundance and feeding of the lesser flamingo. Oecologia (Berl.)
Vareschi E (1979) The ecology of Lake Nakuru (Kenya). II. Biomass and spatial distribution of fish. Oecologia (Berl.) 37:321–335
Vogel St (1978) Organisms that capture currents. Sci Amer 239:108–115
Voipio P (1954) Über die gelbfüßigen Silbermöwen Nordwesteuropas. Acta Soc Pro Fauna et Flora Fennica 71/1:1–56
Voipio P (1968) Zur Verbreitung der Argentatus- und Cachinnans-Möwen. Ornis Fennica 45:73–83
Voipio P (1972) Silbermöwen der Larus argentatus cachinnans-Gruppe als Besiedler des baltischen Raumes. Ann Zool Fennici 9:131–136
Wachter H (1964) Über die Beziehung zwischen Witterung und Buchenmastjahren. Forstarchiv 4:69–78
Wallgren H (1954) Energy metabolism of two species of the genus Emberiza as correlated with distribution and migration. Acta Zool Fenn 84:5–110
Walter H (1949) Einführung in die Pflanzengeographie. III. Grundlagen der Pflanzenverbreitung. Ulmer, Stuttgart
Walter H (1973a) Allgemeine Geobotanik. In: Uni-Taschenbücher, Vol 284. Ulmer, Stuttgart
Walter H (1973b) Vegetation of the earth. In relation to climate and the eco-physiological conditions. Springer, Berlin Heidelberg New York
Waterhouse DF (1974) The biological control of dung. Sci Amer 230/4:100–109
Watson A (1965) A population study of ptarmigan (Lagopus mutus) in Scotland. J Anim Ecol 34:135–172
Watson A, Hewson R, Jenkins D, Parr R (1973) Population densities of mountain hares compared with red grouse on Scottish moors. Oikos 24:225–230
Weeks HP, Kirkatrick ChM (1976) Adaptations of white-tailed deer to naturally occurring sodium deficiencies. J Wildl Managm 40:610–625
Weiß W (1975) Arktis. Urban & Schwarzenberg, Wien München
Wendland V (1975) Dreijähriger Rhythmus im Bestandswechsel der Gelbhalsmaus (Apodemus flavicollis Melch.) Oecologia 20:301–311
Werner D, Roth R (1977) Productivity of diatoms in culture and in marine habitats. Marine Res Indonesia 20:99–113
Whittaker RH (1970) Communities and ecosystems. Macmillan, New York
Wickler W (1973) Mimikry: Nachahmung und Täuschung in der Natur. Fischer Taschenbuch Verlag, Frankfurt
Wieser W (1965) Untersuchungen über die Ernährung und den Gesamtstoffwechsel von Porcellio scaber (Crustacea: Isopoda). Pedobiologia 5:304–331
Wieser W, Zech M (1976) Dehydrogenases as tools in the study of marine sediments. Marine Biology 36:113–122
Wilbert H (1962) Über Festlegung und Einhaltung der mittleren Dichte von Insektenpopulationen. Z Morphol Ökol Tiere 50:576–615
Willmans O (1973) Ökologische Pflanzensoziologie. In: Uni-Taschenbücher, Vol 269. Quelle & Mayer, Heidelberg
Wilson EO (1975) Sociobiology, the new synthesis, Cambridge-Mass., pp 1–697
Wilson EO, Bossert HW (1973) Einführung in die Populationsbiologie. Springer, Berlin Heidelberg New York
Winkler S (1973) Einführung in die Pflanzenökologie. In: Uni-Taschenbücher, Vol 169. Fischer, Stuttgart
Wynne-Edwards VC (1966) Animal dispersion in relation to social behaviour. Edinburgh, pp 1–653
Wyrwoll T (1977) Die Jagdbereitschaft des Habichts (Accipiter gentilis) in Beziehung zum Horstort. J Ornith 118:21–34
Zachariae G (1962) Was leisten Collembolen für den Waldhumus? Soil Organisms, pp 109–124
Zachariae G (1964) Welche Bedeutung haben Enchytraeen im Waldboden? Soil Micromorph, pp 57–68
Zachariae G (1965) Spuren tierischer Tätigkeit im Boden des Buchenwaldes. Forstwissenschaftl Forsch 20:7–68
Zachariae G (1967) Die Streuzersetzung im Köhlgartengebiet. Prog Soil Biology, pp 490–506
Zackrisson O (1976) Vegetation dynamics and land use in the lower reaches of the river Umeälven. Early Norrland 9:10–74
Zackrisson O (1977) Influence of forest fires on the North Swedish boreal forest. Oikos 29:22–32
Zahner R (1959, 1960) Über die Bindung der mitteleuropäischen Calopteryx-Arten (Odonata, Zygoptera) und den Lebensraum des strömenden Was-

sers. I. Der Anteil der Larven an der Biotopbindung. Int Rev ges Hydrobiol 44:51–130. – II. Der Anteil der Imagines an der Biotopbindung. Int Rev ges Hydrobiol 45:101–123
Zebe E (1977) Anaerober Stoffwechsel bei wirbellosen Tieren. Rheinisch-Westfälische Akademie der Wissenschaften. Vorträge Vol 269, pp 51–73
Zethner O (1976) Control experiments on the nun moth (Lymantria monacha L.) by nuclear-polyhedrosis virus in Danish coniferous forests. Angew Ent 81:192–207
Ziegelmeier E (1964) Einwirkungen des kalten Winters 1962/63 auf das Makrobenthos im Ostteil der Deutschen Bucht. Helgol Wiss Meeresunters 10:276–282
Zeigelmeier E (1970) Über Massenvorkommen verschiedener makrobenthaler Wirbelloser während der Wiederbesiedlungsphase nach Schädigungen durch „katastrophale“ Umwelteinflüsse. Helgol Wiss Meeresunters 21:9–20
Ziegler H, Lüttge U (1966) Die Salzdrüsen von Limonium vulgare. I. Mitteilung: Die Feinstruktur. Planta (Berl.) 70:193–206
Zinkler D (1966) Vergleichende Untersuchungen zur Atmungsphysiologie von Collembolen (Apterygota) und anderen Bodenkleinarthropoden. Z vergl Phys 52:99–144
Zinkler D (1968) Vergleichende Untersuchungen zum Wirkungsspektrum der Carbohydrasen von Collembolen (Apterygota). Verh Dtsch Zool Ges Innsbruck, pp 640–644
Zinkler D (1971) Carbohydrasen streubewohnender Collembolen und Oribatiden. Comptes-rendus IV. Colloque Int. Faune du Sol, Dijon 1970. Ann Zool Ecol An nhs, pp 329–334
Zinkler D (1972) Vergleichende Untersuchungen zum Wirkungsspektrum der Carbohydrasen laubstreubewohnender Oribatiden. Verh Dtsch Zool Ges 65. Jahresvers, pp 149–152
Zwölfer H, Ghani MA, Rao VP (1976) Foreign exploration and importation of natural enemies. In: Theory and practice of biological control. Academic Press, New York, pp 189–207
Zwölfer H (1977) Regulationsprozesse in Ökosystemen – ein Beitrag aus dem Bereich der biologischen Schädlingsbekämpfung. 70. Jahrestagung der DZG Erlangen – Abstract

General Books on Subjects Related to Ecology

Cleffmann G (1979) Stoffwechselphysiologie, Bd 791. UTB Ulmer, Stuttgart, p 296
Dawkins R (1977) The selfish gene. Oxford University Press, p 197
Ehrlich P, Ehrlich A, Holdren JP (1977) Ecoscience. Freeman, San Francisco, p 1051
Ellenberg Heinz (1978) Die Vegetation Mitteleuropas mit den Alpen. Ulmer, Stuttgart, p 981
Ewert JP (1976) Neuroethologie. HT 181. Springer, Heidelberg Berlin New York, p 259
Hochachka P, Somero GN (1973) Strategies of biochemical adaptation. Saunders, Philadelphia, p 358
Roeder KR (1968) Nerve cells and insect behavior. Harvard University Press. Cambridge (Mass.)
Schmidt-Nielsen K (1972) How animals work. Cambridge University Press. Cambridge (England), p 124
Schmidt-Nielsen K (1975) Animal physiology. Cambridge University Press. Cambridge (England), p 699
Sengbusch P v (1977) Allgemeine Biologie. 2nd edn. Springer, Berlin Heidelberg New York, p 526
Sperlich D (1973) Populationsgenetik. Stuttgart, p 197
Stern K, Tigerstedt PM (1973) Ökologische Genetik. Stuttgart, p 211
Wilson EO (1975) Sociobiology. Harvard University Press. Cambridge Mass., p 697

Illustration on page 3: Shameless bird (by Erich Ohser; courtesy of M. Klumlies-Ohser)
Illustration on page 107: From Lockley (1953)
Illustration on page 177: From Lohmann (1974)
Illustration on page 267: Three hares with three eares: in the Paderborn Cathedral
Illustration on page 271: The evolution of an ecological model (courtesy of Eastern Deciduous Biome US IBP Analysis of Ecosystems Newsletter, 1972)

Subject Index

Ecological Studies

Analysis and Synthesis

A Selection

Springer-Verlag
Berlin
Heidelberg
New York

Volume 22
R. Guderian

Air Pollution

Phytotoxicity of Acidic Gases and Its Significance in Air Pollution Control

Translated from the German by C. J. Brandt

1977. 40 figures, 4 in color, 26 tables.
VIII, 127 pages
ISBN 3-540-08030-9

Experimental methods are presented and evaluated for determining quantitative relationships between air pollutants and their effects on vegetation. The three most important air pollutants causing injury to vegetation in Central and Western Europe (sulfur dioxide, hydrogen fluoride, and hydrogen chloride) are used as examples. The significance of the reaction of plants for practical air pollution control is discussed.

Volume 25

Microbial Ecology of a Brackish Water Environment

Editor: G. Rheinheimer

With contributions by M. Bölter, K. Gocke, H.-G. Hoppe, J. Lenz, L. A. Meyer-Reil, B. Probst, G. Rheinheimer, J. Schneider, H. Szwerinski, R. Zimmermann

1977. 77 figures, 87 tables. XI, 291 pages
ISBN 3-540-08492-4

The results of a 15-months' investigation in the Kiel Bight (Baltic Sea) are the subject of this volume, containing the measurements of more than 50 hydrographical, chemical, and microbiological parameters. Using methods such as scanning electron microscopy and autoradiography, a comparative analysis of most of the measured data on the bacterial colonization of detritus, and the relations between pollution, production, and remineralization led to the construction of a model for the ecosystem, indicating the energy fluxes. This volume together with Volume 24, A Coastal Marine Ecosystem by Kremer and Nixon (on ecological studies conducted at Narragansett Bay, and system-simulation by a computer model), represent a comprehensive study of coastal marine ecosystems.

Volume 28

Pond Littoral Ecosystems

Structure and Functioning
Methods and Results of Quantitative Ecosystem Research in the Czechoslovakian IBP Wetland Project

Editors: D. Dykyjová, J. Květ

1978. 183 figures, 100 tables. XIV, 464 pages
ISBN 3-540-08569-6

This book contains the results of the Czechoslovakian contribution to the wetland studies of the "International Biological Programme" (1965–1974). A team of ecologists investigated the fishpond littoral ecosystems in two biogeographical regions of Central Europe comparing environmental characteristics and functioning of the biotic components in the Trebon basin, a UNESCO/MAB biosphere reserve in Bohemia (Hercynian region), and in the Lednice fishponds, a State nature reserve in Moravia (Pannonian region). More general information includes structure, productivity, and production processes of the vegetation and its macrophytes and algae, decomposition processes and mineral nutrient regime in fishpond littorals, the role of certain animal populations, and management and conservation of fishpond littoral vegetation and waterfowl.

Volume 29

Vegetation and Production Ecology of an Alaskan Arctic Tundra

Editor: L. L. Tieszen

1978. 217 figures, 115 tables. XVII, 686 pages
ISBN 3-540-90325-9

This is a comprehensive summary of the research conducted during a four-year period as part of the U. S. Tundra Biome Program. It provides an excellent general introduction to the environment and ecology of the Coastal Plain tundra of Northern Alaska, including a floristic analysis of mosses and vascular plants. This is complemented by a thorough description of the vegetation components. The background material provided leads into detailed studies on production and physiology. These studies emphasize mechanistic relationships affecting physiological processes, growth and production in a number of vascular and moss species. Photosynthesis, root growth phenomena, ecosystem CO_2 exchange and water relations are throughly analyzed. The book integrates the modeling approach used in most of these process studies and presents the current state of growth and plant population modeling in other ecosystem processes. Areas where additional research is needed are suggested. This volume provides a solid and comprehensive background for students and researchers interested in tundra and arctic research.

Volume 31
W. Tranquillini

Physiological Ecology of the Alpine Timberline

Tree Existence in High Altitudes with Special Reference to the European Alps

Translated from the German by U. Benecke

1979. 67 figures, 21 tables. XI, 137 pages
ISBN 3-540-09065-7

In all the Earth's mountain areas, nature sets an upper limit to forest growth – the timberline. With the advent of ecophysiological field methods in 1930, the study of this extremely complex phenomenon made rapid strides. Never has there been a more comprehensive analysis of the struggle of trees for survival at their highest altitude of subsistence. Based on decades of research work at various timberline stands in the Austrian Alps, this book explores in-depth how climate changes with increasing altitude, how trees adapt to this increase, and at what point adaptation becomes impossible. The book offers the botanist new insights into the formation of one of the most important vegetation boundaries and provides forestry and agriculture with a sound scientific basis for the preservation and restoration of protective forest in mountain areas.